# THE BUILDING OF AN EMPIRE

# *The Building of an Empire: Italian Land Policy and Practice in Ethiopia 1935–1941*

HAILE M. LAREBO

CLARENDON PRESS · OXFORD

*Oxford University Press, Walton Street, Oxford* OX2 6DP
*Oxford New York*
*Athens Auckland Bangkok Bombay*
*Calcutta Cape Town Dar es Salaam Delhi*
*Florence Hong Kong Istanbul Karachi*
*Kuala Lumpur Madras Madrid Melbourne*
*Mexico City Nairobi Paris Singapore*
*Taipei Tokyo Toronto*
*and associated companies in*
*Berlin Ibadan*

*Oxford is a trade mark of Oxford University Press*

*Published in the United States by*
*Oxford University Press Inc., New York*



*First published 1994*



*British Library Cataloguing in Publication Data*
*Data available*

*Library of Congress Cataloging in Publication Data*
*Larebo, Haile M.*
*The building of an empire : Italian land policy and practice in Ethiopia, 1935-1941 / Haile M. Larebo.*
*p. cm.—(Oxford studies in African affairs)*
*Includes bibliographical references.*
*1. Land use—Government policy—Ethiopia. 2. Ethiopia—Colonization. 3. Italy—Colonies—Africa. I. Title. II. Series.*
*HD979.Z7L37 1994 333.3'163—dc20 93-34531*
*ISBN 0-19-820262-8*

3 5 7 9 10 8 6 4 2

*Printed in Great Britain*
*on acid-free paper by*
*Antony Rowe Ltd*
*Chippenham, Wiltshire*

*To the Memory of Tim Matthews*

# *Preface and Acknowledgements*

THE formidable body of literature dealing with the short-lived Italian occupation of Ethiopia is both Italo-centric and is mainly preoccupied with Italy's aggression against Ethiopia and its impact on international diplomacy. Italy's colonization *per se* is generally ignored or dismissed in a few perfunctory paragraphs. Honourable exceptions to this are the recent works by Grassi and Goglia, Angelo del Boca, and Alberto Sbacchi. Grassi and Goglia's anthology gives an interesting portrait of the model of Italian colonialism as expressed in its ideological framework and in its full practical administrative efficiency. By highlighting its most radical components—direct rule, authoritarianism, racism, and demographic colonization—the authors attempt to dispel the popular Italian view which claims that Italy's brief colonial enterprise was basically different from other imperial powers, for it was both much more benevolent and less exploitative. Even though they totally neglect the Ethiopian context and fail to incorporate and even acknowledge the existence of Ethiopian sources, their work is a welcome departure from the earlier approach. Del Boca and Sbacchi are, in a way, pioneers on the working of Italian colonialism in Ethiopia and its impact on Ethiopian society. These studies are less Italo-centric than most and have the merit of addressing a much wider range of issues raised by the Italian presence in Ethiopia and its impact on the Ethiopian society. But they have serious shortcomings in interpretation and methodology, which are amply addressed in Alessandro Triulzi's review article.[1]

The present work attempts to redress the balance by appraising Italian colonial enterprise—which is now widely condemned—in the terms applicable to its own period rather than the very different criteria commonly applied later. Although a number of subsidiary issues are explored, the major focus is land: for Italian imperialist ambitions from the nineteenth century had been sustained by the need for an outlet to settle Italy's excess population and deploy its surplus capital. In Italian

[1] Review of books on *Italian Colonialism and Ethiopia*, *JAH* 23/2 (1982): 237–43; see also J. Markakis, 'Italian Conquest and Colonisation', *JAH* 281 (1987): 168–9.

eyes, those imperatives justified the conquest of Ethiopia against quasi-universal international opposition.

Unlike the older Italian colonies that were described as a useless collection of sand, Ethiopia, with its varying climate and fertile soil, was portrayed as an El Dorado where Italy's longstanding imperial aspirations could be effectively fulfilled. With the conquest of Ethiopia, Italy claimed to have become one of the 'satisfied' nations. With impressive land-colonization programmes, it aimed at settling Italian colonists with a mission to transform, within a short time, the agricultural sector into the granary of Italy and the Ethiopian Empire into an extension of a *Magna Italia*. The flow of emigrants was to be diverted from the Americas to Ethiopia which, following incorporation into the Italian Empire, was to provide cheap food and raw materials for Italian industry and become a protected market for Italian products. However, the building of the Empire proved extremely difficult, largely because realities on the ground vastly hampered practical results. This book draws the mythology behind these aims and charts in detail the vast chasms that separated intention from execution.

Students of colonial history find many similarities between Italian colonialism and that of other colonial powers in Africa. Colonial systems varied greatly in detail, but their fundamental assumptions and purpose remained the same. Subject people were regarded as an inherently inferior breed. This postulate was supported by the elaboration of a whole range of myths that highlighted the alleged superiority of the White race. In the name of civilizing mission the resources—particularly land and labour—of the subject people were used to serve the interests of the colonial power. The subject people were ruled autocratically by unrepresentative governments and, until they began to organize themselves effectively, lacked the political experience and the mechanism to air their views and affect the political process at the centre. In attaining these objectives, Italian colonialism was certainly no harsher in practice than any other European power. The difference was that Fascism, being a doctrine based on hierarchy and force, was open as to its true principles and intentions. But like other colonial powers, Italy was confronted with enormous practical problems: unprecedented resistance even from a supposedly docile population, general public apathy at home after initial enthusiasm, insufficient support from the home government, and inadequate funds to carry out rapid developmental work. Yet in some respects the Italian case also reveals interesting features that had little or no parallel with the experience of other colonial powers. Italy's primary interest in colonies was to use them as a convenient receptacle for its surplus population and a source of primary goods to the mother country. Agriculture as conducted by other European powers typically took the form of a large farm or ranch employing entirely African labour. The result was a relatively small European population and a restricted market for the local produce. Italy's State, by contrast, sponsored peasant settlement schemes aimed at transforming the Empire into a *Magna Italia*—an extension of the mother country. According to these plans, the peasants would work their own farms with minimal help from the indigenous population and become independent landowners who would form the backbone of an Italian society and transform the Empire into a possession profitable for the motherland. Although on a few occasions France and Portugal and even Britain had flirted with similar enterprises, as a rule this type of settlement policy was, if not essentially foreign, of secondary interest to them and other colonial powers in Africa.

But the practical implementation of these policies, like the policy of any imperial power, was shaped by a complex convergence of forces: the physical landscape, the widely different responses of the local population, the character of the ruling class, and the length of colonial occupation. This point is illustrated clearly by comparing the Italian colonial experience in Ethiopia to those of neighbouring colonial powers, especially the British in Kenya. The Kenyan comparison is especially apt not only because Ethiopia has large areas which are similar in altitude to those in Kenya but, more importantly, because it was in these regions that were deemed to be climatically attractive for the White man that European settlement was largely attempted. Just as Kenya, in the popular British imagination, was the White settler colony par excellence, Ethiopia was perceived to be so in Italian minds. Of course, there are differences of scale. Italy, defined by the advocates of colonialism as a proletarian nation, claimed to possess a large class of workers supposedly prepared to settle as peasant colonists in subtropical conditions. Therefore, unlike in Kenya, where settlement was almost exclusively of a landowning and commercial élite, Italy's main concern remained demographic and the rest was of secondary importance.

Geographically, both Kenya and Ethiopia present dramatic contrasts, with a wide variety of climates and terrains, extending from temperate conditions and elevations of several thousand metres in the highlands to the coastal tropics. Both countries are endowed with great lakes, even though Ethiopia, unlike Kenya, whose rivers are few and, with the exception of Tana, short or ephemeral, possesses a large number of perennial and long rivers, the composition of its soil is richer in mineral salts, and precipitation is, on the whole, more evenly distributed and allowed for greater agricultural potential. Despite all their scenic beauties and fertile zones, both countries had a fundamentally harsh environment whose rugged terrain and unreliable climate presented enormous difficulties for access as well as development. The highlands, where European settlement took shape, though cool, are basically equatorial, and the Italians were confronted with the same technical difficulties that were commonly faced by the settlers in Kenya.

Population patterns did nothing but reinforce these physical obstacles. Like Kenya and most of Africa, Ethiopia has a mosaic of ethnic groups who have distinctive languages, cultures, and social organizations, and varied agricultural practices and technologies that are often influenced by and adapted to the geographical diversity of their habitat. Even though this ethnic diversity often served as a divisive force, facilitating the divide and rule efforts of the colonial power, it presented serious difficulties for administration. The striking contrasts that lurk below the deceptive similarities to the Kenyan condition made this human pattern even more intractable in Ethiopia.

Indeed, the Ethiopian milieu represents a unique political and cultural landscape that has no match elsewhere in Africa. Almost all post-colonial African states, including Kenya, are accidental creations made possible by rivalries between European colonial powers. Although externally imposed and alien to the country, the state and its institutions persisted even after the colonial power left. But the Ethiopian State is an indigenous development and Ethiopia's existence as an independent sovereign nation—albeit with variable frontiers—extends back for millennia. Unlike most of Sub-Saharan Africa, Ethiopia is the home of diverse ancient civilizations with uninterrupted rich literary traditions and splendid monuments left by each of the major religions present for centuries in the region—indigenous religions, Christianity, Islam, and Judaism. Its people had been under an expansionist and dynamic central government, however inefficient to Western eyes, with a substantial intelligentsia and a well-developed administration, maintained by surplus extraction from the peasant cultivators. Well entrenched under complex ancient systems of land tenure, most of the peasants used an agrarian technology which, despite its low level of sophistication, was relatively far more complex than that used in many parts of Sub-Saharan Africa.

Thus Ethiopia presented a very different physical and human texture to the invading Italian forces and their settlement programmes from that encountered by other European colonial powers elsewhere in Africa. Unlike the British in Kenya, for example, the Italians fought a modern war to occupy Ethiopia employing massive air-power, armoured cars, and modern artillery along with over 120,000 men. Once the war was over, they were confronted with a much more strongly organized local society and an intractable land tenure system. Perhaps surprisingly, there was little overt resistance to colonial imposition in Kenya apart from a few skirmishes, and the colonial presence, even though maintained by force, was accepted albeit reluctantly. Serious resistance came only after the Kenyans had experienced colonial rule for many years. Settlement programmes in Kenya were thus carried out in a much more peaceful political environment than in Ethiopia, where the Italians became enmeshed in an almost nationwide opposition whose internationalization put the motherland itself in a very vulnerable position.

Italian occupation was not only precarious but also brief, lasting only five years. Therefore, when Italy was abruptly expelled, its development plans were only at their initial stage. This might make any conclusive judgement and forecast of the feasibility of its programmes and its long-term effect on Ethiopian society difficult and speculative. The occupation came at a time when colonialism, though alive, had diminished in significance as a symbol of world hegemony. The Italian élite itself was well aware of colonies' fading importance. Before it fell back on the easy and unimaginative doctrine of imperialism and racial supremacy, Mussolini's own monthly magazine, *Gerarchia*, came close to tapping one of the most revolutionary forces of our era when it briefly entertained in 1930 the prospects of treating the White and native peoples as absolute equals on the grounds that the age of colonies might be soon over.[2] Understandably, the Ethiopian occupation, unlike most colonial adventures, unleashed unprecedented protests worldwide, especially in Black Africa. Although unsuccessful in ousting the Italians, the international outcry can be credited with heralding a stronger anti-colonial nationalist movement that would sweep across the African continent afterwards.

Italy's own situation as a latecomer to the club of imperialist powers also boded ill for the success of its Ethiopian venture, for Italy possessed neither the experience nor the skill nor the resources to carry out its ambitions effectively. In comparison Britain, with a long history of colonial administration, conducted in a wide variety of social and physical conditions, was well positioned in Kenya to translate its programme into

[2] *Gerarchia* (Oct. 1930), 845–8.

action. But despite strong official support the European entrepreneurs in Kenya, who were far better equipped financially and politically than the penniless Italian peasant or inert and rapacious commercial farmer, were unable either to increase their numbers substantially or to compete effectively against overseas capital and the lower costs of Kenyan peasant production.

Against these formidable obstacles, only the most prejudiced observer could maintain that Italy's programme had any prospect of success. Comparable experiments elsewhere suggest they were doomed to failure. Apart from Kenya, other instructive examples are the French State's colonization programme in Algeria at the end of the nineteenth century and, most importantly, the experiences of Portuguese in Angola and Mozambique in the 1950s. Like Italy, Portugal is a 'proletarian nation'—relatively poor and backward economically but rich in manpower. After initial hesitations, its State colonization programme was applied with some success to its two African possessions to complement Portuguese agriculture and settle its surplus population. Like its French counterpart in Algeria and in spite of all frantic resistance, however, it ended tragically in the throes of self-assertive modern nationalism, the fateful offspring of years of pent-up frustrations, denials, and discrimination. Likewise, Ethiopia could hardly have supported a large European population. Any serious increase in the number of settlers would have likely reduced the opportunities for the Ethiopians and intensified the tensions that would end in a successful struggle of independence.

Thus this book, dealing with an area that has been neglected in the study of African colonial history, should afford new insights into comparative African colonial history. Readers, examining in detail how Italy attempted to build its empire, can conclude for themselves to what extent Italy's imperialist ideas and practices were different from those of other European colonial powers.

Accessible Ethiopian sources were consulted and invaluable information was gained from persons who were closely affected by the Italian land policy itself. In compliance with their wishes, the identity of the sources is not disclosed. But the bulk of this research was conducted in Rome at Archivio Centrale di Stato and, under very difficult conditions, in the widely dispersed and badly organized historical archives of the Comitato per la Documentazione dell'Opera dell'Italia in Africa, successor of the Ministry of Foreign Affairs (1882–1912), the Ministry of Colonies (1912–37), and the Ministry of Italian Africa (1937–43).

Anybody with research experience in Italy will be aware of its peculiar rewards and difficulties, and especially of the problems dealing with the documents of the Comitato. For unexplained reasons, the author was forbidden access to some documents. Although the use of other relevant sources has made these documents redundant for the purpose of the present research, the author shares the view expressed elsewhere that the Ministry's arbitrary approach and lack of clear guidance are harmful to serious scholarly research.[3] The Comitato's recent extensive publications of over forty volumes dealing with Italy's work in Africa cannot fill this vacuum. Both the Comitato, which is a body mainly staffed by 'ex-colonial governors and state functionaries' and its works, which are largely subjective in approach and 'celebratory in intention', cannot be taken as fully dependable historical sources.[4]

The most important data were collected in the Archivio Centrale di Stato in Rome, Archivio Documentario of Istituto Agronomico per L'Oltremare (IAO) in Florence[5] and the Fondazione Einaudi in Turin, where access was gained to the documents of the then Minister of Finance, Conte Paolo Thaon Di Revel. But without the use of learned institutions such as state and private libraries in both Rome and Florence, which are the depositaries of current important periodicals, and the extensive reports and memoirs available at the Public Record Office by the contemporary British officials and their agents, which throw important light on the workings and the difficulties of the Italian regime in Ethiopia, the research would have remained incomplete. Historians of Italian colonialism may object particularly to the use of Italian periodicals and newspapers, which are understandably regarded by many as simple propaganda literature of little or no historical value. It is true that most of their authors were men on the payroll of the Ministry of Propaganda who, along with those employed by the Ministry of Foreign Affairs and Ministry of Italian Africa, diligently published a plethora of articles throughout the occupation. Although these articles took a stereotyped course through platitude to beatitude, ballyhooing the most tenuous progress as a prodigy of achievement, and giving to the plans all the quality of performances, they cannot simply be dismissed. Attentive reading with

[3] Triulzi, Review, 237–8.

[4] R. Rainero, introd. to A. Sbacchi, *Il colonialismo italiano in Etiopia, 1936–1940* (Milan: 1980).

[5] Despite a written recommendation to the author by the Foreign Ministry 'that the most fruitful research could be done at the IAO in Florence,' the author was faced with a number of obstacles from the same authorities whose discordant voices helped only to frustrate his work.

the understanding of the rhetorical garb of the Fascist administration reveals that under this apparently propaganda material lies an untapped wealth of information that was omitted or treated only cursorily by sources commonly regarded as authoritative.

The research would not have been completed without the useful criticism and technical advice, moral support, continuous encouragement, and daily inspiration of many people and institutions. Special thanks goes to the Africa Educational Trust, which supported the author throughout the troubled period of fieldwork, providing financial and moral support and, later in 1991, subsidized another research trip to Rome. My warm thanks go particularly to AET's then director, Tim Matthew, who died at a tragically early age and to whom this book is dedicated; to his successor Kees Maxey, and to his kind, energetic, and understanding staff, particularly Barbara Hart. Without AET's generous help this study could never have been completed. AET's generosity far exceeds my ability to repay it adequately. The Twenty-Seven Foundation, through the London Institute for Historical Research, funded part of the travel, research and support costs. I am deeply indebted to Anthony McAdam, an unfailing friend and source of support since my early days in England. He encouraged me to apply to AET and strongly supported my application. Particular thanks to Philip Nelson, the then UK Counsellor in Rome, whose timely and generous intervention during the difficult period of fieldwork in Italy salvaged my research effort, wrecked by misfortune and muddle-headed, humdrum bureaucrats.

I owe a great debt to Professor Richard Gray who continued to be my inspiration, encouragement, and patient listener even after his formal responsibility as my doctoral supervisor had ended. The completion of this research was partly due to his unfailing attention, indispensable assistance, and advice. Many Italians—friends, private individuals, government officials, public bodies, some of whom were important participants in the events recorded here—gave generously of their time, hospitality, and knowledge. I wish particularly to thank the staff and the management of the Biblioteca Nazionale in Rome and Florence and those of the Istituto Italo-Africano.

In Rome, my debt to Professor Luigi Goglia is greater than any formal acknowledgement could ever express for providing me with friendship, constant intellectual stimulus, and assistance of all kinds. Special thanks are due to Dr Ludovica de Curten and her staff of Archivio Centrale di Stato; and above all to those in charge of the archive of Opera Nazionale di Combattenti, Dr Boccini and Dr Giacozzi, for their patient and gracious assistance and advice in using ONC's uninventoried archive. Dr Enrico Serra, head of Servizio Storico e Documentazione del Archivio Storico of the Ministry of Foreign Affairs, graciously provided me with important published materials; the Superintendent of the archives, Ruffo di Scaletta facilitated access to the archives of the IAO; Mario Gazzini, archivist for the Comitato. I am deeply grateful to all of them.

Many former colonial officers refused to be interviewed, but some were more than helpful. The author is particularly grateful to Dr Attilio Scaglione and Pier Marcello Masotti for sharing with me their reminiscences about the settlers' lives and voracious appetites for land; the ex-settler Gr. Ufficiale Mario Buschi for enlightening me on the difficulties of the colonial Italian State and the role played by some Italians in post-independence Ethiopia; Professor Lanfranco Ricci of Naples University, Dr Salvatore Tedeschi, Professor Loglia, Chancellor of Accademia Nazionale Dei Lincei, and Professor Giampaolo Calchi-Novati, the Director of IPALMO, for their mediation efforts with the Ministry of Foreign Affairs; Dr Giuseppe Rocchetti, former technician of Centro Sperimentale Agrario e Zootecnico of Bolé, and Dr Rolando Guidotti, regent for a short period of Nuovo Ispettorato Generale dell'Agricoltura but whose contact with Ethiopian agriculture went back to 1931, when he was commissioned to study Emperor Haile Sellassie's model farm at Érär.

In Florence, I thank the Director of the IAO for allowing me free xeroxing; the Centro Internazionale Studenti for its succour. In Bologna, Dr Irma Taddia helped me to expand my contacts and broaden my intellectual quest with free gifts of books and personal publications. In Turin, the former journalist and writer on Italian colonialism, Angelo Del Boca, enlightened me on the workings of Italian land policy. From Ravenna, Professor Gian Carlo Stella, wrote me full and informative letters elucidating important and obscure questions and despatched valuable extracts at his expense. I am greatly indebted to all of them.

Many other friends and academics—some of whom I came to know in the course of the research and only a few of whom are named—generously gave invaluable advice, documents, criticism, and help of all kinds that words fail to describe. I am indebted to Professor Christopher Clapham at Lancaster University and the following individuals residing at various locations: Dr Freda Harcourt, Dr David Johnson, Dr Maknun Gamaladin al-Ashami, Dr Muhammad Hassan, Dr Richard Hayward, Professor Jim McCann, Dr Tekeste Negash, Svein Ege, Dr Richard Pankhurst, Dr Tenker Bonger, Dr Osvaldo Rainieri, Dr Alberto Sbacchi, Revd

Vincent Twomeysud, Angela Raven-Roberts and the agronomist and crop-physiologist, Mr Yaqob Edjamo, from whose pedological knowledge I benefited a great deal.

My thanks to Professor David Nicholas, Head of the Department of History at Clemson University, for his time and many useful suggestions; to Steve and Cindy Marks for their valuable friendship and entertaining meals, and my departmental colleagues for their encouragement. It is a must to offer my heartfelt thanks to my all-weather friend and ex-fellow graduate student, Dr Dea Jane Birkett, who, with the rare combination of her eagle eyes and sharp lance of her pen, prevented some of the worst solecisms and obscure semantics slipping into the early draft of this manuscript. On a more personal level, I owe her and Kevin Toolis a great deal more for their exquisite cuisine, engaging companionship, and enduring support.

My deepest gratitude goes to Pauline, a patient and devoted companion, and our two children Iyanna and Manito, for their forbearance, support, and understanding during my long absences. Without their ceaseless encouragement I might never have completed this work successfully.

Finally, I would like to thank my copy-editor, Rowena Anketell, for her invaluable work on this book. My warm tributes go to Sophie MacCallum, Anne Gelling, and Helen Gray of Oxford University Press whose patient and enthusiastic assistance finally brought this work, which originated as a Ph.D. thesis at the school of Oriental and African Studies of the University of London, to a successful conclusion.

H. M. L.

# *Contents*

# *List of Tables*

# *General Notes*

THE question of transcribing Ethiopian names is always a vexing one. My decision involved a compromise, that is, altering spellings which had little resemblance to linguistic reality, and retaining others on account of their familiarity. Bearing this in mind, the following rules are adopted.

## VOWELS

The following characters are used to express the seven Ethiopian intonations.

| | |
|---|---|
| First Order | ä |
| Second Order | u |
| Third Order | i |
| Fourth Order | a |
| Fifth Order | e |
| Sixth Order | é |
| Seventh Order | o |

## CONSONANTS

Whenever a consonant is geminated, as 'b' in 'äbba', the same consonant is repeated twice. For 'explosive', 'glottal', and other consonants to which there is no proper correspondent in the Latin alphabet, the following equivalent is adapted.

| *Latin* | *Equivalent* |
|---|---|
| k | q |
| n | ñ |
| z | ž |
| t | ť |
| c | č |
| s | š |
| p | p̌ |

## NAMES

The Ethiopians have no fixed family names. They use their given name and the name of their father. Thus, Mr Gäbrä and Ms Dobamé, two children of Mr Lafébo (first name) Isa (the personal name of Lafébo's father), will be known as Mr Gäbrä Lafébo and Ms Dobamé Lafébo. After marrying Mr Liränso, Ms Dobamé still keeps her father's name and would be addressed as Mrs Dobamé Lafébo, and not Mrs Liränso, nor Mrs Lafébo.

Ethiopians tend to choose different forms of transliteration in spelling their names and as result many variants are used for the same name. Thus, one finds Häylä-Sellasé being spelt as Haile Sellassie, Haile Selassie, Haile Sellase, Hayla Sellase, Hayle Sellasse, Hailé Sellasé.

## CALENDAR

The Ethiopian calendar year begins on 11 September and lags seven years behind the Gregorian calendar until 31 December, and eight for the rest of the year. The dates of the month lag between ten and seven days behind the Gregorian calendar.

*List of Illustrations*

# *List of Abbreviations*

| | |
|---|---|
| AA | Addis Ababa |
| *AAI* | *Annali dell'Africa Italiana* |
| *AC* | *Agricoltura Coloniale* |
| ACG | Acting Consul General |
| ACP | Affari Civili e Politici ([Directorate for] Civil and Political Affairs) |
| ACS | Archivio Centrale dello Stato (Central State Archive) |
| GRA | Marshal Graziani Files |
| MCP | Ministro della Cultura Popolare (Ministry of Information Files) |
| ONC | Opera Nazionale di Combattenti (Army Veterans' Association National Agency Files) |
| PNC | Partito Nazionale Fascista (National Fascist Party Files) |
| PS | Ministry of Public Security Files |
| SPDR | Segreteria Particolare del Duce (Special Secretariat of Duce Files) |
| A[A]B | Azienda Agraria di Biscioftù (Bishoftu Agricultural Farm) |
| A[A]H | Azienda Agraria di Olettà (Holäta Agricultural Farm) |
| AE | Affari Economici (Economic Affairs) |
| AI | Africa Italiana (Italian Africa) |
| Am | [Regio Governo dell'] Amara ([Royal Governorate of] Amara) |
| AMAR | Archivio di Museo Africano [in Istituto Italo Africano] |
| AO | Africa Orientale (East Africa) |
| AOI | Africa Orientale Italiana (Italian East Africa) |
| AP | Atti Parlamenti (Parliamentary Papers) |
| As | Asmära |
| ASMAE | Archivio storico del Ministero degli Affari Esteri (Foreign Ministry Archives) |
| ASMAI | Archivio storico dell'Africa Italiana (Archives of the Ministry of Italian Africa) |
| AS | Archivio segreto (Secret Archives) |
| ATdR | Archivio Thaon di Revel: Presso la Fondazione Einaudi (Archives of Thaon di Revel at Luigi Einaudi Foundation [Turin]) |
| B | Busta (Box) |
| *BUCE* | *Bollettino Ufficiale della Colonia Eritrea* |
| *BUGA* | *Bollettino Ufficiale del Governo dell'Amara* |
| *BULMAI* | *Bollettino Ufficiale di Legislazione del Minstero dell'Africa Italiana* |
| CAG | Comprensorio del Consorzio Agricolo di Gimma (Jimma Agricultural Consortium) |

| | |
|---|---|
| *CC* | *Consulente Coloniale* |
| CD | Camera dei Deputati (House of Deputies) |
| CD IAO | Centro documentazione: Istituto Agronomico per L'Oltremare (Archives of Overseas Agricultural Institute) |
| *CE* | *Corriere Eritreo* |
| CFA | Confederazione Fascista degli Agricolotori (Fascist Farmers' Union) |
| CFLA | Confederazione Fascista dei Lavoratori dell'Agricoltura (Fascist Agricultural Workers' Union) |
| CFI | Confederazione Fascista degli Industriali (Fascist Industrialists' Union) |
| CFLI | Confederazione Fascista dei Lavoratori dell'Industria (Fascist Confederation of Industrial Workers) |
| CG | Consul General |
| *CI* | *Corriere dell'Impero* |
| CL | Colonizzazione e Lavoro (Office of Colonization and Labour) |
| *CM* | *Corriere Mercantile* |
| Cotetio | Compagnia Nazionale per il Cotone di Etiopia (Ethiopian National Cotton Company) |
| *CS* | *Corriere della Sera* |
| CSFA | Commando Superiore Forze Armate (Armed Forces High Command) |
| DAF | Direttorato di Affari Finanziari (Directorate of Financial Affairs) |
| DCL | Direzione di Colonizzazione e Lavoro (Directorate of Colonization and Labour) |
| DD | Documenti diplomatici (Diplomatic Documents) |
| DG | Direttorato Generale (General Directorate) |
| DG[A]CL | Direzione Generale [degli Affari] della Colonizzazione e del Lavoro (General Directorate of Colonization and Labour) |
| DGG | Decreti Governo/Governatore Generale (Decree of Governorship/Governor-General) |
| DGov | Decreto Governo/Governatore (Decree of Governor/Governorate) |
| DIM | Decreto Interministeriale (Interministerial Decree) |
| DL | Decreto Legge (Decree Law) |
| DM | Decreto Ministeriale (Ministerial Decree) |
| DONC | Direzione Azienda ONC (Directorate of ONC Farm) |
| DRegG | Decreto Reggente il Governo (Decree of the Regent [Governor]) |
| DS | Direttorato Superiore (Central Directorate) |
| AE | Affari Economici (Economic Affairs) |
| AP | Affari Politici (Political Affairs) |
| DSAPGG | Direzione Superiore Affari Politici del Governo Generale ([Central] Superior Directorate of Political Affairs of Governorship-General) |
| EC | Ethiopian Calendar |
| ECAI | Ente per il Cotone dell'Africa Italiana (Italian African Cotton Agency) |
| Eccles. | Ecclesiastical usage |
| ECPE | Ente di Colonizzazione Puglia di Etiopia (Agency for the Puglia Colonization of Ethiopia) |

| | |
|---|---|
| ECRE | Ente di Colonizzazione Romagna di Etiopia (Agency for the Romagna Colonization of Ethiopia) |
| ECVE | Ente di Colonizzazione Veneto di Etiopia (Agency for the Veneto Colonization of Ethiopia) |
| EDR | Ente di Colonizzazione De Rege (De Rege Colonization Board) |
| *EI* | *Espansion Imperiale* |
| EO | Ethiopian Observer |
| EOC | Ethiopian Orthodox Church |
| Er | [Regio Governo dell'] Eritrea ([Royal Governorate of] Eritrea) |
| f | foglio (file) |
| Fl | Florence |
| FO | Public Record Office |
| G | Governatore/Governo (Governor/ate) |
| *GAD* | *Giornale di Agricoltura della Domenica* |
| GG | Governatore/Governo Generale (Governorship-General) |
| *GI* | *Giornale d'Italia* |
| *GM* | *Gazzetta del Mezzogiorno* |
| Gm | Gimma (Jimma) |
| Gn | Gondär |
| *GP* | *Gazzetta del Popolo* |
| GS | [Regio Governo dei] Galla e Sidama ([Royal Governorate of] Galla and Sidamo |
| *GU* | *Gazzetta Ufficiale del Governo Generale dell'AOI* |
| *GUGS* | *Bollettino Ufficiale del Governor Generale dell'AOI e Bollettino del Governo dello Scioa* |
| *GURI* | *Gazzetta Ufficiale del Regno d'Italia* |
| *GV* | *Gazzetta di Venezia* |
| Hr | [Regio Governo del] Harar ([Royal Governorate of] Härär) |
| HSIU | Haile Sellassie I University |
| IA | Ispettorato Agrario (Agricultural Inspectorate Office) |
| IACI | Istituto Agricolo Coloniale Italiano (Italian Colonial Agricultural Institute) |
| IAO | Istituto Agronomico per L'Oltremare |
| *IC* | *Italia Coloniale* |
| ICAI | Società Anonima Impresa Cotoniera Italiana (Italian Cotton Enterprise Corporation) |
| ICI | Istituto Cotoniero Italiano (Italian Cotton Institute) |
| IDC | Istituzione del Distretto Cotoniero . . . (Institution of Cotton District of . . .) |
| IEA | Italian East Africa |
| IFPLAOI | Ispettorato Fascista di Produzione e Lavoro (Fascist Inspectorate for Production and Labour of Italian East Africa) |
| IGA | Ispettorato Generale di Agricoltura/Ispettorato Generale Agrario (General Inspectorate for Agriculture) |

| | |
|---|---|
| *II* | *Impero Italiano* |
| *IL* | *Impero del Lavoro* |
| INFAIL | Istituto Nazionale Fascista per l'Assicurazione contro gli Infortuni del Lavoro (Fascist National Institute for Insurance against Industrial Accidents) |
| INFPS | Istituto Nazionale Fascista per la Previdenza Sociale (Fascist National Institute for Sociale Security) |
| Ing. | Ingegnere |
| *IOM* | *Italia d'Oltremare* |
| *JAH* | *Journal of African History* |
| *JMAH* | *Journal of Modern African History* |
| *JES* | *Journal of Ethiopian Studies* |
| *LI* | *Lunedì dell'Impero* |
| MAE | Ministero degli Affari Esteri (Ministry of Foreign Affairs) |
| MAI | Ministero dell'African Italiana (Ministry of Italian Africa) |
| MC | Ministero delle Colonie (Ministry of Colonies) |
| Mg | Mogadishu |
| Mil. | Military usage |
| MTD | Maria Theresa Thaler |
| MVSN | Milizia Voluntaria di Sicurità Nazionale (Volunteer National Militia) |
| ONC | Opera Nazionale di Combattenti (Army Veterans Association National Agency) |
| P | Pacco |
| *PI* | *Popolo d'Italia* |
| PNF | Partito Nazionale Fascista (National Fascist Party) |
| *PR* | *Popolo di Roma* |
| PSSFA | Principal Secretary of State for Foreign Affairs |
| ql | quintal |
| R | Royal |
| *RC* | *Resto del Carlino* |
| RD | Royal Decree |
| *RDA* | *Rivista di Diritto Agrario* |
| RDL | Royal Decree Law |
| *REAI* | *Rassegna Economica dell'Africa Italiana* |
| *REC* | *Rassegna Economica delle Colonie* |
| RG | Regio Governo |
| RIA | Regio Istituto Agronomico |
| *RISS* | *Rivista Internazionale di Scienze Sociali* |
| Rm | Rome |
| RMC | Regio Ministero delle Colonie (Royal Ministry of Colonies) |
| *RSAI* | *Rassegna Sociale dell'Africa Italiana* |
| SAIS | Società Agricola Italo-Somala (Italo-Somali Farming Company) |
| SAICES | Società Anonima Industriale Commerciale Ethiopia Sud. (South Ethiopian Industrial and Commercial Joint-Stock Co.) |

| | |
|---|---|
| Sc | [Regio Governo dello] Scioa ([Royal Governorate of] Shäwa) |
| SC | Sede Centrale (Central Headquarters) |
| SCCCE | Società per la Coltivazione del Cotone nella Colonia Eritrea (Eritrean Cotton Marketing Company) |
| SIA | Società Imprese Africane (African Development Company) |
| SIOM | Sindacato Italiano d'Oltre Mareb (Italian Syndicate for Beyond Märäb) |
| Sm | [Regio Governo della] Somalia ([Royal Governorate of] Somalia) |
| SM | Stato Maggiore (General Staff) |
| UA | Ufficio Agrario (Agricultural Office) |
| ZN | *Zekrä Nägär* (memoirs written by the court official Balambaras Mahteme Sellassie Wolde Maskal, pub. 1942 EC) |

# I

# *The Evolution of Italian Colonial Policy*

## MISSIONARIES, ADVENTURERS, AND COLONIZERS

The 1935–41 Italian occupation of Ethiopia is rooted in a long and turbulent history that began in the early nineteenth century when both countries were emerging as unified modern states. From this early period, Ethiopia was idealized as a place where Italian colonial aspirations could be realized. Ethiopia was thought to provide a demographic outlet for Italy's surplus population and a valuable market from which Italian industry could draw raw materials and to which it could sell manufactured goods. These same ideas inspired and shaped the agricultural policies and practices of 1935–41 and were the driving force of the Italian colonialist movement which justified the conquest of Ethiopia in 1935, as much as it earlier did the occupation of Eritrea, Libya, and Somalia. It was in these three colonies that the first major attempts were made to translate these ideas into action.

The forerunners of Italian colonialism in Ethiopia were the Italian Catholic missionaries who were scrambling with Islam and Protestantism for the last reserve of disposable souls. They were later followed by scholars, travellers, visionaries, and by those discontented with the anti-imperialist stand of the newly formed Italian state;[1] there were speculators and arms dealers who looked for government subvention.[2] Fascist historians would cherishingly address this heterogeneous group as 'our colonial predecessors'. In background and interest, this group resembled its counterparts in other European countries. As in France and Germany, numerically it was tiny but politically vociferous.

Among the Italian Catholic missionaries, the Lazarist Father Giuseppe Sapeto is credited as the founder of the Catholic missions and the first

[1] The position advocated by the anti-colonialist movement was expressed through phrases such as *politica rinunciataria* (politics of resignation) or *politica delle mani pulite* (politics of clean hands) or *piedi a casa* (firm foot at home).

[2] A. Del Boca, *Gli Italiani in AO*, i: *Dall'unità alla marcia di Roma* (Bari, 1976), 4.

Italian colony in Ethiopia. He came to Ethiopia in 1837 and three decades later acquired for Rubattino Navigation Company the port of Asäb. The mission strengthened with fresh arrivals: in 1839 of De Jacobis, a clergyman, and Montuori, a medical doctor, and seven years later of the Piedmontese Franciscan, Father Massaia, the founder of the Capuchin Order in Ethiopia, and the Ligurian clergyman, Father Stella, of the Congregation of the Turin Mission.

The founding of the Catholic missions coincided with a time of political fragmentation in Ethiopia and intense international activities in the Red Sea and the Nile Valley. It was part of a wider configuration that was taking place in the rest of Africa as a result of the resurgence of renewed European interest in the continent. The missionaries, like the Jesuits under the princes of the seventeenth century, were unconscious and sometimes even unwilling tools of the rival local rulers who, in their unrelenting but cautious bid to gain Western technical assistance, largely saw them as no more than a valuable liaison to the European courts, tolerating their work as long as they were politically useful.[3] However, the local rulers' exploitation of the missions for political objectives should be balanced with the equally important role of the missionaries as agents of European intervention. Even the saintly missionary leaders could not always resist the temptation of believing that the European presence would boast his evangelical work. The case of Massaia, who became one of Menilek's most respected advisers, was not untypical. He maintained contact with the Sardinian kingdom, soliciting it to promote trade with Abyssinia as he assumed that such Italian commercial presence would considerably ease his apostolic activity.[4] But it was his co-religionist, Father Stella, who had as his protégé Däjach Häylu, Governor of Hamasén in Eritrea, that showed a more ambitious proposition and, more than any other missionary, played an important role in the history of Italy's colonialist movement.

Stella considered colonies as centres of settlement and urged the Sardinian court to start such a scheme at Hamasén, which he thought an ideal location where Europeans would easily feel at home. He painted an enticing portrait of the area as the most fertile and the richest in all Abyssinia, where 'there is available extensive unoccupied land, for their

[3] For a more detailed discussion of this aspect see D. Crummey, *Priests and Politicians: Protestant and Catholic Missionaries in Orthodox Ethiopia, 1838–1868* (Oxford, 1972).

[4] AMAR, misc. f4, Cardinal Massaia to Conte di Cavour, 1 Feb. 1858.

old owners have either emigrated or are dead.'[5] In his former mission station, Bogos, where he was once 'a moral dictator and amphitryonic judge' of seventeen Bilén villages,[6] Stella set up in 1867 the first Italian agricultural colony on a 1,000 sq-km concession. It was situated 1,200 m above sea level and within six km from Käränn, the main town and took the name of Colonia Italo-Africana di Sciottel, or Italo-African Shotäl Colony. Though predominantly Italian, the settlers were a mix of nationalities, all united by the prospects of easy gain.[7] According to the plan, the new immigrants were expected to assert themselves as incontestable masters of the land and develop it into a flourishing metropolitan commercial enterprise under the Italian protectorate.[8] Work began in earnest with cotton cultivation and animal breeding. But the initial enthusiasm was short-lived and the scheme was a ludicrous failure. After a series of quarrels, most of the party with capital left, disgruntled with the whole business. Then infighting, a hostile climate, lack of Italian government support and capital took its toll and by 1870 nothing was left of the colony but rubble.[9]

If the plan failed in practice, the legacy of Stella's vision of a colony as a centre of population settlement lingered, sustained by geographical and exploration societies[10] who were financed partly by the State and partly by public donations. In their effort to win the hearts and minds of the

[5] AMAR, misc. f4, Padre Stella to Cavour, 3 Mar. 1859; CDDD, XV, *Libro Verde: Etiopia*, (Rm, 1890), doc. 7, pp. 15–16; R. Rainero, *I primi tentativi di colonizzazione agricola e di popolamento dell'Eritrea, 1890–1895* (Milan, 1960), 38–9.

[6] G. Lejean, *Voyage aux deux Nils, 1860–1864* (Paris, 1865), 154.

[7] Almost all of them came from Egypt and, according to the contract, they were divided into two clear-cut groups, capital investors, known as *coloni capitalisti cooperatori*, and manual workers [cf. Del Boca, *Dall'unità*, 19–22]; MAE *L'Italia in Africa i/2. L'avvaloramento e la colonizzazione* (Rm, 1970), 33.

[8] AMAR, f4, Stella to Cavour.

[9] O. Antinori, *Viaggio fra i Bogos* (Rm, 1887), 95.

[10] Exploration and geographical societies mushroomed in most of Italy's commercial cities. The most important were: Società Geografica Italiana of Fl (1867), whose first president Cristoforo Negri prompted the Italian missionaries in Ethiopia to promote metropolitan interests in the territory of their mission stations; Società di Esplorazioni Commerciali in Africa of Milan, which had on its executive board high-ranking figures of Lombardian business who played a crucial role in economic and political penetration of the AI colonies and later Ethiopia; Club Africano di Napoli which, founded in 1880, in the immediate aftermath of the occupation of Asäb by Rubattino Navigation Company, transformed itself into Società Africana di Italia. Other societies worked in the same direction in most Italian commercial cities: Società Coloniale di Studi in Fl, Associazione di Geografia Commerciale in Bari, Comitato Per le Esplorazioni in Africa in Turin, Società D'Esplorazione of Genoa. Cf. Del Boca, *Dall'unità*, 51–5; R. Ciasca, *Storia coloniale dell'Italia contemporanea* (Milan, 1940), 35.

Italian public, the societies organized expeditions, widely publicized their cause in journals and at public meetings.[11] But their attempt to justify their ambitions, like other European expansionists, as a search for markets and raw materials, and as a civilizing mission in the Roman tradition, fell on deaf ears. The Italy that had emerged as a single nation in 1861 was not only politically weak and divided but economically backward. For a country not sufficiently industrialized to acquire raw materials and produce substantial manufactured goods, these arguments meant little.

However, Italy's emigration problem made the issue of the demographic myth nationally emotive. Between 1861 and 1911 the population grew from 25.7 million to 35.9 million, increasing the density from 87 persons to 123 persons per sq km. Geography offered little immediate solution to this growth; almost four-fifths of the peninsula was mountainous. Economically, Italy was a poor country and industrialization, even in the north, was advancing slowly and extensive development took place only at the turn of the century. In what came to be known as the *Mezzo Giorno*, the Southern region, the social structure and land tenure systems further compounded the misery of the peasantry. In these circumstances, emigration was the easiest solution, reaching the proportions of a major exodus by the turn of the century, with regional shifts and variations according to economic tide. An upsurge in Southern emigration coincided with the world agricultural slump of the mid-1880s, while previously the bulk of emigrants had come from the North.[12]

A tense intellectual battle gathered around the emigration issue. Some felt it was a natural phenomenon offering long-term benefits: socially, emigration served as a safety-valve for domestic unrest, as the nation rid itself of discontented elements. Economically, the remittances from overseas were important in helping Italy's foreign balance of payments. The opponents discounted both arguments as irrelevant. In their view remittances could not offset the economic losses of the investment in the upbringing and education of the emigrant, and dependence on other

[11] Leading journals included M. Camperio's *L'Esploratore*, 'a journal of journey and commercial geography that has nothing to do with "abstract science" ' cited by Del Boca in *Dall'unità*, 52; and *Africa Italiana* of the Società Africana d'Italia.

[12] C. M. Cipollla, 'Four Centuries of Italian Demographic development', in D. V. Glass and D. E. C. Eversley (eds.), *Population in History* (Chicago, 1965), 576–87; C. Gini, 'Il fattore demografico nella politica coloniale', *AAI* 4 (Sept. 1941), 811; J. S. McDonald, 'Italy's Rural Social Structure and Emigration', *Occidente*, 22 (Sept.–Oct. 1956), 437–56; A. Capanna, 'Economic Problems and Reconstruction in Italy,' *International Labour Review*, 62 (June 1951), 607–52; M. Rossi-Doria, 'Land Tenure System and Class in Southern Italy,' *American Historical Review*, 64 (Oct. 1958), 46–53.

countries to dispose of social malcontents hurt national pride. The military saw emigration as a serious drain on manpower needed for national defence; the big landowners envisioned labour shortages and a rise in agricultural wages. Articles were published giving graphic descriptions of the wretched and hopeless plight of the emigrants in the USA and South America and publicizing the episodes of outrageous treatment of emigrants.[13]

The expeditions to Ethiopia highlight the close link between the trinity of politics, science, and business—a classic combination in the development of colonialism elsewhere in Africa. It is not possible in this context to recount the exploits of all the expeditions and their leaders. But mention should be made of the Great Expedition of 1875, led by the sexagenarian aristocrat and military careerist Orazio Antinori and organized by Società Geografica Italiana. On a 95-ha plot granted by King Menilek at Lét-Maräfiya, near his capital Ankobär, the mission set up the first modern agricultural station which was recognized by the International Association of Brussels and until 1897 served a two-pronged function of a study centre for the members of the Società Geografica and an intelligence-gathering base for the Italian political authorities.[14]

Antinori saw Lét-Maräfiya as a springboard for effective Italian penetration of Abyssinia and expansion into the rich southern Galla provinces. Although he ironically held his munificent Shäwan hosts in contempt, regarding them as morally devious, he sensibly proposed that the expansion of Italian peasant families in Shäwa could be achieved 'by acclimatizing itself with the country, intermarrying with the natives, learning the language'.[15] These liberal ideas were far in advance of the spirit of Fascism which forbade mixed-race relationships and, by a complex body of discriminatory laws, upheld the racial status of the colonists.

The scientific missions came to a halt with Italy's ignominious defeat at Adwa and even the Società Geografica was ordered to close its station. Although the expeditions resumed after Italo-Ethiopian relations

[13] For an authoritative account on emigration see V. Briani, *L'emigrazione italiana ieri e oggi* (Rm, 1957); A. A. Castagno, 'The Development of Expansionist Concept in Italy, 1861–1896', Ph.D. thesis (Columbia University, 1957), 737; F. Manzotti, *La polemica sull 'emigrazione nell' Italia unita fino alla prima guerra mondiale* (Milan, 1962); G. Dore, *La democrazia italiana e l'emigrazione in America* (Brescia, 1964).

[14] Antinori's death in 1882 was followed by a rapid management change. He was succeeded by Pietro Antonelli, the medical doctor Vincenzo Ragazzi, the naturalist Leopaldo Traversi, and finally in 1894 by engineer Luigi Cappucci who was in charge until the station was closed on the eve of Adwa in 1895. Cf. MAE, *Avvaloramento*, 334.

[15] L. Traversi, *Let Marefià* (Milan, 1931), 64–5.

improved in 1903, the efforts of the colonialist movement to develop a *coscienza coloniale* or 'colonial consciousness' proved futile as the Italian public failed to cultivate a lasting taste for colonialism. The brief peaks of popular enthusiasm for the Libyan war in 1911 soon foundered on long troughs of apathy, hostility, and indifference, as they did later in 1935 with the conquest of Ethiopia.[16]

Nevertheless, the work of these expansionists offers a useful and stimulating insight into the psyche and preconceptions of the colonial pioneers. But more significantly, their prolific writings, dealing with almost the same vigour, and often in a fair amount of detail, on a wide variety of subjects ranging from climate to botany, from ornithology to ethnography, from geography to history, provide a valuable contribution to the sparse knowledge of contemporary Ethiopia. Understandably, as works written in an atmosphere polluted by aggressive colonial climate and prompted primarily by political, patriotic, or idealistic motives, they are marred with tendentious judgements. Ethiopia is presented as a country with unlimited economic potential but inhabited by idle and immoral people indulging in abhorrent practices of slavery and oppression and administered by savage rulers.[17]

Most significantly, it was the findings and propaganda of these missions which contributed to shape Italy's imperial policy towards Ethiopia. It was also largely on the claims made by these 'pioneers' that Fascist Italy would try to carry out its task of empire-building and the programme of demographic colonization.

## THE ROLE OF THE ITALIAN STATE

When assessing the attitude of the Italian State towards the Ethiopian enterprise there are three distinguishable stages: the first stage, 1861–81; the second, 1882–96; the third, 1897–1935. Before 1861 the material and intellectual resources of the court of the Sardinian kingdom were almost exclusively absorbed by more serious and pressing domestic and international problems.

The official policy in the period between 1861–81 was to encourage 'commercial colonies' that 'could offer Italy outlets for its industrial

[16] C. G. Segrè, *Fourth Shore: Italian Colonization of Libya* (Chicago, 1974), 33; ACS GRA 44/36/2, Furio Lantini to Petretti (Vice-GG AOI), Rm, 24 Sept. 1937.

[17] G. Bianchi, *Alla terra dei Galla* (Milan, 1884), 158, 175, 181; Traversi, *Let Marefià*, 64–5.

products and an opportunity to accomplish works of civilization'.[18] The early eighties inaugurated a more active Italian colonial enterprise which in 1882 unfolded in the decision by P. S. Mancini to purchase Asäb from the bankrupt Rubattino Navigation Company and his authorization to occupy Meš̌wa. After the Berlin Conference (1884–5), Italy's hitherto cautious colonial diplomacy became bold. The scramble for Africa coincided with the beginning of mass southern emigration upon which hinged the appeal of the new policy. Demographic arguments were evoked to validate the occupation of Meš̌wa, in addition to strategic considerations. As Mancini said, the occupation was part of an indispensable programme to direct migration 'into hospitable lands under the Italian flag rather than letting it be dispersed on the face of the earth'.[19] For the visionary Crispi, Ethiopia was one such land whose vast cultivable areas 'will offer an outlet in the near future to that overflowing Italian fecundity which now goes to other civilized countries . . . [and] is lost to the motherland'.[20]

But both Mancini and Crispi failed to rally the public behind their expansionist programmes, and their erratic policy lacked a strong sense of conviction. Mancini resigned following the Dogalé debacle in 1887, when the Ethiopian General, Ras Alula, put to rout about a 500-strong Italian task force on a mission to relieve a besieged garrison. Crispi's government was also forced out of office after Italians suffered a humiliating defeat at the battle of Adwa in March 1896.

With the Adwa defeat the ambitious plan of demographic settlement was laid to one side until resurrected by Fascism. Adwa also marked a historical watershed between aggressive imperialism that strived for the imposition of direct European political control and peaceful economic penetration resolved to operate within the restrictive social and political parameters of an independent African state. Thus territorial ambition was replaced with competition to economic prominence with concessions as its centrepiece. Through concessions Italy aimed at institutionalizing Rome's economic supremacy over a large part of Ethiopia and transforming the Solomonic Empire into an economic dependency and then, when the first favourable occasion arose, into a protectorate.

The new Italian programme of peaceful penetration owed much to the

[18] J. L. Miege, *L'imperialismo coloniale italiano dal 1870 ai giorni nostri* (Milan, 1976), 16–19; E. de Leone, 'Le prime ricerche di una colonia e la esplorazione geografica politica ed economica', *L'Italia in Africa*, ii (Rm, 1955): 2.

[19] *Discorsi Parlamentari*, viii. 162–90 cited by Del Boca, *Dall'unità*, 182.

[20] *Scritti e discorsi politici, 1848–1890* (Turin, n.d.), 738.

statesmanship and diplomatic skill of its Minister in Addis Ababa, Major Federico Ciccodicola. As he was of the view that both the Ethiopian rulers and the people had an instinctive distrust of representatives or companies supported by any neighbouring colonial powers, he warned of any open identification of the Italian government with business interests. As a result, he laid a guideline that the concessions should be demanded by private individuals of Italian origin with apparent business interest or companies of multinational character with business connection with Italy and yet quietly supported by the Italian government. Until the occupation of Ethiopia in 1935, this policy was pursued, with varying degrees of commitment, by Ciccodicola's consular successors, strongly backed by the successive governors of Eritrea and, later, of Somalia.[21] Ciccodicola explained the significance:

> This new political strategy, cammouflaged with industrial interest, has greater advantages than the platonic desire of possession contained in the old concept of *zone of influence* for it has a practical and concrete foundation based upon a documented and real understanding with the Emperor who willingly grants . . . , for a fixed period of time, the exploration of resources of a quite extensive area, excluded to any similar foreign influence. It offers the opportunity to secure an exclusive political and commercial influence in the richest region of the territory for a relatively brief period but sufficient enough to the energies and activities of . . . capital to create such political and trade interests as to render any immediate foreign activity impossible.[22]

Broadly speaking there were three types of concessions: mining, agriculture, and commerce. Italy's attempt to outbid Great Britain and France, the two adjoining colonial powers contending to tap as much as possible the vast (imagined or real) riches of Ethiopia, in these three fields proved disastrous. The concessions as an instrument to political penetration had the backing of the State but not of capital.[23] The companies on whom the Italian government relied lacked the spirit of entrepreneurship, adequate capital, and necessary skills. Appalled by the heavy expenditure and far-reaching responsibilities that these initiatives entailed, the business class expected the Italian government to bear the costs.

The classic example was the conduct of two companies: SIOM and SCCCE. Formed in Asmära in September 1903 to exploit the mineral

[21] ASMAE 51/1: Martini to MAE, As, 29 June 1902; MAE to Lavelli De Capitani, Rm 24 Apr. 1909; MAE to SIOM, Rm, 12 Sept. 1905; 'Estratto per S.E. il Ministro dal rapporto del 21. 7. 1911 della R. Legazione in Etiopia'.

[22] ASMAE 51/1 [Ciccodicola] Rapports del Rappresentante in AA to MAE, AA 11 Apr. 1901.

[23] ASMAI 51/1, Presidente SIOM to MAE, Milan 18 May 1905.

resources of Tegray region in a concession obtained by Ciccodicola, SIOM was an outcome of a joint venture spearheaded by the Eritrean Governor Martini, the Società Italiana per il Commercio delle Colonie and the Società Eritrea per le Miniere D'Oro.[24] Believing that it offered Italy an economic foothold that would enable its long-term plan of gradual and peaceful penetration into Ethiopia to come to fruition, the officialdom saw this concession as a historical breakthrough and urged fast and discreet action.[25]

SIOM aspired to attain an impressive variety of objectives which went beyond the scope of the initial concession. It included, in addition to mineral exploration and extraction, trade in local resources, agricultural exploitation, road and railway construction, and the development of a communication network. The plan was to make the concession self-financing and as part of this design, SIOM also endeavoured to incorporate into its concession the distant territories adjoining the Lake Tana region with the aim of building a railway link stretching from Sätit to Gondär in order to exploit the rich agricultural resources of these areas. The driving force behind the later plan was the Eritrean Governor, Marchese Salvago Raggi, whose paramount concern was to pre-empt the encroachment into the region of any foreign interest, notably British, whose quest for concession to build a dam on Lake Tana had remained a perennial preoccupation.[26] Raggi's scheme seemed to be on the verge of success when the Ethiopian government asked SIOM to despatch its representative to Addis Ababa in order to discuss the proposal. But by then SIOM, unable to raise the necessary capital, was on its dramatic journey into bankruptcy, a course started within a year of its formation. After one highly publicized but muddled expedition to the area,[27] followed

[24] It took a long bargaining between the two rival Eritrean-based companies. SIOM was led by Count Felice Scheibler who made his fortune in the overseas trade and was one of the major shareholders of the Società du Minier du Wollega, a mining company based in south-west Ethiopia; Società per le Miniere d'Oro, headed by the engineer Talamo, was involved in mining in Eritrea.

[25] 'The economic and political advantages of the concession should not be lost due to our carelessness and delay. By acting fast and forcefully we have to belie the popular adage claiming that "good and fast cannot go together".' (ASMAE 51/1: Martini to MAE, As 11 June 1901; MAE to Maggiore Ciccodicola, Rm 12 June 1901.

[26] ASMAE 51/1 Marchese Salvago Raggi to MAE, As 26 May 1908 and 3 Aug. 1908; G Er to MAE, As 2 June 1908; MAE to Conte Colli di Felizzano, 8 July 1908. Bahru Zewde, 'The Fumbling Debut of British Capital in Ethiopia', in S. Rubenson (ed.), *Proceedings of the VIIth International Conference of Ethiopian Studies* (AA University, 1984), 332.

[27] The expedition, ten people strong including six miners, was organized by the MAE in conjunction with the Eritrean Governor and the Italian Minister in AA. The members of the expedition received special treatment during their sea journey from Italy 'as a matter of encouragement' and as something 'totally exceptional'. Included with the variety of goods

by years of existence on paper only, SIOM was forced into receivership within less than a decade of its creation.[28]

Undaunted by SIOM's insolvency and determined to pursue his grand design, Raggi persuaded SCCCE to step into the footsteps of the bankrupt company. Since 1904 SCCCE's major investment in Eritrea was cotton farming. To make it profitable, the company was engaged in a monopolistic trade with the surrounding districts of neighbouring Ethiopia to whose population the company supplied agricultural tools and cotton seeds, administered through their leaders. This cross-border trade, conducted in connivance with the Italian political authorities and without Ethiopian government official approval, fitted Italy's strategy of peaceful penetration and Raggi's own scheme of gradual absorption of part of the Ethiopian northern frontier regions into the Eritrean highland.[29] But Eritrean cotton remained unprofitable and SCCCE's cross-border illegal trade monopoly stopped, undermined by resurgence of the British interest in the Nile Valley area and boycotted by the dissatisfied leaders of the Ethiopian frontier, who resented the company's unauthorized intrusion in their local affairs and stinginess.[30] Thus when approached by Raggi, SCCCE was on the brink of financial crisis and in accepting the proposal, hoped not only to legitimize its controversial cross-border operation but also more importantly, though unrealistically, for a greater State subsidy.[31] However, when asked to despatch its agent to Addis Ababa for the negotiation of the planned extension of the original concession, SCCCE took a narrow view of the paramount interest of its shareholders, making the despatch conditional upon the government's financial support:

If, in fact, His Excellency, for the benefit of all and to oppose the requests made by other foreigners, likes to make use of us to obtain the due protection of the

shipped were a large quantity of small arms and 'gifts for the native chiefs of the territories *en route*'. (ASMAE 51/1: MAE to Società Navigazione Italiana, Rm 10 Sept. 1903; MAE to SIOM, Rm 18 Sept. 1903; Talamo to MAE, Rm 19 Sept. 1903; Scheibler to MAE, Milan, 8 Oct. 1903.)

[28] ASMAE 51/1: Conte Colli di Felizzano to MAE, AA 17 Aug. 1908; Legazione AA to MAE, AA 24 Nov. 1908; MAE to SIOM, Rm. 24 Apr. 1909.

[29] Expansion of the Eritrean border into part of Ethiopia is unambiguously spelt out in one of Raggi's undespatched reports to the MC: 'Eritrea's great shortcoming is that of being too small, and, given that the mountainous Abyssinian highlands do not have good agricultural prospects, an enlargment of the colony towards Tegray would not be a bad idea', cited by Del Boca, *Dall'unità*, 838.

[30] ASMAE 51/1: SCCCE to MAE, Milan 17 July 1908 and 27 July 1908; Legazione AA to MAE, AA 8 Sept. 1908.

[31] ASMAE 51/1: SCCCE to Commissario regionale Agordat, Agordat 25 May 1908; Salvago Raggi to MAE, As 26 May 1908.

Italian interest, we are disposed to do whatever His Excellency likes to tell us, provided the expenses, well known to us and to you and which are to be met in Addis Ababa, are sustained by the government.[32]

Thus the Italian policy of using the business class as pawns for its political chess-game had only frustrating results. SCCCE's conduct was deplored as 'extremely inert and indecisive'.[33] And yet the Italian government had neither the will nor the resources to invest in the ventures.[34]

The policy of economic penetration was part of Italy's overall strategy of destabilization of Ethiopia and was backed up, particularly after the first decade of the twentieth century, with the so-called 'politics of attraction of the periphery', or *politica periferica*, aimed at winning the goodwill of the Ethiopian population, especially those on the frontiers of their two colonies, Eritrea and Somalia. Italy attempted to advance this partly by distribution of largesse to local chiefs, and partly by establishing schools, clinics, and commercial agencies in key Ethiopian towns as means of information-gathering, espionage, and political infiltration.[35] Among these, the attempt to draw the Ethiopian export trade into Italian economic orbit seemed the most promising prospect as it was reasonably thought that land-locked Ethiopia would conveniently agree to divert a substantial part of its overseas trade to the adjoining Italian colonies. However, even during the most ideal condition of Italo-Ethiopian relations, Ethiopian commercial traffic through Italy's colonial frontiers remained negligible. Contributing to this was not only the modest nature of the volume of the Ethiopian trade *per se*, but the ports of Italian colonies experienced strong competition by better equipped and strategically well-placed French-ruled Djibuti in the East and Anglo-Egyptian trading posts in the West. Italian Somaliland's share of Ethiopian foreign trade was remarkable only for its insignificance and the Eritrean route pulled about 10 per cent compared to 20 per cent of Sudan and 70 per cent of Djibuti.[36]

The outcome of Italy's overall plans was abject failure in its objectives but the legacy lingered, to be inherited by Fascist Italy. Fascism added

[32] ASMAE 51/1, Gino Lavelli De Capitani (Consigliere della SCCCE) to Marchese Salvago Raggi, As 6 May 1909.

[33] ASMAE 51/1, MAE to Legazione AA, Rm 27 May 1909.

[34] ASMAE 51/1: SCCCE to MAE, 17 July 1908; MAE to Legazione AA, Rm 27 May 1909; SIOM to MAE, Milan 31 July 1908; Marchese Salvago Raggi to MAE, As 3 Aug. 1908. For the Fascist period see L. Federzoni, *Venti anni di azione coloniale* (Milan, 1926), 63; id., *AO: Il posto al sole* (Bologna, 1936), 153–4, 156–9.

[35] Del Boca, *Dall'unità*, 838–9.

[36] D. Teferra, *Social History and Theoretical Analyses of the Economy of Ethiopia* (New York, 1990), 34.

little new. As a good socialist, Mussolini had condemned the Italian liberal government for its atrocities against colonial peoples and squandering the country's wealth on prestige imperialism, diverting attention from pressing problems at home. But once in power, Fascism, using the same arguments of his liberal predecessors, vigorously campaigned for the necessity for the Empire. The difference was mainly only of style and degree. Decoyed by the sense of its fanciful mission of restoring the power and glory of the ancient Roman Empire, Fascist Italy lacked the cautious diplomacy that characterized liberal Italy in its confrontation with great powers and, after an uncertain start, its policy of the destabilization of Ethiopia became increasingly aggressive. Parallel to this went peaceful diplomacy which reached its climax in August 1928 with the signing of a twenty year Peace and Friendship treaty aimed to trap Ethiopia into becoming an economic protectorate of Italy. But Ethiopia's effort to diversify its foreign contacts convinced Fascism that more forcible measures were necessary.[37] Within Ethiopia subversion aimed at weakening the central power and strengthening the centrifugal forces became tenacious, while in Italy the Fascist propaganda machinery was mobilized to promote collective imperial consciousness and diffuse awareness of a revived imperial Rome among the Italian masses.

Although their policies towards Ethiopia had received a series of setbacks, in their colonies the Italians did not fail to put some of their main colonial agricultural policies into practice. In formulating these policies, be it in Libya, Somalia, or Ethiopia, Eritrea was often the classic point of reference.[38]

## ERITREA AS A MODEL COLONY 'CAPITALIST' FARMING VERSUS 'DEMOGRAPHIC' SETTLEMENT

Later nicknamed 'La Colonia Primogenita' (first-born colony), Eritrea was planned as a model self-sufficient agricultural colony where Italy's imperial aspirations could be realized. It was to serve as a centre of

[37] D. Mack Smith, *Mussolini's Roman Empire* (Harmondsworth, 1979), 59.

[38] Segrè, *Fourth Shore*, 12; R. Hess, *Italian Colonialism in Somalia* (Chicago, 1966); I. Taddia, *L'Eritrea—colonia, 1890–1952* (Milan, 1986); S. F. Nadel, 'Land Tenure on the Eritrean Plateau', *Africa*, 16/1–2 (1946), 1–22, 99–109; R. Pankhurst, 'Italian Settlement Policy in Eritrea and its Repercussions, 1889–1896', in J. Butler (ed.), *Boston University Papers on African History*, i (Boston, Mass., 1964), 121–56.

settlement and source of raw materials. Demographic colonization aimed to direct to the colony the agricultural workforce emigrating overseas whereas capitalist colonization was tied up with the deployment of surplus private capital. The controversy over these two schemes, which later dominated Italy's erratic colonization programme in Libya and Ethiopia, also plagued Eritrea's agricultural development.

*Demographic Colonization*

Demographic settlement is associated with Leopaldo Franchetti who in June 1890 was put in charge of Italian colonization in Eritrea, and subsequently became the head of the Settlement Office, or Ufficio per la Colonizzazione.[39] A parliamentarian born at Livorno in Florence to a wealthy Jewish family, Franchetti seemed to possess the skill and experience that the programme required.[40] He belonged to the school of thought which considered colonies an outlet for a surplus population[41] and wanted to make Eritrea the showpiece of his ideas. He maintained that, in contrast to the desolate port of Mešwa, the Eritrean highlands, with their cool and healthy climate, their altitude ranging from 1,400 to 2,300 m, and their sparse population, offered an ideal place for European settlement.

Franchetti opposed capitalist colonization based on privately financed large estates; this, he believed, would leave most of the country undeveloped and degenerate into land speculation and latifundism—a rural phenomenon with which he was familiar in his studies of southern Italy. His proposed peasant or 'demographic' colonization envisaged small family-sized concessions, supported, if necessary, by the State.[42] In his view, peasant settlements would create a large, stable, and productive rural population that would defend the colony militarily and soon make itself economically self-sufficient. Private initiative, he argued, would confine the Italian presence to the military at great expense to the nation, and

[39] Rainero, *Primi tentativi*, 32, 62, 149–51, 159–62. For a more detailed biography, see the introd. to Franchetti, *Mezzogiorno e Colonie* (Fl, 1950), written by his collaborator and student Umberto Zanotti-Bianco.

[40] He was a philanthropist and expert on the agricultural and social problems of rural south Italy, and co-author with Sidney Sonnino of a celebrated report on the agricultural conditions in Sicily. He also edited, together with Sonnino, the *Rassegna Settimanale*, a journal concerned with social and agricultural problems (cf. Rainero, *Primi tentativi*, 32).

[41] This view he, as an energetic propagandist, expounded in the Chamber of Deputies, in journals, such as the *Nuova Antologia*, a journal of opinion where his own pamphlet 'L'Italia e la sua Colonia Eritrea' first appeared as an article (cf. 'L'Italia e la sua Colonia Africana', *Nuova Antologia*, 1 June 1891) and also reprinted in *Mezzogiorno e Colonie*.

[42] 'L'avvenire della nostra colonia', *Nuova Antologia* (Apr. 1895), 614, 622–3.

reduce the colony to an unproductive desert where 'a little wealth would flourish on public misery'.[43]

But Franchetti's plan met with serious opposition. Crispi himself, unconvinced by the exclusion of private capital, regarded it as a noble but fanciful vision.[44] A Royal Commission of Enquiry, set up in 1891 to investigate into allegations of colonial maladministration and settlement possibilities in Eritrea, strongly disagreed with him.[45] Although the Commission shared Franchetti's conclusion on the colony's potential as 'an outlet for Italy's partial emigration' and eventual financial self-sufficiency, describing it as a 'virgin and fertile land' awaiting Italian labour, it also asserted in no uncertain terms that both private capitalist farming and State-sponsored peasant colonization were complementary and interdependent, and neither could stand or progress without the other.[46]

Notwithstanding the controversy, in the latter part of 1890 Franchetti opened an agricultural experimental station just outside Asmära. No fewer than ninety-six different types of seeds were sown. In the following year two additional stations were established; one in Akhälä Guzay at the Gura'e district and the other at Godofälassi, in Säraye. Satisfactory results in all these cases suggested the practicality of White settlement, even though at later stages the methods he used as well as the results obtained and the procedures in the selection of the area were strongly contested.[47]

Colonization began at the end of 1893 when ten Lombard, Sicilian, and Venetian families settled in stone houses built in advance at Addi-Wugri, a few miles from Godofälassi.[48] Another five families followed shortly afterwards.[49] Seeing 'the promised land', the peasants were impressed.[50]

[43] *Nuova Antologia* (Apr. 1895), 622–3; Rainero, *Primi tentativi*, 72–3, 125.

[44] Rainero, *Primi tentativi*, 58.

[45] Ibid. 85; *Relazione generale della R. Commissione d'Inchiesta sulla colonia Eritrea* (Rm, 1891), 178, 186, 189; Pankhurst, 'Italian Settlement', 127.

[46] Rainero, *Primi tentativi*, 83, 85–6.

[47] [E. Cagnassi], *I nostri errori: Tredici anni in Eritrea* (Turin: 1898), 133–5; about the experiments themselves see B. Melli, *La colonia Eritrea* (Parma, 1899), 52–3, 73–4.

[48] The families consisted of 29 men, 15 women, and 17 children. Of these, 7 families were Lombard (Magentino), 2 Sicilian (Pedara), and 1 Venetian (Frioli). Cf. Rainero, *Primi tentativi*, 123; Del Boca, *Dall'unità*, 517.

[49] Similar efforts at settling the landless peasantry were made by others. One was promoted by the Association for Aid of Catholic Missionaries (Associazione per Soccorrere i Missionari Cattolici) of Senator Alessandro Rossi and the Apostolic Prefecture of Eritrea who in early Jan. 1896 brought in 16 Venetian families, 128 strong. Cf. Del Boca, *Dall' unità*, 615.

[50] Looking at the land, the families cheered and kissed it and sang their country songs. The following year the settlement was officially inaugurated taking the name of Umberto I.

Each family was given 20–5 ha of land, agricultural tools, cattle, and provisions necessary to tide them over until the next crop.[51] The families were expected to repay these loans, estimated at L.4,000, over five years at 3 per cent interest. Provided that they cultivated the land uninterruptedly during these years with their own means, they were promised ownership.[52]

In his report of April 1894 to Parliament, Franchetti asserted that demographic colonization was firmly rooted in highland Eritrean soils and expressed his strong belief that the country would absorb within a few years a large proportion of Italian emigrants.[53] But contrary to official expectations, there was no great wave of immigration. By the end of 1896 a total of thirty-two families were settled. By early 1898 all but one of them had left.[54] When in 1904 a member of the Commission of Agricultural Labourers from Romagna sent to investigate settlement possibilities, ascended the slopes, he indignantly exclaimed 'Romagna shall never come here.'[55]

What went wrong with this attempt to establish an emigrants' paradise in Africa? Life was arduous, wages low, communication difficult, and there was no easy access to market to sell goods. Most settlers were harassed by hailstorms, locusts, and the lingering rinderpest that decimated livestock.[56]

Mistakes in planning and poor selection of farmers compounded the problem. In some cases shortage of supplies was acute. Some of the peasants were factory workers from Milan with no experience in farming, old persons too inflexible to change their lifestyle, and families from rival regions in Italy.

The policy was further frustrated by continuous opposition by successive Eritrean governors. Both Barattieri and his predecessor, General Gandolfi, undermined Franchetti. General Barattieri was explicit in his determination to fight proletarian colonization 'with all available means . . . convinced once and for all that the system cannot be maintained as an

[51] The animals consisted of 8 oxen, 10 chickens, and 1 pig.

[52] Del Boca, *Dall'unità*, 517.

[53] Franchetti seems to have put Italian immigration on an average at about 143,000 persons per year. On the basis of this estimate more than 5,700,000 individuals would have settled in Eritrea in a span of only two generations or over a forty-year period. In both accounts his assessment was flawed. The scale of Italian exodus grew rapidly. During the four-year period between 1887 and 1891, 717,000 Italians settled abroad. Although in the subsequent fifteen years this figure had trebled, the number of settlers in Eritrea remained insignificant. Cf. ibid. 479.

[54] Ibid. 753; Taddia, *Eritrea—colonia*, 218.

[55] F. Coletti, *Dell'immigrazione italiana*, ii (Rm, 1911), 137.

[56] [Cagnassi], *I nostri errori*, 134–6.

official policy at the expense of the State'.[57] His decision to start his own colonization programme within the framework of the Royal Commission of Enquiry's recommendation granting land only to those peasants or capitalists who had adequate capital and technical skill, led in February 1895 to the resignation of Franchetti. Barattieri immediately seized the opportunity to abolish the Office of Colonization, personally assuming charge of the colonization programme. One of the stations was then suppressed. Franchetti vigorously protested to the Ministry of Foreign Affairs against the 'works of disorganization and destruction of his colonization efforts'. He was given a polite rebuff, clearly demonstrating that Rome had torpedoed his plans and opted for Barattieri.[58]

The move marked the end of the myth of Eritrea as a settlement colony and the triumphal assertion of capitalist colonization. With Franchetti's resignation, the state subsidy to attract the colonists was cut off. The prospective settler family was invited to spend about L.2,500 or more with a very uncertain future ahead. On this an anonymous critic commented: 'What farmer in Italy possessing such a sum would like to emigrate? And if he did, would he wish to risk his life and savings in a country where the results of European farming are still unknown? Would he not prefer to go elsewhere?'[59]

But the *coup de grâce* came with the Bahta Hagos uprising. The events leading to it were symptomatic of the basic incompatibility of interests between the colonizers and the colonized. The point at issue was Italy's confused and misguided land policy. Land was appropriated for colonization following the guidelines set out by the Royal Commission which recommended that settlement be directed towards better State lands, demanding that land legislation should be enacted with the ultimate goal being 'to facilitate the colonization and agricultural progress, in other words, the easy transfer of lands into the hands of the Italians, and, among the Italians, of those who could best cultivate it'.[60] This policy aimed at a gradual erosion of the traditional land system to serve the settlers' interest. As a result, in order to 'remove a great obstacle to colonization', private property among indigenes was encouraged at the expense of prevailing communal or village ownership, as the latter hindered the sale of long-term concessions because of the need to secure unanimous consensus of the community.[61]

[57] Rainero, *Primi tentativi*, 127. [58] Ibid. 232.

[59] Pankhurst, 'Italian Settlement', 150. [60] *Relazione generale*, 178–86.

[61] A. Omodeo, *et al.*, *La colonia Eritrea: Condizioni e problemi* (Rm, 1913), 16; Pankhurst, 'Italian Settlement', 131.

The Commission's recommendations on land policy did not reflect Eritrean customary law nor the Italian civil code. It stressed that existing land laws and rights tended to be 'incompatible with a rapid increase of agricultural and scientific colonization' and should therefore be dispensed with. It rejected the doctrinaire dogmatism that respected 'all local laws and customs relating to land' as much as it did the rigid application of the Italian civil code. For the creation of State lands arguments were concocted on views based on extensive numbers of examples from, and incorrect interpretations of, Ethiopian traditional law. The policy rested on three major assumptions: first, that ownership of land in Ethiopia was traditionally vested in the sovereign who could allocate it or appropriate at will; secondly, that the State lands were not owned by their cultivators and those abandoning their lands forfeited to the State all rights of tenure; thirdly, that the needs of the Italian immigrants and the progress of agriculture being paramount, this warranted the placing of vast estates at their disposal.[62]

These assumptions were an outcome of gross misinterpretation of traditional legal systems and a self-serving distortion of facts. Undue emphasis on the sovereign's right failed to balance equally important claims by actual landowners.[63] State lands were devised to supply the sovereign and its entourage with certain necessities but they were inhabited by the tillers of the soil who remained on the land through the succession of lords, and for generations had ensured its productivity. The Commission deliberately failed to see the circumstances in which lands were abandoned. Natural and political disasters had often been the cause for population dislocation. Land seemed plentiful only because the Eritrean highlanders had temporarily abandoned the area after a series of local wars, and the great famine and epidemics (1888–92). To the Commission such areas constituted abandoned lands irrespective of whether the owners might still be alive. But once the situation improved, the villages were reoccupied and the Eritreans resumed their cultivation, clashing with the colonists.[64]

[62] *Relazione generale*, 156; Pankhurst, 'Italian Settlement', 128–9.

[63] Land tenure in Eritrean highlands was governed by principles identical to those paramount in the ancient Abyssinia which is discussed in Ch. 2.

[64] T. Bent, *The Sacred City of the Ethiopians*, (London, 1896), 11–12. This was the conclusion reached by Società Italiana per il Progresso delle Scienze, a study mission, which looked into the Eritrean highland uprising in 1913. Cf. Omodeo *et al.*, *Eritrea*, 16, 17 n., 47–9. The same point was emphasized by Agnese, the Director of Colonial Affairs in the MAE, at a congress organized by Istituto Coloniale Italiano in 1911. (Cf. Istituto Coloniale Italiano, *Atti del II Congresso degli Italiani all 'Estero*, ii 1 (1911): 482–3.)

Franchetti was alarmed by what he called 'disorderly extension of native cultivations' which threatened the success of his initiative and, in his report of 1893–4, demanded Crispi's 'swift, economical and efficacious action' to stop the arbitrary occupation by 'natives' of 'lands reserved by the government for Italian colonization'.[65] Crispi acted accordingly and ordered General Barattieri 'to watch out that native colonization should not bar the way for our domestic colonization nor that exaggerated scruples deter us from transforming into public domain lands which are effectively abandoned, *res nullius*, to the mercy of the first occupant.'[66]

As a result, the expropriation of prime agricultural lands forged ahead. Out of Eritrea's estimated total productive land of 648,938 ha, by the end of 1893 19,020 ha were made State domain. In 1894 alone Barattieri added 280,039 ha. By the end of 1895 the total figure had risen to 412,892 ha—almost two-thirds of the productive land.

Expropriation was carried out amid mounting discontent and simmering rebellion.[67] It was resented particularly by the Christians, and in December 1894 led to the revolt of a catholic convert and most trusted Italian agent, Däjach Bahta Hagos, the Governor of Akhälä Guzay.[68] The insurrection spilt over into Ethiopia, culminating a year later in the Italo-Ethiopian war in which the Italians suffered a humiliating defeat at the battle of Adwa in March 1896.[69] The victory caused havoc amongst the settlers and 'nearly all the Italians immediately ran away to the coast'. From there all except one repatriated to Italy.[70]

*Per se* the Bahta Hagos's revolt was an insignificant localized episode in the otherwise sporadic highland Eritrean resistance to Italian rule. It had none of the elements of a mass movement attributed to it by some.[71] It was swiftly crushed and easily contained, and the cause of revolt came into clear perspective only at a later stage.[72] What gave significance to the

[65] AP CD, *Appendice alla relazione annuale sulla colonia Eritrea, 28-4-1894* (Rm, 1894), 17.

[66] *La Prima Guerra d'Africa* (Milan, 1914), 273.

[67] Rainero, *Primi tentativi*, 197–202.

[68] His epigram 'from the bite of the black snake one recovers, but the bite of the white snake is fatal', which he used as a rallying ground, epitomized the deep ill-feeling towards the settlers. (Cf. Tekeste Negash, *No Medicine for the Bite of a White Snake* (Uppsala, 1986), 37–45.)

[69] Ibid. 43–4; id., *Italian Colonialism in Eritrea, 1882–1941* (Uppsala, 1987), 33–5; H. Marcus, *The Life and Times of Menelik II: Ethiopia 1844–1913* (Oxford, 1974), 154–8; G. F. H. Berkley, *The Campaign of Adowa and the Rise of Menelik* (London, 1935), 62–4.

[70] Segrè, *Fourth Shore*, 15.

[71] Cf. Tekeste Negash, *No Medecine*, 43–5; id., *Italian Colonialism*, 124–5.

[72] Although few allusions were made to the land alienation as the cause of the rebellion, until the study of Alberto Pollera the motives were not clearly explained. Cf. E. Ardemani,

revolt was its Ethiopian dimension. Not only it did take place at a time when Italo-Ethiopian relations were at their lowest ebb but it also helped to precipitate the Adwa débâcle. Adwa, however, was not fought over Italy's misguided land policy in Eritrea but over Ethiopia's fate as an independent Black African state. But its results had an immediate and long-lasting effect not only on Italian policy in Eritrea but also elsewhere: Adwa remained a constant reminder of Italy's crushing defeat by a Black African power, and at later dates the Bahta Hagos's revolt would be evoked by dissenting voices to counter both settlers' voracious appetite for land and adventurous state land policies.

The civilian administration that replaced the military regime in the immediate aftermath of Adwa rejected the idea of Eritrea as a settlement colony and abrogated most of the decrees relating to public domains.[73] Most of the land appropriated by the government for demographic settlement was restored to its original owners.[74]

## *Capitalist Farming*

The resignation of Franchetti marked the triumphal assertion of *colonizzazione libera*—a capitalist form of colonization based on market forces and involving only those who were technically and financially fit. The decree of April 1895 excluded from settlement those who did not possess the start-up capital estimated at L.2,500–3,500.[75]

Following the three traditional Ethiopian agricultural classifications of land—the lowlands or *qolla* (below 1,400 m), the midlands or *wäynä däga* (1,400–2,400 m), and highlands or *däga* (2,400–4,600 m)[76]—the Italians promoted a particular form of colonization. The trend reflected general Italian agricultural policy and had its justification in climate and demography. The densely populated highlands—both *däga* and *wäynä däga*—had a healthy climate and a complex and well advanced agricultural system. They were reserved for small-scale and medium-sized farms devoted in *wäynä däga* to cereal farming, and in *däga* to cattle-breeding

*Tre pagine gloriose nella storia militare-civile-religiosa della colonia Eritrea* (Rm, 1901), 90; Pollera, *Il regime della proprietà terriera in Etiopia e nella colonia Eritrea* (Rm, 1913), 67.

[73] The Land Act of 1906 (RD 31-1-1909, no. 378) declared that the existing land rights of the Eritreans would be respected. That of 1926 (RD 7-2-1926, no. 269), excluded any concession of land to the settlers in the highland areas.

[74] Pankhurst 'Italian Settlement', 153; Del Boca, *Dall'unità*, 837.

[75] Rainero, *Primi tentativi*, 216–19.

[76] The three climatic zones reflected different type of flora grown in them and approximately corresponded to the Mexican classification of *tierra fria*, *tierra templada*, *tierra caliente*.

and pasture. Small-scale farms of 20–30 ha were given to those who had a start-up capital of L.50 per ha. The concessionaire was expected to set up his residence inside the farm. Medium-sized farms were 30–300 ha and were granted to the farmers with a minimum of L.100 per ha. The scarcely populated, agriculturally less advanced, and reputedly unhealthy lowlands were the realm of large concessions which, using Eritrean labour and Italian technical expertise and capital, engaged in cash-crop farming. The farms consisted of between 300 ha and 10,000 ha.[77]

All three types of concession were granted on a thirty-year lease which, with the exception of large-scale farms, could eventually be turned into ownership. The key feature of the concessions was their mutual interdependence. According to the Italian plan, within Eritrea, the highland concessions had, as their priority, to first feed the settler population and then export the surplus; the lowland concessionaires, whose target was cash-crop for export, were to rely on the highlands for their subsistence. But the revenue from their export was expected to finance the expansion of highland concessions. In the wider context of the metropolis, the highland concessions had as their ultimate objective the absorption of metropolitan emigration and the lowlands its surplus capital.[78]

To attract the capital, the State offered cheap land and tax advantages. In return, the concessionaire had to develop the land or face revocation. The model of development required the use of advanced technology. The concessionaire had to man the farms personally and no renting or sharecropping was permitted. By setting these conditions, the colonial authorities aimed to retain control over the land and discourage speculators and colonists without capital or suitable agricultural experience.[79]

Initially the government remained committed to the notion of a populating colonization and laid great emphasis on the small and medium investor in the highlands. But, contrary to the government's expectations, the concessions proved to be uneconomical and the areas politically sensitive. The majority of concessionaires possessed neither the will, skills, nor capital required, and, with a few exceptions, the technology they used was indistinguishable from that of the local farmers. Moreover, not only were they completely ignorant of local conditions and potential for agriculture but few of them had much agricultural background or practical experience. The difference was that settlers monopolized the best lands and benefited from State aid. Yet most of the concessionaires found leasing their concessions to the Eritreans, i.e. the same people whose

[77] Taddia, *Eritrea—colonia*, 221–2. [78] Ibid. 231.
[79] Ibid. 222; Rainero, *Primi tentativi*, 216–19.

lands were expropriated to be given free to the settlers, much more lucrative than working them personally. Even though the government intervened vigorously in 1907 by repatriating most of the absentee landlords, this practice persisted, with virtually all the farmers leasing their concessions to the Eritrean peasants for a share of between a third and a quarter of the crop.[80]

The poor productivity, particularly of the small concessions, and the sad realization that Eritrea no longer constituted the emigrants' haven, led in 1926 to a major reassessment and subsequent policy shift towards medium-sized and large-scale farming.[81] At the same time, however dimly, the government became aware of the need of modernizing traditional Eritrean agriculture. Until then any such attempts were opposed by the concessionaires and the authorities who feared competition for land and markets. In the 1930s Eritreans were eligible for government aid. But this initiative remained no more than a timid gesture, involving a few specific cases. To provide substantial aid to the peasantry would mean allowing them to produce more, threatening the colonists continually. Wary of any possible disruption both in production and the political situation, the metropolitan government maintained a policy of minimum interference in Eritrean economic and social organization.[82]

The policy shift away from the small farmer to medium and large farms failed to attract rich investors. Settler concessions remained static from 1907 to 1931 when economic depression left a number of large-scale farms bankrupt (see Table 1). With their concessions starved of capital and themselves lacking entrepreneurship, the concessionaries were ready to give up farming whenever a better opportunity presented itself. Thus during the Italo-Ethiopian war many concessions were abandoned as their owners found the war economy much more lucrative; others joined the government administration which was badly in need of labour.[83]

One of the discouraging factors was the harsh fluctuating Eritrean

[80] Taddia, *Eritrea—colonia*, 237; R. Salis Serotolis, *L'ordinamento fondiario Eritreo* (Padua, 1932), 91.

[81] CD IAO 1965, UA Er, Attività agricola in Eritrea dal 1923 al 1931; CD IAO 2004, A. Maugini, Avvaloramento agrario dell' Er e della Sm, 1940; A. Maugini, 'La valorizzazione agricola della colonia Eritrea', *REC* 19/3–4 (Mar.–Apr. 1931), 365–79; I. Baldrati, 'Lo sviluppo dell' agricoltura in Er nei cinquanta anni di occupazione italiana', *REC* 7/1 (1933), 43–53.

[82] The Eritreans sold their labour for low wages to supplement their income and pay their taxes which was a major source of government revenue. Cf. Tekeste Negash, *No Medecine*, 22–36. However, the number of Eritreans employed in the agricultural sector was modest. In 1937 of the 1,911 total workforce employed in 124 farms, there were 20 Italians and 1,891 Eritreans—1,119 permanent and 772 casual; cf. Taddia, *Eritrea—colonia*, 252.

[83] Taddia, *Eritrea—colonia*, 235, 238; Segrè, *Fourth Store*, 15.

TABLE 1. Land concession areas in Eritrea (ha)

| Type of tenure | Zone | 1907 | 1930 | 1932 | 1939 |
|---|---|---|---|---|---|
| Concession | Highland | 6,483 | 1,251 | 1,251 | 329 |
| | Lowland | 4,569 | 4,159 | 359 | 1,287 |
| Ownership | Highland | — | 4,654 | 4,654 | 4,150 |
| | Lowland | — | 67 | 67 | 282 |
| TOTAL | | 11,052 | 10,131 | 6,331 | 6,048 |

*Source:* Taddia, *Eritrea—colonia*, 236, 237.

climate.[84] In freak years the farmers had to contract unbearable debts against which the government offered no protection. From 1927 to 1932 locusts destroyed the harvest for five consecutive years. In 1927 wheat production slumped from 60,000 ql to 5,000 ql within one year, and other cereals were almost totally destroyed.[85]

The vagary of climate and dearth of capital were partly responsible for lack of advocacy for a monoculture, one of the most remarkable features of Eritrean settler agriculture. The Italians had prided themselves on quoting this as proof that Italian colonialism was not exploitative and that their colonial policies were geared towards the development of a colony's internal economic needs. But the truth was that monoculture demanded greater financial commitment at the initial stages and a stable labour market, and Eritrea offered neither. With a meagre state subsidy, large-scale production was unaffordable. Labour was volatile and the concessionaires often had to make recourse to overseas markets.[86]

After the Adwa defeat, the government, perhaps wary of any possible political backlash, adopted the posture of hanging on rather than really developing the colony. Out of its skeleton budget, expenditure in agriculture amounted to less than one per cent, while defence and administration swallowed the rest.[87]

[84] A. Maugini, *Flora ed economia agraria degli indigeni* (Rm, 1931), 116–17.

[85] I. Baldrati, *Mostra delle attività economiche della colonia Eritrea* (As, 1932), 15.

[86] Del Boca, *Gli Italiani in AO, ii. La conquista dell'Impero* (Bari, 1979), 32; Taddia, *Eritrea—colonia*, 312, 321.

[87] See Table 2. This rate should be compared with that of Libya where government subsidy amounted to between 12 and 26%. Cf. P. Lombardi, 'La colonizzazione agraria in Libia durante il periodo del fascismo' paper presented to 'Seminario sulla Libia: Storia e Revoluzione', Rm 27–9 Jan. 1981, cited by Taddia, *Eritrea—colonia*, 307.

TABLE 2. Italian colonies' nine year budget, 1921/2–1929/30 (L. m)

| Colony | Revenue | | Expenditure | |
|---|---|---|---|---|
| | Internal | State subsidy | Administration | Defence |
| Libya | 1,087 | 3,189 | 1,753 | 3,380 |
| Somalia | 319 | 439 | 1,034 | 168 |
| Eritrea | 261 | 246 | 782 | 155 |

*Source:* Taddia, *Eritrea—colonia*, 306.

However, within the constraints imposed upon it, the colonial administration financed and managed developmental infrastructures that it considered to be conducive to agricultural development. Perhaps the most significant was the building of a rail and road network which could be used as much for the rapid deployment of the colonial army as for the exploitation of the peripheral areas. Between 1900 and 1930 up to L.100,000,000 had been invested in this sector.

Another important undertaking was the exclusion of non-Italian capital. This was one of the unique features of Italian colonialism. But this protectionist policy left the colony hostage to an indifferent and weak metropolitan capital, thus hindering the exploitation of agricultural resources. Whenever it took place, private investment was erratic and timid, with an eye more to milking quick profits and government concessions than on risk-taking and the burden of investment that would allow long-term rational exploitation. Unlike other Italian colonies, particularly Libya, Eritrea had no agricultural credit facilities. Government intervention in the lowland concessions was sporadic, while in the highlands it was conspicuous only by its absence. In the 1930s steps were taken to set up credit institutions but they were no more than token gestures, either too narrowly restricted to particular groups and crops or totally inadequate.[88]

One exception was Täsänäy, near the Sudanese border, where in 1923 the State intervened with massive investment on a cotton farm pioneered by SCCCE, which had fallen into bankruptcy after ten difficult years. A

[88] The assistance was predominantly given to cultivators of coffee on the eastern slope (*pendici orientali*) and cotton. For a more detailed discussion on credit institutions and their limitations see Taddia, *Eritrea—colonia*, 284–90.

dam was built on Gash River with a potential to irrigate up to 16,000 ha.[89] After a vain attempt to involve Italian capital, the government had little option but to run it as a State farm. Named Azienda Agricola Statale di Tessenei, it was predominantly run using a sharecropping system with Africans, similar to the Gezira cotton scheme.[90] The State provided small irrigated plots of between 1 and 4 ha, ploughed mechanically and ready for sowing. It distributed seeds and cash advances at three fixed intervals—sowing, weeding, and harvest. It provided health care and assisted in the building of dwellings. The company insisted that the peasant raise food crops on half of his land and a cash crop of cotton on the other half. The peasant did all the manual work from sowing to harvest until the cotton was deposited in the company's store. In exchange he received a daily ration for his maintenance. At harvest, cereal crops were partitioned equally between the company and the peasant, while the peasant's share of cotton was purchased by the farm.[91] Described by some as 'a cathedral in the desert', the enterprise was a disappointment and exposed the failure of state entrepreneurship in Eritrea.[92] Out of a planned 16,000 ha, the State was able to develop less than 3,000 ha with a production fluctuating between 3 and 10 ql per ha and gradually diminishing. Rather than an economic miracle, Täsänäy's name remained associated in the

[89] The initiative is associated with Governor Gasparini, who, like De Vecchi and Luigi di Savoia, then engaged in Somalia in building two similar projects—Azienda Agraria Governative di Genale (Genale State Farm) and Villagio Duca d'Abruzzi (Villabruzzi)—wanted to leave his personal imprint. Like these two schemes, Täsänäy Farm was motivated more by the Fascists' quest for prestige and glory than economic considerations. On the two schemes cf. G. Scassellati-Sforzolini, *La SAIS in Somalia* (Fl, 1926); Del Boca: *Conquista*, 78–87; *Gli Italiani in AO: La Caduta dell'Impero*, iii (Bari, 1982), 214–16; Hess, Italian Colonialism, 163–6.

[90] For Gezira scheme see A. Gaitskell, *Gezira: A Story of Development in the Sudan* (London, 1959); A. W. Abdel Rahim, 'An Economic History of the Gezira Scheme, 1900–1956', Ph.D. (Manchester, 1968); T. Barnet, 'The Gezira Scheme: Production of Cotton and the Reproduction of Underdevelopment', in I. Oxaal *et al.* (eds.), *Beyond the Sociology of Development* (London, 1975), 183–207.

[91] A similar method was followed in Somalia. Cf. Hess, *Italian Colonialism*, 164–5.

[92] See Table 3. This is how one visitor describes it: 'In Italy there is much talk of the Täsänäy scheme. Magnificent photographs show us dams and canals: no doubt these are wonderful works even from a panoramic point of view. But, after seeing the pictures, one has to look at the balance book. By 1930 Täsänäy Farm will need another L.3,000,000 to repair the damages incurred during these twenty months. So, to cultivate 3,000 ha of land a total of between L.35,000,000–36,000,000 is needed', R. Martinelli, *Sud. Rapporto di un viaggio in Eritrea e in Etiopia* (Fl, 1930), 256–7. The damages were caused by flooding of the river between 1927–8. This colossal figure was three times more than the colony's ordinary revenue. The concession went from one crisis to another. In 1926 and 1928 an attack by *Helio Thrips Indicus* halved the crop. Then at the end of 1928 the world cotton price slump left it badly debt-ridden. Cf. Del Boca, *Conquista*, 34; Taddia, *Eritrea—colonia*, 328.

TABLE 3. Täsänäy cotton farm: areas under cultivation

| Year | Category of Farm | Farmed area (ha) | Crop (ql) |
|---|---|---|---|
| 1925–6 | State | 700 | * |
| 1926–7 | State | 1,500 | 3,000 |
| 1927–8 | State | 1,000 | 6,000 |
| 1928–9 | State | 1,200 | 3,500 |
| 1929–30 | State | 1,600 | 7,433 |
| 1931–2 | State | * | * |
| 1932–3 | Private | 1,230 | * |
| 1933–4 | Private | 1,480 | * |
| 1934–5 | Private | 2,147 | * |
| 1935–6 | Private | * | 3,360 |
| 1936–7 | Private | * | 3,185 |
| 1937–8 | Private | * | 2,801 |
| 1938–9 | Private | * | 4,225 |

*Note:* * indicates that no figures are available.

*Source:* MAI, 'La valorizzazione agraria e colonizzazione', *AAI* 3 (1939), 234–5; 'Notiziario agricolo commerciale', *AC* 34/11 (1940), 482; Taddia, *Eritrea—colonia*, 323, 327.

popular mind with the death toll by malaria that decimated the labour force.[93]

As an outlet for emigration, the colony was a disappointment. Between 1890 and 1905 an estimated total of over 3,500,000 Italians emigrated.[94] And yet the Italian population in Eritrea totalled only 3,949. In 1931 it had increased to 4,182 and to over 46,000 at the end of 1939.[95] In 1913 out of 1,617 adult male Italians in the colony, only 62 classified themselves as agriculturalists, as against 834 in the military, 349 in industry, and 219 in commerce. In 1939 the farming population totalled 112, of which 98 were owner farmers. With the slow decline of the area under cultivation, the dream of Eritrea as a small farmers' paradise also evaporated. In 1907

[93] G. Marescalchi, *Eritrea* (Milan, 1935), 124.

[94] Del Boca, *Dall'unità*, 479; Segrè almost trebles this figure (*Fourth Shore*, 14).

[95] See Table 4; *VII Censimento generale della popolazione*, 21-4-1931, Rm 1935, p. 34; Pankhurst, 'Italian Settlement', 135–6.

TABLE 4. Eritrea as a demographic outlet, 1940

| Place | Male | Female | Total |
|---|---|---|---|
| Addi Kuala | 121 | 29 | 150 |
| Addi Qäyeh | 59 | 22 | 81 |
| Addi Wugri | 499 | 104 | 603 |
| Aquirdät | 175 | 25 | 200 |
| Asmära | — | — | 30,000 |
| Asäb | 1,492 | 89 | 1,581 |
| Däqämeharä | 5,439 | 905 | 6,344 |
| Ghindae | 122 | 33 | 155 |
| Käran | 400 | 202 | 602 |
| Mešwa | 5,633 | 304 | 5,937 |
| Sägänäyeti | 598 | 107 | 705 |
| TOTAL | 14,538 | 1,820 | 46,358 |

*Source:* 'Il popolo italiano ha creato col suo sangue l'Impero e lo feconda col suo lavoro', *Popolo Fascita* (Salerno), 25 Feb. 1939; 'Quanti connazionali vivono nell'Impero', *GI*, 25 Feb. 1939.

29,553 ha were set aside for colonization. By 1913 the total area under cultivation consisted of 1,146 ha, increasing to 6,048 ha in 1939.[96]

Agricultural self-sufficiency also remained untenable. Eritrea was only a model colony in supplying colonial soldiers to aid Italy's imperial expansion and consolidation in Somalia, Libya, and Ethiopia.[97] Although the

[96] Omodeo *et al.*, *Colonia Eritrea*; Pankhurst, 'Italian Settlement', 155; Taddia, *Eritrea—colonia* 237. Other Italian colonies fared better. In 1939 in Somalia, where only capitalist colonization was practised, out of 64,936 ha of land under concession, 33,604 were cultivated. According to 1933 statistics, there were 115 concessionaires. Of these, SAIS owned 25,000 ha at Villabruzzi, the 100 concessionaires of Genale State Farm had 20,142 ha, and fourteen private farms 15,483 ha. SAIS had developed 10,000 ha, reserving the rest for pasture; the Genale State Farm 18,000 ha; and the fourteen farmers about 10% of the total area. Cf. Hess, *Italian Colonialism*, 163–6. In Libya, where both settlement schemes and capitalist enterprise operated, in 1937 there were 840 farms, 384 still in concession and 411 owned outright. Out of 364,723 ha of land under concession, in 1940 222,386 ha were cultivated. During this same period 8,782 agricultural families were settled, numbering 29,876 individuals. Cf. E. Massi, 'Economia dell'AI', *RISS* 48/11 (May 1940), 429–33; 'Notiziario agricolo Commerciale: AOI', *AC* 32/9 (1938) 426. Segrè, *Fourth Shore*, 97–100.

[97] Eritrea assumed this role in 1912, which reached its climax with the invasion of Ethiopia towards which the colony supplied over 60,000 soldiers. The invasion drastically

illusion of transforming Eritrea into the White man's paradise lingered on even after the conquest of Ethiopia,[98] the colony became neither the privileged field of financial magnates nor the ideal destination for a prospective Italian emigrant.

Contrary to early optimistic forecasts, settlement possibilities in Italy's African colonies remained patently untenable. And contrary to expectations, and despite a considerable expenditure of finance and effort, the colonies were more of a liability than an economic asset to the mother country. Yet such a history of continuous failure did not cause Italy to alter its colonial aspirations, nor was it deflected from targeting Ethiopia as a place where such aspirations could be fulfilled. To achieve this, Ethiopia had to be peacefully subdued. This had become a remote possibility, particularly under Haile Sellassie. The policies of subversion which intensified from 1929 and peaceful economic penetration proved unworkable. Refusing to acknowledge that the era of colonial conquests had passed, Italy embarked, under the slogan of 'Expand or Explode', on conquering Ethiopia.[99]

altered the demographic map of Eritrea, for the needs of the war opened up unprecedented opportunities. With large-scale immigration, the Italian population grew fast in order to take advantage of this transitory and heightened economic activity (see Table 4).

[98] The latest attempts at colonization by Governor De Feo were opposed by Graziani. Cf. ACS GRA 43/34/22: Graziani to Lessona, AA 2 Nov. 1937; Lessona to Graziani, Rm 4 Nov. 1937.

[99] R. De Felice, *Mussolini il Duce: Gli anni del consenso, 1926–1936* (Turin, 1974), 374; 'Fame coloniale', *Echi e Commenti*, 25 Jan. 1931.

# 2

# *Ethiopian Land Tenure and its Development before 1935*

The main justification for Italy's ambition to dominate Ethiopia rested on the proposition that traditional Ethiopian society was incapable of achieving economic progress by itself. Full-scale reform was necessary and this could only take place if Italy assumed responsibility for the transformation in order to make its untapped natural wealth available to the general benefit of mankind. Severe criticism was directed against what was known as the *gäbbar* system. Interpreted as a mechanism by which the ruling class appropriated the surplus of the peasant's produce and labour, the *gäbbar* system had always been an emotive issue. Both progressive Ethiopian opinion and foreign writers had alleged an appalling exploitation through the *gäbbar* by parasitic landlords and church officials, providing Italy with material to exploit.[1]

The *gäbbar* must be looked at from the perspective of the country's land tenure system. Politically, Ethiopia may be a nation. But from a sociological, ethnicocultural, and geographical point of view it is a conglomerate of societies producing highly differentiated forms of ownership and utilization of land. Land tenure not only varies from one ethnic group to another, but within each group, from place to place.

A broad distinction can be made between the land tenure systems in the North and those in the South. The North is the heartland of ancient Abyssinia which, with the help of European military hardware, imposed its political model on the southern areas.

[1] ASMAE 2/2/1, Paternò to MAE, AA 20 June 1931 where he describes the Ethiopians as 'inherently corrupt, decadent, and reactionary, steeped in slavery and feudalism'; G. C. Baravelli [M. Missiroli], *The Last Stronghold of Slavery: What Abyssinia Is* (Rm, 1935), 41 where he reports the statement of the wife of the British ex-Minister for Foreign Affairs, Lady Simon, in support of the need of foreign intervention to establish a new state in Ethiopia.

## NORTHERN LAND TENURE PATTERNS

Ancient Abyssinia refers to the geographical regions comprising the north and north central highlands of modern Ethiopia: the Eritrean plateau; Bägémder; Gojjam; parts of Tegray, Shäwa, and Wällo. Ethnically it is inhabited by two major groups, the Amharas and the Tegréans, who belonged to the EOC and, since the establishment of the modern empire state by Emperor Menilek II (1889–1913) in the last quarter of the nineteenth century, remained economically and politically dominant. Land provided the social and economic basis for public administration and national defence. State employees, both civilian and military, were granted land in lieu of salary. Socially, land was the most valued asset, conferring status, endowing membership of social organizations, and giving access to political office. An Amharic proverb says: 'To be landless is to be sub-human.'[2] All those who did not own land, such as merchants, artisans, and religious minorities were excluded from political life. Concern for land possession and rights were jealously defended and backed by physical force.[3]

The variety of functions and services attached to land gave rise to an elaborate system of land tenure and land tribute deeply rooted in tradition and sanctioned by biblical myth.[4] There were two basic forms of land ownership, based on two fundamental principles, distinct in theory but often overlapping in concrete situations. The first principle was of eminent domain under which the State owned all the land and distributed it as it pleased. This generated *gult* tenures.[5] Coexisting with the dominion

[2] Cf. Mesfin Wolde Mariam, in the introd. to 'Some Aspects of Land Ownership in Ethiopia', paper presented to Seminar of Ethiopian Studies, HSIU, 1965; id., *Rural Vulnerability to Famine in Ethiopia, 1958–1977* (London, 1986), 76–7.

[3] This discussion is based on a vast literature indicated in the Biblio. The major studies are: Gebre-Weld Ingida Werq, 'Ethiopia's Traditional System of Land Tenure and Taxation', *EO* 5/4 (1962): 302–9; ZN; Mahteme Sellassie Wolde Mascal, 'The Land System of Ethiopia', *EO* 1/9 (Oct. 1957), 283–301; A. Hoben, *Land Tenure among the Amhara of Ethiopia: The Dynamics of Cognatic Descent* (Chicago, 1973); F. Bauer, 'Land, Leadership and Legitimacy among the Inderta Tigray of Ethiopia', Ph.D. thesis (University of Rochester, 1972); Nadel, 'Land Tenure'; M. M. Moreno, 'Il regime terriero abissino nel GS', *REAI* 25/10 (Oct. 1937): 1496–1508; E. Brotto, *Il regime delle terre nel Governo del Harar* (AA, 1939).

[4] R. Pankhurst, *State and Land in Ethiopian Society* (AA, 1966), 55; Addis Hiwet, *Ethiopia: From Autocracy to Revolution* (London, 1975), A. D'Abbadie, *Douze ans de séjour dans la Haute-Éthiopie (Abyssinie)* (Paris, 1868), 101, 130; Hoben, *Land Tenure*, 82–5; M. Perham, *The Government of Ethiopia* (London, 1969), 281.

[5] The technical language used refers mainly to Amharic unless stated otherwise. The etymology does not necessarily correspond to Tegreña.

right was the principle of first occupancy which gave rise to *rest* tenures. From these tenures a variety of landholdings evolved, broadly classified as church, private, government, and kinship or communal.

Under *rest* tenure, the 'land of a parent was divided equally among all his or her biological children without regard to seniority or sex'.[6] In addition to actual ownership of land, it also gave the right to claim a share of land by tracing kinship through either parent to a recognized and often fictitious original occupant or founding father.[7] With few exceptions, descent was ambilineal and traceable over a number of generations, allowing one to obtain plots scattered over a wide area. The success of a claim hinged upon a number of factors, such as the claimant's political skill and ability to influence the elders in charge of allocation and the local judges, his social standing in the community, his financial capability to prolong litigation, the availability of land on the area of his claim, and the importance of the claimant.[8]

Technically, *rest* referred not to right of ownership but usage—*ius utendi et abutendi*—within the restrictions imposed by the collective interest of the kinship group in whom was vested the reversionary right.[9] *Rest* could be enjoyed for life, leased, and at death passed on to siblings. But it could not be sold, mortgaged, or forfeited through absence and the dictum 'rest bäshi amätu läbaläbétu'[10] seems to reinforce this common belief. *Rest* lands were subordinate to the State's paramountcy which, under the principle of eminent domain, could confiscate in case of rebellion or tax evasion or inherit in the absence of a successor. However, such a principle was rarely invoked.[11] *Rest* made chances for social mobility greater and assured

[6] A. Hoben, The Role of Ambilineal Descent Groups in Gojjam Amhara Social Organization', Ph.D. thesis (UC, Berkeley, Calif., 1963), 43.

[7] In Tegreña the point of reference is *enda*, a group of families claiming to have descended from a single ancestor. The most important element in the claim is biological descent from either parent. Thus, children born out of wedlock (*deqala*) suffer no disability and have equal claims.

[8] J. Markakis, 'Review of *Land tenure among the Amhara of Ethiopia: The Dynamics of Cognatic descent* by Allan Hoben', *JMAH* 12/2 (1974): 341–2.

[9] P. Schwab, 'Rebellion in Ethiopia', *East Africa Journal*, 6/11 (Nov. 1969), 29.

[10] '*Rest* returns to its owner even after a thousand years.'

[11] Indeed people talk of *rest* as a 'fundamental' and 'sacred' possession and the proverb—'as land is fixed by nature and cannot be moved, equally the *rest*-holder cannot be moved from *rest*'—clearly denotes that *rest* provided almost absolute security to its holder. This is contrary to the absolute power of the State, reported in western travellers' account, such as A. Lobo's: 'The King's authority is so unlimited that no man can in this country be called with Justice Proprietor of anything, nor doth any man when he Sows his Field know that he shall Reap it' (*A Voyage to Abyssinia*, trans. Mr Le Grand (1735), 263) and more recently the Duchesne-Fournet mission according to whom the 'authority of the Emperor is unlimited . . . The people are his and his empire is his property' (J. Duchesne-Fournet, *Mission*

security of tenure. As it allowed access to land to all freeborn Abyssinians, it minimized landlessness and tenancy.

Where it existed in the North, tenancy was associated only with either submerged caste groups, religious minorities—commonly considered foreigners—or young *rest*-holders, known as *restäña*, who sought more land.[12] The status of the landlord and the tenant were not necessarily equal, but the tenancy was seen as a joint venture by two or more people who combined their resources of land, oxen, labour, and seed for mutual advantage. Each shared the output in proportion to their input. The contract was entered by a landlord who had more land than he could work or who, for some reason (old age, ill-health, lack of plough-oxen), was incapable of working. If the tenant was a 'foreigner', the landlord had to make sure that no one among his relations or fellow villagers was interested in the work.

Among the considerable number of tenancy systems, the most widespread were *erbo arash*, *sisso arash*, and *ekkul arash*, translated respectively as 'tiller of one-fourth', 'tiller of one-third', 'tiller of one-half'. The form adopted was largely determined by the degree of fertility of the land and the quantity of labour required. For this purpose, the land was usually divided into four categories: *mäṫefo märét*, *mäna märét*, *dähena märét*, and *maläfiya märét*—'arid land', 'semi-arid land', 'semi-fertile land', 'fertile land'.[13]

*en Éthiopie 1901–1903*, i (Paris, 1908–9), 261). This view seems to derive from a mistaken or literary interpretation of popular proverb 'the land belongs to the king'. As a group of Eritrean elders expressed, rather than 'a juridical concept' the statement intends to 'affirm that the earth belongs to the king in the same way as the heavens belong to God. We allude to this statement when we wish to enhance the power of the State, but do not thereby intend to refer to the ownership of the fields. [It] . . . refers only to that kind of command relating to the imposition of taxation on land and prevention of power abuse and violence. But no one can take away our lands; the State awards rank, office, and *gulti*, and can take them away, but it cannot deprive us of our lands except in case of confiscation resulting from such crimes as we may commit.' (C. Conti Rossini, *Principii di diritto consuetudinario dell'Eritrea* (Rm, 1916), 115–16.)

[12] Abyssinian culture equated craftsmen with outcasts and manual labour done for others with low status. As the Tegreña proverb says—'aslamay addi [awdi] yäbellu, sämay andi yäbellu' (as the sky has no pillars, the Muslim has no land [threshing-floor])—the Muslims were the largest religious group excluded from *rest* ownership in the Christian areas. Emperor Yohannes IV's attempt at their forceful conversion to Christianity opened access to *rest* for many Muslims, several of whom were described as 'mäalti krestiyan läyti aslam' (Christians by day but Muslims by night). Cf. R. Perini, *Di quà del Maréb* (Fl, 1905), 344.

[13] CD IAO Sc 3025, T. Moreschini, Principali contratti agrari indigeni dello Sc (AA, 1939), 3–4; J. Cohen and D. Weintraub, *Land and Peasants in Imperial Ethiopia: The Social Background to Revolution* (Assen, 1975), 53. Sharecropping appears to have been a universal practice throughout the rest of Ethiopian regions and, notwithstanding some concessions to

In all three tenancies, the cultivator either resided inside the plot—in which case he became *ťisäña*—or lived somewhere else carrying out the work as *ťämaj* or *mofär zämach*. Unlike *ťisäña*, *ťämaj* generally had no fixed period of contract and maintained his freedom to work elsewhere. In *erbo arash*, the landlord, in addition to grazing-land (maqomiya) to be used as pasture for his cattle by the cultivator, provided him with reclaimed land. The cultivator contributed plough animals, agricultural implements, seeds, and labour. At harvest, as in all three forms, land tax or *asrat*, valued at one-tenth of produce, was levied. One-fourth of the remainder was allocated to the landlord, while the tenant retained the rest plus the straw and the chaff, with a right to graze his cattle on stubble. *Sisso arash*, the most widespread, was identical to *erbo arash* except that the landlord was paid one-third of the remainder.

*Ekkul arash* varied according to the type and scale of input provided by the parties. The landlord, in addition to reclaimed land, provided a plough with one or two oxen and with part or all of the seed. The produce was divided equally on a fifty-fifty basis. Unlike the other two rental forms, in *ekkul arash* often no grazing-land was provided and the tenants owning cattle had to make separate rental arrangements often with the same landlord. The benefit was that the tenant was not obliged to provide services tied to the grazing-land, such as making available his labour to the landlord five days per year. But to maintain good relations with the landowner, the tenant provided supplementary services, assisting him in housework at family celebrations and in house repairs, and providing him with eggs and chickens on special occasions. The assistance was usually reciprocated.[14] Normally the landlord gave burnous or other garments to the tenant's family. When the tenancy was between an ordinary man and a man with authority and power, the tenant often became one of the landowner's retinue, visiting him frequently with gifts, escorting him on journeys and to church services. When such extra-economic relationship became paramount, the landlord's share tended to be less than the customary amount. For the honour and support he provided to the landlord,

local variations, almost everywhere it operated on the same principles. (Cf. UA, Am, 'Aspetti generali e zootecnici del Lago Tana', *AC* 32/6 (1938), 262–3; E. Conforti, 'Cenni sulla regione dei Guraghe', *AC* 33/7 (1939), 419; also 'La regione dei Guraghe', *AC* 35/6 (1941), 243–4; V. Pierrucci, 'Impressioni agrarie sull'Aussa', *AC* 34/4 (1940), 165–6; G. Piani, 'L'Agricoltura indigena nel G Hr e i mezzi per farla progredire', *AC* 33/7 (1941), 404.)

[14] Cohen and Weintraub, *Land and Peasants*, 50–5. Cf. CD IAO Sc 3025, Moreschini, Principali Contratti, p. 9; Hoben, *Land Tenure*, 138.

the tenant received help in court litigation, political protection, and an occasional gift.[15]

An important variation of *rest* is village tenure, described variously as *shäna*, *dässa*, *čegurafgottet*. These were restricted to the Eritrean highlands, particularly Hamasén and Akhälä Guzay and part of Tegray province. Under this tenure, land was collectively owned by the village and eligibility was based on residence rather than descent. Need to accommodate new members and correct inequality and fragmentation necessitated distribution, often by lot, every five or twenty-five years, with each member of the village being entitled to an equal share of the land.[16]

The *restäña* were proud of their system and the laws regulating it. But close examination reveals *rest*'s in-built flaws: it encouraged innumerable claims and counterclaims that led to long-drawn and costly court disputes and made ownership insecure as a successful claimant could take it away at any time. Village tenure attempted to eliminate the grave injustices embodied in *rest*, but periodic redistributions aggravated insecurity and tended to minimize the incentive for proper care of the land. Constant division and subdivision in *rest* fostered excessive fragmentation and diminution of plots. The *restäña* clung to all their privileges. Even under village ownership, they maintained narrow and rigid economic and political privileges. The village chiefs, known as *čeqa-shum*, were elected only from their ranks. As a result, a large section of the population was permanently excluded from political life and kept in a state of dependence and economic insecurity.[17] Attempts to reform were restricted to particular areas.[18]

Owing to the gradual imposition of *gult* over *rest*, the two systems had always lived in a state of tension, actual or potential. Etymologically *gult*

[15] Such combination of servitude and tenancy normally occurs when the landowner was a *gult*-holder (cf. Hoben, 'Descent Groups', 182; Nadel, 'Land Tenure'; 16/1: 17).

[16] Bauer, 'Land, Leadership'; Cohen and Weintraub, *Land and Peasants*, 33–4; Ambaye Zekarias *Land Tenure in Eritrea* (Ethiopia) (AA, 1966). But the most authoritative research is by the anthropologist Nadel, who worked in the Eritrean highland under the British military administration. Cf. Nadel, 'Land Tenure', 16/1: 11–15.

[17] Nadel, 'Land Tenure', 16/2: 108; Dessalegn Rahmato, *Agrarian Reform in Ethiopia*, (Trenton, NJ, 1985), 18–19; Perham, *Government*, 291–2; Cohen and Weintraub, *Land and Peasants*, 47–50.

[18] The most important intrusions were the edicts attributed to Emperor Yohannes IV and his General, Ras Alula, who sanctioned legal ownership tantamount to *rest* after forty years' undisputed occupancy. According to Nadel this edict, 'crystallized in a period of considerable immigrations and military colonization', enabled many landless 'foreign' squatters to buy their entry into the 'jealously guarded ranks' of the *restäña* (Nadel, 'Land Tenure', 16/1: 11, 18).

simply means 'grant',[19] but popular opinion put the *gult*-holder into a distinct and powerful class. In fact, *gult* indicated a territorial unit of administration where the State renounced part or all of its fiscal rights in favour of a *gult*-ruler.[20] *Gult* was also a major mechanism of surplus appropriation—labour and produce—and also the only way whereby a large estate could be accumulated. *Gult* estates were classified according to the status of the grantee, the specific services attached to them and the time length imposed for their enjoyment. The broadest distinction, however, was between *gult* destined for religious dispensation (*bétä-kehenät*) and secular administration (*bétä-mängest*).

The *gult*-holder, whether secular or religious, was a lord but not a landowner except in his own personal *hudad*—State land attached to his office. He was not involved in the process of production. Peasants owning the land produced what they liked and in whatever way they liked; the *gult*-holder only dispatched his agents at harvest time to collect his dues.[21]

*Gult* rights were granted by the State to its favourites, unsalaried local government officials, religious institutions, local gentry, and others. With such grants it secured support and loyalty from politically influential groups or rewarded meritorious servants while it discouraged opposition by threats of dispossession. Theoretically, *gult* was not hereditary, but in practice many *gults* transmogrified into some kind of *rest*, giving rise to a hybrid form known as *restä-gult*. But the State retained the ultimate right to modify, upgrade, abolish, or transfer the original grant.

In order to counter future emergencies, reward deserving individuals, and cater for its own needs, the State also kept at its disposal a vast amount of land, commonly known as *hudad*. Composed largely of uncultivated land, *hudad* constantly fluctuated in size, swelling with fresh intake of lands confiscated for tax evasion, high treason, or rebellion, and contracting with further apportionment. Its name changed depending on the entitlement and destination. Thus, the *hudad* allocated for public administrators, civil servants, and military personnel in lieu of salary became *madärya* or *restä-gult*, depending on whether it was a temporary or permanent grant;[22] those providing supplies for the imperial court (*gebbi*)

[19] From the Ge'ez word *gwällätä* (to donate), *gwelt* indicates 'an officially sanctioned reward'.

[20] Pankhurst *State and Land*, 29.

[21] The *gult*-holder or his agents visited the crop three times—during harvest, piling, and threshing—after being notified by the peasant. Cf. ZN, 333–44 where there are shown detailed regulations sanctioned by the decree of *Ṫeqemt* 7, 1914 EC.

[22] G. W. B. Huntingford, *The Land Charters of Northern Ethiopia* (AA, 1965); Taddesse Tamrat, *Church and State in Ethiopia, 1270–1527* (Oxford, 1972), 100–3.

were known as *ganä-gäb* (pot- or king-bound), *mad-bét* (kitchen), and *wärä-gänu* (herdsman) and were directly administered by central government.[23]

What did *gult* actually entail to its holder? Two clearly distinct but overlapping rights were involved. For taxation purposes, the landowning class can be divided into *gäbbar*—who were *rest*-holders on ground of descent—and *tekläña* (settlers)—who obtained their land through the goodwill of the State.[24] All land was burdened with two forms of taxes: *asrat* (tithe)—a form of land tax—and *geber*[25]—a form of rate. Originally, *asrat* was paid to the central government, in one-tenth of produce per *gasha*. The bulk of *geber* was paid in labour or services. Both *gäbbar* and *tekläña* paid the same amount of *asrat*, but how they fulfilled their *geber* differed substantially.

For the *gäbbar*, *geber* was mainly paid in statutary labour with a small payment in kind, varying according to locality: regions renowned for a particular type of produce, such as honey, white *ťéf*, or horses had their *geber* commuted with one of these items. Others had a fixed annual payment in *gämäta čäw* (salt bar). The statutary labour, normally one-third of a gäbbar's working time, involved assistance in construction of fortifications, buildings, and storage bins, and maintenance of major trails as well as the cultivation of *hudad* for the benefit of political appointees or the imperial court. In monetary terms, the total tax (*asrat* and *geber*) amounted almost to one-fifth of the produce.[26]

The *geber* of the *tekläña* was commuted with a specific service attached to the land, and was comparatively less onerous than that paid on *gäbbar*

[23] ZN, 124–5. R. Pankhurst, *Economic History of Ethiopia, 1800–1935* (AA, 1968), 148.

[24] *Tekläña* commonly refers to military, as virtually all *gult*-holders were originally either military men or linked to some form of military service. Analysis of ownership reveals that the holders included fusiliers, artillerymen, avanguardists, quartermasters, spear- and gunpowder-manufacturers, drummers, makers and transporters of the imperial tent, baggage-transporters, muleteers, cavalrymen and horse-breeders, animal game suppliers, foragers, prison-wardens, tailors, kitchen-gardeners, postmen, land surveyors. The services were to be provided both in peace and war time. These lands were originally given on a temporary basis but through the course of time most of them took on a permanent character and have all the features of *rest*. Here the term *tekläña* is used to indicate all categories of people who obtained their land in the form of *gult* rather than descent.

[25] In Amharic *geber* is a generic form and means 'banquet, tax' and does not specifically relate to this second form of tax. But like *asrat*, the term *geber* is Ge'ez—so not Amharic—and means 'work, labour', thus specifically denoting the nature of the taxation.

[26] e.g. in Bishoftu area on a revenue of 1,000 *dawulla* produce from one *gasha* land, valued at 400 MTD, the estimated total tax is 20.50% of the produce with the cash value of 82 MTD. This was composed of 40 MTD *asrat*, and 42 MTD *geber* (the estimate of 12 MTD for 1 *gundo mar*, 12 MTD for 3 *dawulla* cereal, and 12 MTD for corvée). Cf. CD IAO 794, Centurione P. Ciocca, Elementi di diritto fondiario e tributario nello Scioa, n.d., pp. 65–74. For other regional variations see Pankhurst *Economic History*, 511–19.

land. As the obligations were on land rather than person, the holder lacking the specific skill attached to the land had the duty either to relinquish it or discharge it by a proxy. If, for example, a woman inherited land which had to supply soldiers or clergymen, it was her duty to raise and equip them.[27]

At the lowest administrative level, the *gult*-holder, whether religious or secular, was almost the sole link between the provincial government and the local *gäbbar*. He administered justice and settled local disputes, subject to appeal to higher authority. He was also responsible for maintenance of civil order, communication of government decrees, organization of statutary labour, and the collection of *asrat* and *geber*. He made proclamations concerning public work. He also ensured that the *gäbbar* provided government officials and guests in transit with food, drink, and lodging as might be required. In his work he was assisted by the village chief, *čeqa-shum*, who was elected annually by, and from among, the local landowning class. If he was a provincial governor, he appointed all the functionaries under him and conferred military titles up to the rank below his own.[28] In return, he was expected to maintain a personal army and provide military assistance to the central government whenever required.[29]

The most important benefit accruing to the *gult*-holder was the *geber* of the *gäbbar* under his jurisdiction which he used in the way he deemed fit for the common good. He was normally exempt from *geber* on his own

[27] The normal practice was to hire against payment or rent part of the land to a third party who had the skill.

[28] In Shäwa e.g. the lowest position in the political hierarchy was occupied by the *čeqa-shum*, a village headman, annually elected among the *gäbbar*. Next came *mälkäña* whose jurisdiction extended over a district involving a number of more or less extensive *gasha* (*gasha* of *gäbbar*, *gendäbäl*, *sämon*, *madärya*) which constituted his jurisdiction or *mälkäñennät*. To this bulk may be added one or more *hudad*—*hudad* of the *mälkäña*, given for his personal use as a temporary (*madärya*) or permanent (*rest*) grant. Note that *hudad* in *rest* form is numerically limited. The *mälkäñennät* became a *gult* if a total or partial exemption from *äsrat* payment, and free usufruct of market and court fees was attached to it. *Mälkäña* were accountable to *mesläné*, a governor of a vast territory. Above *mesläné*, a distinction can be made between land directly under the central government, such as *ganägäb*, *wärägänu* (they were so because of imperial disposition or their proximity to the royal capital), and territories controlled by great regional chiefs. *Ganägäb* countries were governed by a *wämbär* of the Ministry of the Imperial Court, answerable to the Minister himself, and the others by a *shaläqa*, directly responsible to the Emperor. Cf. CD IAO 794, Ciocca, Elementi di diritto, pp. 29–32; ZN, 124.

[29] D. Donham and Wendy James (eds.), *The Southern Marches of Imperial Ethiopia* (Cambridge, 1986), 9; CD IAO 794, Ciocca, Elementi di diritto, 29; Nadel, 'Land Tenure', 16/1: 4; Hoben *Land Tenure*, 77–8.

personal *hudad*[30] which he administered often aided by his representative, *wäkil.*[31] He might also be allowed to keep market and court fees for himself. *Hudad*, however, consisted of undeveloped land and, unless there were peasants to cultivate it, was worthless. So, an entrepreneurial *gult*-holder might exploit it in a number of ways: using statutary labour of the *gäbbar*, or under some form of rental arrangement, or leasing it for an extended period. In exceptional cases, particularly when it was a *restä-gult*, it was sold or settled with a landless person who became his *gäbbar*. This latter often was much more profitable as the *gult*-holder—transformed into a landlord—in addition to proceeds from sale secured a lifetime income in the form of *geber* or *asrat* or both.[32]

Ancient Abyssinia's social organization rested on the ideological underpinnings that its Christianity, as expressed in the teachings of the EOC, provided. Catering for the spiritual needs of its faithful, education and welfare formed part of the Church's public duties, with the State defraying the associated costs, largely discharged by granting land necessary for the purpose and often supplemented by additional subsidy (cf. Table 5). This was done in a quite familiar fashion. Whenever a new church was built, the *liqä-kahnat*, regional representative of the *abun*, the Church's highest authority in the land, fixed the number of the officiating clergy, and the government allocated the land and, until this was effected, paid the clergy from its own treasury. Allocation was commensurate with the importance of the church, and the number and the status of its clergy. Normally, in each *gäťär* (village) church a portion amounting to a maximum of one-third of available total land is set aside in this fashion.

Even though *sämon*, a quite ambiguous term, is generally used to denote this class of land, the nomenclature varied from place to place. The broadest distinction one can make is that between land allotted as

[30] *Gult*-holders paid *asrat* on their personal *gult* but not necessarily *geber*. Their service was accounted for the latter. Those exempted from *asrat* made a token payment which had largely symbolic significance.

[31] Hoben, *Land Tenure*, 77. It should be emphasized that the right of the *gult-gäž i* (*gult*-ruler) included land only if the land was given to him as *hudad*, or as a source of personal subsistence; or *restä-gult* in which case he had the power to evict and repossess his land. *Rest*-holders could be evicted only for failure to pay tax. In this case the land was given to a better candidate promising to ensure future revenue.

[32] Sale was rare and, depending on the nature of the land, it had to be approved by the State. The *gult*-holder's right to *äsrat* and *geber*, if his land entailed him to it, remained unaffected by eventual transactions. Theoretically, those owning such land became the *gäbbar* of the *gult*-holder for their taxes were paid to him. This process largely took place in the South where massive sale of *gult* lands gave rise to *yä-mälkäña gäbbar* or *gäbbar* of the *mälkäña*.

TABLE 5. Bishoftu's EOC population and revenues in the 1930s

| Area | No. of churches | No. of clergy | Sämon land[a] | Revenues | | | | | | |
|---|---|---|---|---|---|---|---|---|---|---|
| | | | | Landholder | | | | State | | |
| | | | | Cereal | | | Cash (MTD) | Cereal | | Cash (MTD) |
| | | | | Dawulla | Qunna | Wood load | | Dawulla | Qunna | |
| Aqaqi | 13 | 160 | 134.75 | 227 | 109 | 777 | 27 | 8 | — | 45.00 |
| Ada | 13 | 160 | 122.50 | 187 | 147 | 1,213 | 65 | 111 | 12 | 139.00 |
| Chaffadänsa | 8 | 98 | 64.25 | 80 | 53 | 504 | 28 | 11 | — | 56.00 |
| Moggio | 18 | 233 | 180.50 | 314 | 91 | 1,474 | 142 | 15 | — | 220.00 |
| Balci | 40 | 458 | 453.75 | 225 | 222 | 1,963 | 88 | 152 | 36 | 866.50 |
| Zuquala[b] | 8 | 205 | 182.00 | 671 | 20 | 78 | 30 | 802 | 10 | 3,022.00 |
| TOTAL | 100 | 1,311 | 1,137.75 | 1,704 | 642 | 6,009 | 380 | 1,099 | 58 | 4,348.50 |

[a] The list does not include *gult* lands. The largest holder was Zuquala *gädam*. The two churches of the *gädam*, St Abbò and Wämbärä Maryam, seem to have had about 746 *gasha* between them.

[b] The figure includes the two churches of the Zuquala *gädam* where about 800 individuals lived (With DGG 18-6-1937 no. 79276, and 8-8-1938 no. 118321/A-3-B/6 the Italian government set at 200 the maximum number allowed to live in the two churches of Zuquala *gädam*. As part of the financial settlement, the abbot was acknowledged as a *gult*-holder. He was allocated an annuity of L.13,041 while the *gädam* received L.24,000 per annum.)

*Source:* CD IAO 791, Commissariato di G. di Biscioftù, La Chiesa Etiopica nel R. Commissariato di Biscioftù, Moggio 31 Dec. 1938 (Totals as in the original).

endowment fund and personal emolument. The first, referring to lands reserved for the Church as an institution, was administered by a *gäbäz* (Church administrator) who allocated part to his dependents—four persons per *gasha*—and leased or rented the rest, with the revenue supporting the poor.[33] The second group, which comprised a vast category of land, was allocated to individual servants of the Church who exploited it as they liked.[34] The owners of these lands either worked them personally or rented or leased them but they were not allowed to sell, mortgage, or exchange in other ways. They could be inherited by both sexes and, except in a few areas, could not be revoked even if the grantee was defrocked or incapacitated as long as the service was duly performed by a proxy.

The allotments are made in both *rest* and *gult* forms. The latter, however, is largely associated with the two higher institutions, *gädam* (monastery), and *däbr* (abbey) and their inhabitants as well as other important ecclesiastical dignitaries, both secular and religious.[35] The overall administration of *däbr* land was in the hands of *aläqa* who normally bore a special title that was unique to each *däbr*. *Gädam* land was under the *mäggabi*.[36] Unlike its secular counterparts, which could be temporary and

[33] Under this group came the lands known as *yä-tabot* (the Holy Arlc) and *yä-qurban* (Eucharist).

[34] This land is normally divided in three categories: (1) *Yä-sämonäña märét*, land reserved for the weekly mass celebrant priests and deacons—normally five per week—included: (*a*) *yä-qéssenna* (priesthood); (*b*) *yä-diqunna* (diaconate); (*c*) *yä-haymanotä-äbäw [änbabi]* ([reader] of Book of our Fathers). (2) *Yä-däbtära märét*, land allocated to the church musical group, included: (*a*) *yä-däbtära* or *yämäwwädes* (musician); (*b*) *yä-šomä-deggua* (Lenten music); (*c*) *yä-säatat Quämi* (horologer); (*d*) *yä-zäynägges* (Psalm reader). The two categories are also known under a summary title of *yä-asqädash-yä-aswädash märet* meaning land of celebrants and musicians. (3) *Church wardens* included land belonging to those providing supportive functions: (*a*) *yä-antafi* (upholsterer); (*b*) *yä-atabi* or *yä-wälway* (washer/cleaner) (*c*) *yä-aqabit* (keeper); (*d*) *yä-däwway* (bell-ringer); (*e*) *yä-guäzguaž* (person in charge of covering church floor with dry grass); (*f*) *yä-qärafi* (woodcutter). It seems appropriate to put in this same category lands reserved for some purposeful animals such as *yä-demmät märét* (cat). For some variations in Church land nomenclature in other regions; see E. Bartolozzi, 'Principali aspetti della vita economica delle genti del lago Tana', *AC* 36/6 (1942), 151.

[35] In *gädam* and *däbr* institutions, the ecclesiastical population is much larger and more stratified than the *gätär* church which is *reflected* in the amount of land and the variation of their nomenclature. For more detailed information on the three church institutions—*gätär*, *däbr*, and *gädam*—see Haile M. Larebo, 'The EOC', in P. Ramet (ed.), *Eastern Christianity and Politics in the Twentieth Century*, i (Durham, NC 1988), 381–3; id., 'The EOC and Politics in the Twentieth Century', *Northeast African Studies*, 9/3 (1987), 1–15.

[36] The practical distribution of the *gädam* land between the monks varied depending on whether the *gädam* was coenobitic (*yä-andennät*), or idiorrhtmic (*yä-qurrit*). Detailed studies on church lands are few, making description and analysis difficult. Although limited

revocable, Church *gults* were, at least in theory, permanent and irrevocable, because the Church is a legal entity that cannot die and, as Fetha Nägäst rules, 'what goes into the Church shall not come out of it', for 'what has been given to God cannot be taken away'.[37]

Church lands were meant to support the Church's activities, its clergy, and those who provided various liturgical and related services. Depending on their status, they were exploited in the same manner as an ordinary *rest* or *gult* with the Church acting in much the same way as a kinship group or secular *gult*-holder. Church land, however, was considered to be more attractive, for the holders owed their *geber* to the Church, which was judged as less onerous.[38]

## THE FORMATION OF SOUTHERN TENURES

The prevalent characteristic of the southern Ethiopian land tenure system was widespread private ownership. Its emergence was inextricably linked with Abyssinian expansion during the last quarter of the nineteenth century. Conquest caught up the southern societies in varied stages of social formations: politically, they ranged from centralized kingships to segmented communities; economically, although a tiny minority were linked with international trade network, the vast majority were settled agriculturalists; the rest were either pastoralists or, like forest societies, combined farming with hunting and gathering. Peripheral variations notwithstanding, the traditional land tenure in most of these societies was, like in ancient Abyssinia, communal, based on lineage system, with the peasantry enjoying only usufructuary rights over the land.

to a specific area, the following are the most important: CD IAO 791, Commissariato di G. di Biscioftù, La Chiesa Etiopica nel R. Commissariato di Biscioftù, Moggio 31 Dec. 1938; G. Villari, 'I "gulti" della regione di Axum', *REAI* 26/9 (Sept. 1938), 1430–44; Dagne Haile Gabriel, "The Gebzenna Charter, 1894', *JES* 9/2 (1972), 67–80.

[37] Fethä Nägäst is a traditional civil and religious code. See 'Fetha Nagast' [Law of Kings], trans. Paulos Tzadua (AA, mimeograph) 498; Berhanou Abbebe, *Évolution de la propriété foncière au choa* (Paris, 1971), 140. Ownership had been often unclear and the source of continuous dispute between the cultivators who considered themselves as *restañas* and the Church who regarded them simply as sharecroppers, cf. Villari, 'Gulti' 1434–6.

[38] The *geber* consisted in construction and repair of the church building, supply of firewood, and some fees payable in cash or kind. However, as the fiscal reforms, tortuously pursued by the Ethiopian government from 1942 to 1967 in order to alleviate the plight of the peasantry, left most of the Church's land unaffected, the peasants cultivating these lands remained under heavier burden than those farming State lands.

The conventional view was that the invaders expropriated land indiscriminately. But the State was careful not to disturb the vernacular structure partly because it had neither the administrative manpower to govern these vast areas nor the technical and material resources to exploit them; and partly because of the cultural heterogeneity of the conquered peoples and the need to stifle opposition from the traditional authorities. Certainly serious abuses and brutalities took place. It was also true that the State claimed all conquered lands as its own, but practical assertion of this claim depended upon the level of resistance to conquest, the degree of cultivation of the area and the productivity of the land, as well as climatic conditions.[39] Response to conquest was as diverse as the societies themselves, ranging from armed resistance to peaceful submission.

Conquest was brutal where the local leaders fought the new invaders. The leaders and their followers were killed, deported, or sold as slaves, their land confiscated and declared state property. But the masses were left in their land which the state aimed to exploit by imposing what has been called the Abyssinian model of administration. This chiefly consisted in massive reorientation of the existing economic and political relations by introducing *gult* grants and maintaining such a system by force. Where suitable for colonization, land was divided into two categories: inhabited land already under cultivation, *yä-gäbbar märét*; and undeveloped areas, sometimes described as *yä-säqäla märét*, which comprised the whole gamut of lands considered by the Abyssinians as underpopulated and underutilized: land confiscated from the resistance leaders, rebels, and those who fell into political disgrace; pasture; or forest.[40]

In inhabited areas *gult* grants involved reducing the local population, some of whom had previously lived in relatively unstratified societies, into tribute-paying peasants, or *gäbbar*. As in the North, the *gäbbar* system affected neither the original status of, nor the existing transactions in, landownership; it only demanded the transfer of tributes due to the central administration to the *gult*-holders, who mainly were the settler soldiers. Land, measured by either *qällad* or 'by eye' and organized in *gasha*, was classified as fertile, semi-fertile, or poor. In inhabited areas,

[39] Both Enrico Brotto, adviser to the G Hr during the occupation, and Martino Mario Moreno, an official of the MAI, reject the view of massive and arbitrary expropriation. Brotto did his field-research in Hr, Arsi, and Balé, and Moreno in the G GS. Their view is supported by a number of other writers. This belies Italian propaganda which capitalized on this issue.

[40] For a detailed account of the consequences of conquest cf. Donham and James, *Southern Marches*; C. W. McClellan, 'Reaction to Ethiopian Expansionism: The Case of Darassa, 1895–1935', Ph.D. thesis (Michigan State University, 1988).

having developed a roster of tribute payers, a number of lesser *gäbbar* were regrouped together to form a full *gäbbar*—normally four per *gasha*.[41] Then, a number of full *gäbbar* or (if uncultivated) *gasha* lands, or both were allocated in lieu of salary to a wide range of people and institutions on the basis of their merit, rank, social status, or importance:[42] northern officials who came to administer the areas;[43] ecclesiastical institutions and dignitaries who were given the task of expanding the Christian religion; the local rulers who, stripped of their original power and invested with an exotic title, *balabbat*, were incorporated into the new polity as the low-level functionaries, entrusted with the task of developing the local government system and acting as intermediaries between their former subjects and the new incumbents.[44] Surplus *gasha* lands were either given gratis or sold for a ridiculously low price as an incentive to private developers, including central and provincial élites, loyal to the crown, and peasants moving from the north under demographic pressure.[45]

Those areas which peacefully submitted to Abyssinian expansion were spared from the imposition of *gäbbar* system and maintained some degree of internal autonomy, albeit surrounded by imperial agents who acted as spies. As a price for their administrative, financial, and military autonomy, they paid a fixed tribute (*qurť geber*) to upkeep the royal court. There were six provinces,[46] all of them fairly commercialized with five of them located in the West near the gold rich tributaries of the Abbay River. But over time tributes became heavier, and increasing centralization eroded their autonomy to the extent that at the time of the Italian occupation,

[41] A full *gäbbar*, described also as *quťer gäbbar*, was one able to pay the full amount of tax laid on a *gasha* (cf. ZN, 133–4).

[42] As in the North, the lands took a variety of names depending on the status of the grantee, the purpose they were meant to serve, and the fiscal obligation laid upon them. Thus lands given to the State officials in lieu of their salary were known as *madärya* and as such the grant was a temporary one; church institutions and the leading clergymen took land in the form of *sämon* or *sämon gult*. For more detailed information on the variety of land nomenclatures see ZN, 109–28; J. Mantel-Niecko, *The Role of Land Tenure in the System of Ethiopian Imperial Government in Modern Times* (Warsaw, 1980); Cohen and Weintraub, *Land and Peasants*, 36–47.

[43] *Ras* and *däjach* took thousands of *gäbbar*, *fitawurari* 300, and *qäñazmach* 150; the soldiers, according to their seniority, 10, 15, and 20. The State fixed the number and the *balabbat* allocated the land and the people.

[44] J. Markakis, *Ethiopia: Anatomy of a Traditional Polity* (Oxford, 1974), 115–16; Cohen and Weintraub, *Land and Peasants*, 35–6, 38; Donham and James, *Southern Marches*, 39.

[45] Cohen and Weintraub, *Land and Peasants*, 35–6.

[46] These were Jimma, Wälläga Näqämti, Wälläga Qälläm, Béni-Shangul, Guba, and Awsa (cf. ZN, 165).

only the rulers of Näqämti, Béni Shangul, and Ausa retained more or less their original relationship with the centre.[47]

Central control remained equally tenuous in the lowland areas inhabited by hunters, pastoralists, and shifting cultivators. In these areas, the *gäbbar* system was never instituted partly because the population was much harder to control and partly because the land was not suitable for colonization by highland agriculturalists. Initial extraction consisted in periodical raids for both cattle and slaves, and exploitation of important catches such as ivory and leopard-skins, by the nearby settlers. But with the increasing assertion of the central government and depletion of these commodities, the raids gradually regularized into tax collection, particularly when local leaders were brought into the realm of central administration. With the establishment of *balabbat* a fixed rate was imposed on cattle.[48]

## DEVELOPMENTS UP TO 1935

The most visible feature of the Abyssinian expansion was to homogenize, at least theoretically, the situation of the peasantry in the North and South. And in both areas land had become more than a simple commodity exchangeable in a market economy: it had multiple attributes that reflected the composition of social and political organizations. The relation between land and man bonded men together in political units. Rights over land use and landownership, rather than being absolute or exclusive, remained often multiple and overlapping. Not all people held these land rights. There were three distinct strata in society: the landed, the landless, and slaves.

Slaves were widely used for domestic purposes especially in the houses of the rich, but much less in agriculture, with notable exception of Abba Jiffar II's estates in Jimma. Anyhow the lucrative trade in slaves that, stimulated by the Abyssinian conquest, devastated much of the southern territories, was on the wane due to exertion by international pressure and

[47] Donham and James, *Southern Marches*, 37–8, 51–68; for other studies cf. H. S. Lewis, *A Galla Monarchy: Jimma Abba Gifar. Ethiopia, 1830–1932*, (Madison, Wis., 1965); P. P. Garretson, 'Shaykh Hamdan Abu Shok (1898–1938) and the Administration of Guba', in J. Tubiana (ed.), *Modern Ethiopia from the Accession of Menelik II to the Present* (Rotterdam, 1980); Maknun Gamaledin Ashami, 'The Political Economy of the Afar Region of Ethiopia: A Dynamic Periphery' Ph.D. Thesis (Cambridge, 1985).

[48] Donham and James, *Southern Marches*, 42, 148–71, 219–45.

domestic forces which in 1932 culminated in the vigorous restructuring of the anti-slavery bureau established in the previous decade. The landless, if they were not engaged in trade or traditional handicraft, were linked to land through a variety of sharecropping systems, or practised both. Among the landed, the most important classes were *gäbbar* and *tekläña* both supporting the Ethiopian State in a way that involved mutual responsibilities: provisioning and labour services on one hand, administrative, judicial, and military obligations on the other. The elaborate systems of land tenure and taxation with all their provincial variations represented the different means by which the State ultimately appropriated or rewarded services. But the backbone of the economy was the *gäbbar*, as it was his labour that materially sustained the State apparatus.[49]

Despite the inferior social position, it is wrong to equate the *gäbbar* with slaves or serfs. He was not legally bound to the *gult*-holder as, for example, the slave was. The *gäbbar* owned his lands but, unlike the serfs, he could leave and set up his homestead elsewhere. It was a fluid term and refers to no distinct social class. A study conducted at Balci in Bishoftu indicated that the Emperor Haile Sellassie himself was one of the *gäbbar of* the church of Somsa Mädhané-Aläm at Näč Dengay where he held a land burdened with *gäbbar*-associated services.[50] As tribute was based upon land and not person, anybody could be a *gäbbar* as long as satisfactory arrangements were made to provide the services imposed on the land.[51]

Considering the regime's decentralized structure and the absence of a money economy, the difficulties in transport and communication, and taking a full account of the material and manpower resources at its disposal, the *gäbbar* system efficiently financed local government and remitted residual revenue to the central treasury. The State was able to administer a vast country without resorting to loans from European powers or international finance at a time when technologically more advanced African and Asian countries had been occupied or reduced to satellites of a remote foreign country under the pretext of avoiding bankruptcy. But it did so at the expense of the *gäbbar* themselves who gained little or no material benefit from the taxes they paid.[52] Indeed, the system, while enriching the centre and its large machinery, depleted the local resources, deprived the peasantry of a fair share of its produce, and exploited its labour for financing the development of social and economic infrastructure.

[49] Perham, *Government*, 278. [50] CD IAO 791, La Chiesa Etiopica, 97.

[51] Perham, *Government*, 278; Markakis, *Ethiopia*, 341–2.

[52] ASMAI 54/21, Informazioni commerciali: Finanze Etiopiche, n.d; H. G. Marcus, *Haile Sellassie I: The Formative Years, 1892–1936* (Berkeley, Calif., 1987), 41.

But this need not be dramatized into describing the system as ruthlessly exploitative or barbaric.

Like any taxpayer, the *gäbbar* owed dues to the central government or his representative, in kind and labour. The relationship between the *gäbbar* and the *gult*-holder tended to be patrimonial—personal in nature and diffuse in content, varying according to the character and disposition of individual *gult*-holder. In a country where the central authority was remote, the politics often fluid, and in situations where the *gäbbar* happened to be in a weak position, this gave opportunities to an unscrupulous *gult*-holder for arbitrary extortion. But despite the wide scope of his authority, a *gult*-holder could not afford to consistently antagonize his *gäbbar* who, at least in theory, could always leave and set up his homestead elsewhere or, if he was daring and had sufficient resources, appeal to central government.[53]

In the popular mind, the evils of *gäbbar* were largely associated with the landlessness and harsh treatment experienced by southern peasantry. But no governmental decree existed that made its situation different from that of its northern counterpart, nor were the *rest* guarantees that existed in the North eliminated in the South. The *gäbbar* system functioned differently in the South because of a combination of cultural differences and the inroads of modernization. Northern peasantry, though well aware of the inadequacies of the system,[54] prided itself as part and parcel of its organization. But not so for the recently conquered South where its imposition was seen as an unwarranted intrusion in land that had been freely enjoyed. The brutal force used in the process helped only to highlight its alien character. In the North *gäbbar* and *gult*-holder belonged to the same ethnic group, shared the same religion, and were often united by kinship ties. In the South, the northern settlers interpreted their military superiority culturally and viewed their position in terms of a 'civilizing mission'. The northern peasantry, through its long exposure, had developed an appropriate defence-mechanism to resist the insatiable exactions of the rapacious *gult*-holder. The southerners lacked such valuable experience; the language barrier and the cultural differences only worsened their plight.[55]

[53] Perham, *Government*, 279; Hoben, *Land Tenure*, 79; Donham and James, *Southern Marches*, 185–6.

[54] The following verse reveals the awareness of the peasantry of the exploitative nature of the system: 'If land tax we must pay, / So must the monkey, / For is it not the same land, / That it scratches with its hand.' Quoted by Mesfin Wolde Mariam, *Rural Vulnerability*, 25; Hoben, *Land Tenure*, 81.

[55] Donham and James, *Southern Marches*, 8.

The relatively rapid advance of protocapitalist agriculture in the South helped only to increasingly exacerbate the condition of the *gäbbar*. In most of the North, political turmoil, growing ecological degradation, and poor communication kept the region almost agriculturally static. Many northerners sought to overcome their economic destitution, or supplement the shortfalls, as soldiers of fortune, or migrant labourers to neighbouring colonial territories, or moving to the South. Improved communications, relative tranquility, and in particular the construction of Addis Ababa–Djibuti railway revolutionized much of the fertile southern countryside. The metamorphosis had made the South by the late 1920s into the Ethiopian government's 'large colony of exploitation—meant in the most rigorous sense of the term.'[56] The exploitation was made possible by the emergence of a significant number of fairly modern agricultural farms, remarkably absent in the North. These developments witnessed the rise of coffee as the area's prime agricultural resource, the flourishing of new commercial towns competing with, or replacing, the old military garrisons, and rapid transformation of land into a marketable commodity. As speculation on land intensified, undeveloped areas were put under cultivation. In Arsi, for example, the *gult*-holders sold their *hudad* as *rest* to both Amharas and Oromos who were squeezed out by commercial developments in Addis Ababa and became the *gäbbar* of the gult-holders. Identical trends were in operation in Härärgé.[57]

One of the most visible features of these developments was the absence of mechanization. With some rare exceptions, the basic means of production even for most of the advanced modern farms remained traditional, relying largely on animal traction, ploughshare, and human labour. Rather than in the application of machine or new technologies to farming, the changes largely consisted in putting more land under plough, in rationalizing production by better manuring, use of better crops, diversification, irrigation, and labour-intensification. A development bank, known as La Société Nationale d'Éthiopie pour le Développement de l'Agriculture et du Commerce, had existed since 1908 to assist in agricultural and commercial development. But the public entertained a wholesome suspicion and, despite restructuring in 1928, its importance in the agricultural economy seems insignificant.[58]

The developments in southern agriculture had a drastic effect on its

[56] E. Cerulli, 'Le popolazioni ed i capi dell'Etiopia sud-occidentale', in ASMAI 54/34/137, Zoli to MAE, Rm 8 Jan. 1927.

[57] Brotto, *Regime*, 58–60, 95–7, 101; Moreno, '*Regime*', 1504, 1506–8.

[58] Pankhurst, *Economic History*, 208.

*gäbbar* in two ways. The settler increasingly saw him as a resource to be exploited for his own benefit and made constant and heavy demands on his labour.[59] With increasing monopolization of his labour by the *gult*-holder, the *gäbbar* devoted less and less time to the care and cultivation of his own plots and the State lost desperately needed revenue. As a result, several *gäbbar* were forced to sell their land and turn into share-croppers. On the other hand, the expansion of cultivable land put an increasing pressure on grazing land, causing serious shortage of traction animals. In places such as Arsi, as pasture became scarce, the collapse of the traditional pastoral economy had the effect of forcibly transforming the once-proud cattle-herdsmen into tenant cultivators.[60]

The changing role of the settler at the periphery was matched with the growing assertion of authority at the centre whose continuously expanding needs and priorities dictated a shift in the country's economic organization. The process, put into motion by Lej Iyyasu, Menilek's grandson and successor (1911–16), gained momentum under Ras Täfäri Mäkannen—heir to the throne from 1916, *negus* from 1928, and crowned Emperor Haile Sellassie in 1930—culminating in the enactment of the 1931 Constitution that acknowledged the monarchy as the sole embodiment of all power in the land. By his successful action, Emperor Haile Sellassie had made hollow Italy's peripheral policy which, with its twofold action of bribing the provincial leaders with money, and the masses with schools, hospitals, and clinics while pretending to maintain friendly relations with the regime in Addis Ababa, aimed to subvert Addis Ababa's efforts at centralization.[61] According to one Italian official, Emperor Haile Sellassie's

[59] Somehow the plight of the southern peasant may not be different from that of the Ankobär, most eloquently talked by Asbä Häylu. The article captures the human dimension by its moving and graphic description on the conditions of the *gäbbar*. Cf. Berhanenna Sälam, 21 July 1927, reprod. at length in English by Addis Hiwet, *Ethiopia*, 71–3.

[60] Brotto, *Regime*, 92–7.

[61] *Politica periferica* counted much on the rebellion of the two scions of old aristocratic families who were opposed to Täfäri's reforms: Empress Zäwditu's husband, Ras Gugsa Wolé of Bägémder, who belittled the western-educated young Ethiopians who ran the government; Ras Häylu of Gojjam, an acutely business-minded and arrogant snob who publicly scorned them as 'people from nothing who were only simple stablemen during Menilek's time'. But it was a great disappointment. Gugsa's revolt was the nadir of the policy but it was easily crushed. Ras Häylu's rebellion was skilfully pre-empted; and the half-million bullets rushed by the Italian adventurer Raimondo Franchetti were too late to be of any help to him after his arrest. Contrary to the expectation, there was no massive upheaval of Gojjamese who felt rather relieved by his arrest as they hated his penny-pinching ways. Cf. ASMAE: 3/1/1, Raimondo Franchetti, Situazione Etiopica dopo l'incoronazione del Negus, AA Dec. 1930; ASMAE 7/1/7, Scamacca to R. MAE, AA 2 [illeg.] Nov. 1932; ASMAE 7/1/5, Spese politiche (1924–33) contains accounts of money spent on the Ethiopian notables and spies. Marcus, *Haile Sellassie*, 92–5, 120.

modernization policy was so impressive that it was bound 'to lead Ethiopia through a phase of rapid evolution to form a country different from the traditional Abyssinia to which we have up to now been accustomed.'[62]

Economically, centralization was largely precipitated by fiscal needs of the State. Various measures that were enacted since Iyyasu II aimed to boost revenues of the central treasury by releasing the productive potential of the rural society and by 1930s the State was able to affect the local agricultural economics to a degree impossible two decades earlier. The intervention had far-reaching repercussions on the life of the *gäbbar*, and this was particularly so under Emperor Haile Sellassie I. Economic imperatives apart, Haile Sellassie was also pressured by the ardent campaign of the burgeoning progressive young Ethiopian intellectuals and international criticism against the *gäbbar* system. He was of the view that the *gult* system had outlived its usefulness, and his agrarian reforms, set in motion when he was still a young and ambitious governor of Härärgé province and vigorously pursued during his regency, progressively eroded the *gult*-holders' power and institutions on which they depended for their support.[63] The immediate victims became those economically minded provincial lords whose contribution to the centre was not generous and who made *gäbbar* life unbearable. Not surprisingly, the complaints of the *gäbbar* were used as a pretext to eliminate the landlords. If a reasonable number of peasants complained about arbitrary treatment, the offending governor was called to Addis Ababa.[64]

Towards the end of 1920s taxation was rationalized by the introduction of receipts, making the provincial lords accountable for any loss of taxes. But it was the decree of 1929 that seriously eroded the prerogatives of the *gult*-holder: it abolished special public works, and restricted others to periods of agricultural recess. Most importantly, labour on *hudad* land was reduced to only three days per year with the option open to the *gäbbar* to commute it with payment in cash or in kind. Having emphasized that the

[62] ASMAI 54/31/124, Lessona to Ministro, Rm 27 July 1931. The Italians were not anyhow refrained by such reality. The call was rather to strengthen the policy. Hence the Eritrean Governor, Astuto, wrote: 'If the policy of Täfäri since 1917 was to transform [Ethiopia] from a feudal to a modern state, the peripheral policy should aim to obstruct this policy by fomenting—with all precautions required to avoid our game being publicly exposed—the ego of these chiefs encouraging them to passive or active resistance.' (ASMAE 2/4, Astuto to Ministro, As 29 July 1931.)

[63] Haile Sellassie I, *My Life and Ethiopia's Progress, 1892–1937*, trans. and ed. E. Ullendorff (Oxford, 1976), 38–41, 73.

[64] Cases in point are those of Däjach Balcha Abba Näfso, one of Menilek's ablest commanders and the hero of Adwa, and Ras Häylu Cf. Haile Sellassie I, *My Life*, 151–2, 204–5; Marcus, *Haile Sellasie*, 120.

peasant is the lifeblood of the country, the decree warned the *gult*-ruler: 'Don't go near his doorway and never dare to ask him to carry, or till, or fence, or offer you gifts, or transport the impedimenta for you.'[65] More fundamental changes took place in the 1930s, with the introduction of a standardized system of land measurement and survey by an independent body. Tax incentives were given to the developers of the uncultivated areas allowing them a three-year graduated tax exemption. With a 1931 decree prohibiting State employees from obtaining land, the government again aimed to curb the growing tendency to land speculation by the northern settlers who took no part in the production process and, therefore, counteract the divorce of the southern *gäbbar* especially from his land.[66] A high point in such ongoing agrarian reform was reached with the decree of September 1934 which, in its attempt to stamp out the abuses of the *gult*-owners, encouraged tax payment in cash rather than in kind and labour, and drastically curtailed personal services.[67] Owing to inadequate manpower and finance, the application of most of these measures proved slow and costly. Moreover, while they were welcomed in most of the South, in the traditional North they were invariably resisted. The resistance often took the form of élite-led local and regional rebellions. These did not stop the central power from asserting its authority but the successive military campaigns conducted to suppress them had the effect of putting serious stress on local resources and production.[68]

However, attempts to educate the peasantry using modern methods of cultivation were not successful. In 1931 the Emperor's proposal to exploit the country's agriculture with the formation of joint Italo-Ethiopian ventures did not materialize. The Emperor wanted to set up farms similar to his Érär Farm, one of several model farms established since the early decades of the twentieth century by the Ethiopian and foreign concessionaires.[69] Consisting of 1,000 ha, Érär Farm was established in 1925 and run by an Italian adventurer Pastorelli who by 1931 had developed about 100 ha.[70]

[65] ZN, 137. [66] Brotto, *Regime*, 58, 95–7. [67] ZN, 128–130, 138–9.

[68] In part of Härärgé, they were most effectively applied. ZN, 128–30, 138–41; Mahteme Sellassie Wolde Mascal, 'Land System', 296–7; CD IAO 794, Ciocca, Elementi, 91–5; Brotto, *Regime*, 55–6, 58–61.

[69] Pankhurst, *Economic History*, 208–9.

[70] Dissatisfied with his employer and instigated by Fascist agents, Pastorelli contacted Mussolini urging his intervention and played some influence in the resurgence of ideas of peasant settlement in Italy. He wrote embellished letters, describing Ethiopia's agricultural potential as enormous. In a letter to his cousin Di Vittorio, presented to Mussolini, he particularly pointed out that Ethiopia 'is the country where Italy should concentrate its energies for its compelling needs of demographic expansion and that our Duce should devise the most appropriate means for our people and capital to find here an immediate and

When approached by the Ethiopian government the Italian authorities were put in a predicament. They saw the move to modernize Ethiopian agriculture as thwarting Italy's longstanding design. The scheme was acknowledged 'as a useful instrument to penetrate the country economically and politically'. But as the difficulties of finding Italian capital or willing settlers proved insurmountable, the authorities clung to their strategy of peripheral politics.[71] As one official pointed out, for Italy's policy of destabilization, agricultural collaboration with the Emperor would not be a good investment, as are schools and clinics which, 'apart from being an end in themselves, will be used one day when a definite settlement with Abyssinia comes (and which I believe is quite imminent) as a bargaining chip for our specific legitimate interest'.[72]

vast field of action', Pastorelli to Di Vittorio, Dirè Daoua 30 Nov. 1932, in ASMAE 8/4/1, Di Vittorio Angelo to Benito Mussolini: Brevi cenni illustrativi sull'opera colonizzatrice svolta da Pastorelli Arcangelo, Rm 24 July 1934.

[71] ASMAE 2/2: Paterno' to MAE, AA 4 Apr. 1931; Grandi to RMC, Rm 14 Apr. 1931; De Bono to MAE, Rm 23 Apr. 1931.

[72] ASMAE 2/2, Guariglia to RMC, Rm 24 Apr. 1931.

# 3
# *The Contours of Land Policy*

## THE ADMINISTRATIVE AND ECONOMIC ENVIRONMENT

Italy's land policy developed within the framework of the colonial administration—a cumbersome, undermanned, and underfinanced machine with tangled lines of authority. Overlapping military and civilian jurisdiction and influence wielded by the government and National Fascist Party officials, who played a far greater role in Ethiopia than in Italy, left the administration in a parlous state.[1]

The Italian administration was based on the theory that colonies were an extension of the motherland to be populated by Italian settlers and exploited by Italian capital. The objective was to transform the colony into a region of *Magna Italia*. The chief architect of this policy in Ethiopia was the first Minister of Colonies, Lessona, a man with a relatively sound educational background and marked literary acumen, but an outspokenly self-styled Fascist with boundless self-confidence and obsession for discipline and authority.[2] The policy was characterized by excessive centralization and the exclusion of foreign capital, and failed to enlist the co-operation of the subject population.

In order to conquer Ethiopia, the total dismemberment of the existing administrative structure of the Empire became necessary. Everything was hurriedly reorganized *ex-novo* cutting across religious and ethnic divisions. The most outstanding feature was the total disappearance of Ethiopia as a geographical unit, the name surviving only in the king's subsidiary title of emperor of Ethiopia. The frontiers were dismantled and the

[1] ACS GRA 46/41/9, Petretti to Graziani, AA 13 May 1937; FO371/22021/J1224/40/1, CG Bird to Lambert, AA 12 Feb. 1938; Del Boca, *Caduta*, 137–58; D. Fossa, 'L'intervento del partito nel governo dell'Impero', *REAI* 20/3 (Mar. 1939), 255–62; P. M. Masotti, *Ricordi d'Etiopia di un funzionario coloniale* (Milan, 1981), 48.

[2] Born in Sept. 1891 and an orphan of a talented lawyer who taught at the University of Pisa, Lessona had a rigid education in a military school (cf. ACS SPDR 87/W/R/1, [Secret Report on] Alessandro Lessona, 27 Dec. 1927; Del Boca, *Caduta*, 39).

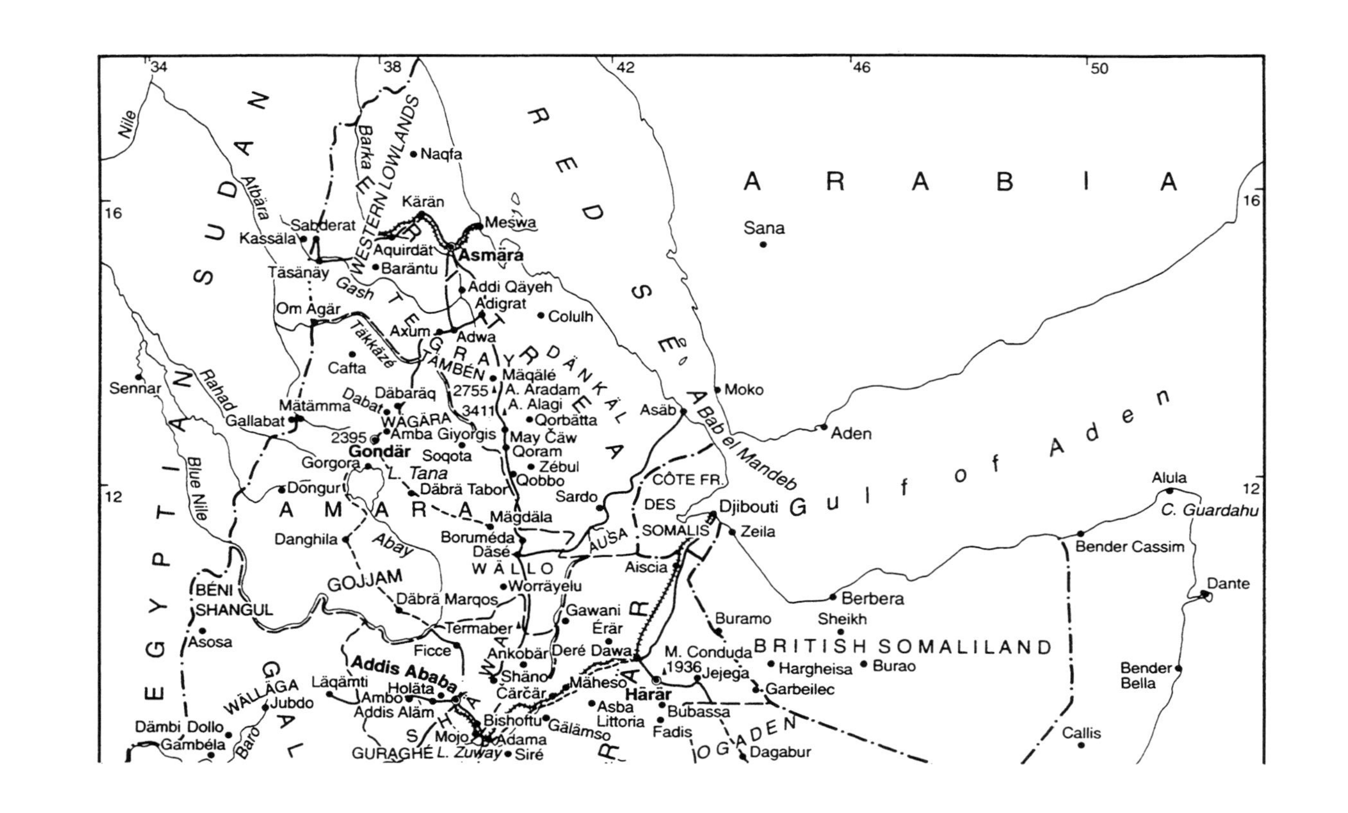

ARABIA
RED SEA
Gulf of Aden
SUDAN
EGYPTIAN SUDAN
ERITREA
TIGRAY
DÄNKAL
AMARA
GOJJAM
WÄLLO
HARAR
SHAWA
GALLA
OGADEN
BRITISH SOMALILAND
CÔTE FR. DES SOMALIS
WESTERN LOWLANDS
BÉNI SHANGUL
WÄLLÄGA
GURAGHÉ
TAMBÉN
WAGARA
AUSA
Nile
Atbära
Barka
Gash
Täkkäzé
Rahad
Blue Nile
Abay
Baro
Dabat
L. Tana
L. Zuway
Bab el Mandeb
C. Guardahu
Naqfa
Kärän
Meswa
Asmära
Aquirdät
Barãntu
Sabderat
Kassäla
Täsänäy
Om Agär
Addi Qäyeh
Adigrat
Colulh
Axum
Adwa
Cafta
Mäqälé
2755
A. Aradam
3411
A. Alagi
Qorbätta
May Čäw
Qoram
Zébul
Qobbo
Däbaräq
Mätämma
Gallabat
Amba Giyorgis
2395
Gondär
Soqota
Gorgora
Dongur
Däbrä Tabor
Mägdäla
Sardo
Asäb
Moko
Aden
Sana
Sennar
Danghila
Boruméda
Däsé
Worräyelu
Däbrä Marqos
Termaber
Gawani
Érär
Deré Dawa
Ankobär
Shäno
Cärcär
Mäheso
Härär
Asba
Littoria
Bubassa
Fadis
Gälämso
Bishoftu
Adama
Mojo
Siré
Ficce
Addis Ababa
Holäta
Ambo
Addis Aläm
Läqämti
Jubdo
Asosa
Dämbi Dollo
Gambéla
Aiscia
Djibouti
Zeila
Berbera
Buramo
Sheikh
M. Conduda
1936
Jejega
Hargheisa
Burao
Garbeilec
Dagabur
Alula
Bender Cassim
Dante
Bender Bella
Callis
34
38
42
46
50
16
12

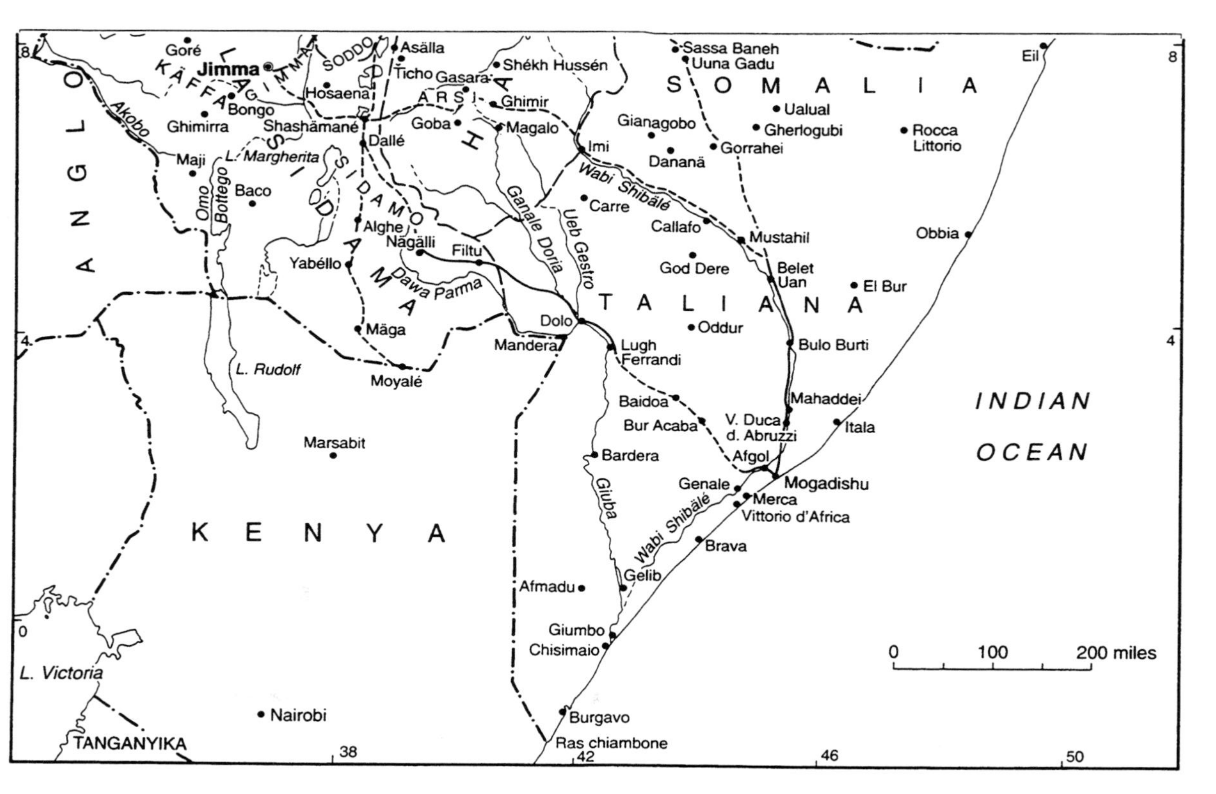


FIG. 1. Map of IEA Empire

Ethiopian Empire was merged with the two former colonies Eritrea and Somalia to form an Italian East Africa (IEA) or Africa Orientale Italiana (AOI). This was divided initially into five, and later in November 1938 into six, semi-autonomous governorates which 'behaved as if they were more or less five States unwillingly joined together by a capital mini-state'.[3] Of these, Ethiopia proper consisted of four territories—Addis Ababa (later changed to Shäwa) in the centre, Amara in the north and north-west, Galla and Sidama (GS) in the west and south-west, and Härär[4] in the east and south-east with their capitals Addis Ababa, Gondär, Jimma, and Härär respectively. The two governorates of Eritrea and Somalia were expanded by adding the present province of Tegray to the first and the Ogaden region to the second. Each governorate was then divided into a dozen districts known as *commissariato* and these in turn into a varying number of subdistricts and sub-subdistricts *residenza* and *vice-residenza*. The districts and both residencies and vice-residencies never had a definite shape and their number and boundaries fluctuated.[5]

Each governorate was administered by a governor who resided in the respective capital and was subject, theoretically at least, to the Governor-General in Addis Ababa who bore the title of the Governor-General of IEA and the Viceroy of the King of Italy and Emperor of Ethiopia. The major incumbents of this supreme post were Marshal Rodolfo Graziani and Duke of Aosta. Graziani was preceded by the brief occupation of Marshal Badoglio. Albeit described by some foreign journalists as an amiable Christian gentleman,[6] Graziani was a bloodthirsty general and an untutored and bumptious ruffian who ruled ruthlessly with an iron hand until the excesses of his 'systematic policy of terror and extermination' rebounded on him and an attempt was made to assassinate him. After this he was forcibly replaced by the Duke of Aosta, a temperamentally moderate, but politically indecisive and uninspiring, man with no previous administrative experience, whose well-known Fascist sympathy remained one of his less attractive characteristics.[7] Indeed, as the first civilian

[3] ATdR 24/108, Relazione viaggio AOI, Rm, 1937, p. 2; Villa Santa *et al.*, *Amedeo Duca D'Aosta* (Rm, 1954), 214.

[4] Am included most of the Ethiopian heartland such as Bägémder, Wällo, Gojjam, and northern Shäwa; GS consisted of southern and southwestern provinces of Arsi, Sidama-Boräna, Gämu-Gofa, Käffa, Illubabor, and Walläga, whereas Hr incorporated Balé and part of Härärgé provinces.

[5] F. Quaranta, *Ethiopia: An Empire in the Making* (London, 1939), 1–5.

[6] E. Waugh, *Waugh in Abyssinia* (London, 1936), 230.

[7] ACS SPDR 87/W/R/1, Testa Temistocle [Prefect of Udine] to Arturo, Stresa 25 July 1937.

governor, the Duke made spasmodic gestures to correct the worst of his military predecessor's mistakes, but characteristically racial discrimination under his rule intensified nor did harsh policy methods come to an end.[8] The governorships changed hands even more rapidly than the viceroyalty, within a year or two, and of the six original governors only the governor of Härär, General Nasi, remained in the country until the end. The personal characteristics of these governors are hard to define but all, imbued with the innate sense of superiority, were remarkable for their taste for authority, strong paternalism, and racism which were accepted as dogmas beyond any dispute.

As an overseas province of Italy, and also in line with the Fascist administrative doctrine that advocated a centralized authoritarian rule, the Empire was under the direct control of Rome to whom even the smallest internal matter had to be referred back.[9] Therefore, the Viceroy took his instructions from the Ministry of Colonies (MC), lately changed into Ministry of Italian Africa or Ministero dell'Africa Italiana (MAI), whose occupants, until Duke Aosta's incumbency, were Lessona, and then Terruzzi, a man with a reputation for mediocrity, presumption, money-grabbing, and whirlwind romances, loathed and despised by most settlers as one of the hangers-on.[10] The governors were assisted by an executive council, a secretariat, and a gamut of technical and administrative departments staffed by officers drawn from the MAI or loaned by other ministries in Rome. For all practical purposes, these offices and organizations were almost an exact replica of those set up inside the MAI itself. At the head of each district there was a commissioner (*commissario*) and at that of subdistrict and sub-subdistrict an officer or *residente* and *vice-residente* respectively. Most of the officers were ex-army officials. Many of them had no administrative skill and their colonial experience had been largely gained commanding native troops. Attempts to replace them with civilian staff were hampered by shortage of applicants, for whose recruitment the Ministry relied on newspaper advertisements. Within the Empire these posts, and particularly those of *residente*, were treasured, for they held sway in administrative, judicial, and political matters within their

[8] D. Mack Smith, *Mussolini's Roman Empire* (Harmondsworth, 1979), 81.

[9] FO371/22020/J395/40/1, Philipps to FO, Cairo 1 Feb. 1938.

[10] Del Boca, *Caduta*, 39; Sbacchi, 'Italian Colonialism' 96. By 1940 he had become extremely unpopular among the settler community. Despite the official precaution to avoid disturbances on his visit of Feb. 1940 by rounding up possible troublemakers, there were shouts, and handbills posted with 'Evviva Ministro' (Up with the Minister) were altered to 'Abbasso Ministro' (Down with the Minister). (See FO371/24635/J887/18/1, ACG Gibbs to Cavendish-Bentinck, AA 21 Feb. 1940.)

areas. The administration, however, was demanding, covering a vast territory, further exacerbated by language barriers which often required the employment of three or four interpreters at once.[11]

Perhaps quite rightly the Italian administration in Ethiopia was described as being riddled with opposing personalities, corruption, and inefficiency, and inept and narrow-minded officialdom.[12] In a situation where control was exerted from a distance this could not conceivably have been avoided. Despite all the emphasis upon the high-sounding Fascist ethos of hierarchy and discipline, the unruly and disobedient behaviour of the officialdom remained one of the Duke of Aosta's complaints. He blamed the MAI for dealing him 'out with some queer fish, many of whom should be kept in sealed boxes and only released when their services were needed'.[13] Civil service salaries, though almost fivefold those normal in Italy, were seldom regular which made graft unavoidable and the Duke of Aosta's often-reported characterization of the administrative personnel as 50 per cent inept and 25 per cent thieves was not perhaps exaggerated.[14]

An inevitable corollary of Italy's colonial policy was the exclusion of the subject population from all forms of power-sharing. This was no different from in other Italian colonies. In the Ethiopian case, there was a brief flirtation with indirect rule under the governorship of General Badoglio, using leading chiefs as 'intermediary organs' and 'giving some concessions to the old order'.[15] Badoglio's approach rested on an ideology associated with elements of the old colonial school, dubbed indigenophiles. This view contradicted the Fascist style of government which judged any power-sharing with the feudal lords as unacceptable. Mussolini imposed direct rule from Rome, epitomized by Lessona in his motto 'no power-sharing with the *ras*'.[16] This alienated many traditional leaders who had

[11] Abysmal performance by the colonial officers led to the formation of the Fascist Academy for AI (Accademia Fascista dell'AI) where university graduates selected for a colonial career were expected to attend a two-year specialized course. Cf. Quaranta, *Ethiopia,* 4; Masotti, *Ricordi,* 49–50, 55–6, 160.

[12] ACS SPDR B44/242/R/39, Farinacci to Presidente, Cremona 25 Dec. 1938; Villa Santa *et al.*, *Amedeo*, 219–20; FO371/22021/J2300/40/1, CG Bird to FO, AA 10 May 1938.

[13] FO371/22021/J1214/40/1, CG Bird to FO, AA 16 Feb. 1938.

[14] Villa Santa *et al.*, *Amedeo*, 219–20; G. Ciano, *Diario, 1937–1938* (Bologna, 1948), 194.

[15] R. Pankhurst, 'Economic Verdict on the Italian Occupation of Ethiopia, 1936–1941', *EO* 5/4 (1971): 72.

[16] ACS GRA 45/41/4, Lessona to GG AOI, Rm 5 Aug. 1936, p. 1; Lessona, *Memorie* (Fl, 1958), 296; ACS SPDR 87/W/R/1 La, Davide Fossa to Lessona, AA 5 Oct. 1937; Masotti, *Ricordi*, 89–91.

collaborated during the conquest with the hope of gaining wider territorial control and a share of the power lost under Emperor Haile Sellassie. Instead they found themselves stripped overnight even of the little power and privilege they had left. The salaries and titles given for their collaboration did not satisfy them.[17]

As the chief architect of the policy, Lessona made it clear that the aim of his native policy or *politica indigena* was to subordinate the indigenous population's interests to metropolitan politics. He did not see this as being incompatible with the humanitarian objectives that Italy, as a 'civilizing nation', claimed towards its Ethiopian subjects.[18] Even though no comprehensive policy on how to bring civilization to the Ethiopians was formulated, it was self-evident that subscription to a civilizing mission did not mean granting the Ethiopians the same rights as the Italians nor paying any special attention to their wellbeing. It certainly meant the establishment of the bastion of the Italian supremacy, benign not tyrannical, in which the Ethiopians would be relegated to the role of mere grateful and passive subservients. A leading newspaper succinctly summed up what all this was about in practice: '[economic] collaboration between the two races such as no power has offered or will ever offer; but separation such as no other country had been able to apply, not on the grounds of principle, but for inaction, hypocrasy and fear'. Regarded as far-sighted and corresponding to a higher concept of civilization, the policy was seen as steering the middle road between the Anglo-Saxon policy of the colour bar, rejected as too extreme since it went too far in dividing the administrators and the subject peoples, and the assimilationist policy of the French, criticized as too soft since it led to the destruction of the superiority of the governing race.[19] In contrast, Italian 'native' policy was regarded as 'exquisitely humane, profoundly realistic, with an essentially civilizing purpose, and politically necessary and important'.[20]

[17] ACS GRA 40/33/14: Regente Governo Redini to DS AP GG, As 14 June 1937; Lessona to Graziani, Bengasi 14 Mar. 1937. Sbacchi, *Ethiopia*, 129–40; Pankhurst, 'Economic Verdict', 81.

[18] Lessona, *Memorie*, 298. In an interview with the author on 24 May 1984, Lessona still claimed: 'We went there to improve the life of these poor people [*quella povera gente*] but things did not go as planned. As you can see they are still suffering.' He was referring to the 1984 Ethiopian famine.

[19] FO371/22021/J3016/40/1, Lord Perth to Visc. Halifax, Rm 29 July 1939 where he reports Terruzzi's brief to the weekly newspaper *Azione Coloniale* on Italian administration in Ethiopia.

[20] *GI*, 3 July 1938; FO371/22021/J2677/40/1, Lord Perth to FO, Rm 6 July 1938.

Naturally, the policy culminated in laws that set out rigid rules of separate social and residential development based on race.[21] Wherever Italy had effective control, particularly in urban and settlement areas, these laws were fairly rigidly applied. The policy entailed massive forced removal of Ethiopians from their residences to new quarters. A form of apartheid was in the making, with the settlers and the Ethiopians living in geographically distinct quarters, each of them with their own social amenities and exclusive way of life. The two communities were forbidden to have any social contact with each other except of course when the imperatives of the labour-exchange market required it. In large urban centres Ethiopian ghettos developed which Europeans needed a permit to enter.[22]

Unlike the surrounding colonial countries, before the war Ethiopia had a growing and a well-articulated intellectual élite, many of whom had attended schools and universities overseas. Fearing perhaps that their acclaimed racial superiority could be badly exposed or their administration challenged, the Italians systematically eliminated the educated Ethiopians, as well as those suspected of having any meaningful education, and this was to have the long-term effect of setting back the development of the country for decades. Plans were also underway to restrict the education of Ethiopians to elementary level and, unless the needs of the settler farms made a necessary requirement, to exclude them from practical skill training. Most of the pre-war educational establishments and health facilities were already devoted to the exclusive use of the Italians. During the occupation the Ethiopians had unprecedented employment opportunities, but they were relegated to the most demeaning jobs and explicitly excluded from participation in any sector of the economy where they might compete with Italians.[23]

Attempts to isolate the Ethiopian population culturally and physically from outside contact led to the expulsion and ruthless treatment of the non-Italian missionaries as their thoughts and ideas regarding the natives

[21] CD IAO Misc. 1038, Alberto Pollera, Europei e indigeni nella valorizzazione e nell'economia dell'Impero, As 7 Oct. 1938, pp. 2–5.

[22] M. Moreno, 'Politica di razza e politica coloniale', *AAI* 2/2 (1937): 456–67; FO371/22021/J3439/40/1, D. M. Riches [Acting Consul] to HM CG, Hr 12 Aug. 1938. A. Berretta, *Con Amedeo D'Aosta in AOI in pace e in guerra* (Milan, 1952), 139.

[23] ACS GRA 46/41/9, Petretti to Graziani, AA 13 May 1937; FO371/23376/J574/41/1, CG Bird to PSSFA, AA 2 Jan. 1939. For a more detailed discussion see R. Pankhurst, 'Fascist Racial Policies in Ethiopia 1922–1941', *EO* 12 (1969): 270–86; *Universalità Fascista* (Nov. 1937): 7; L. Goglia, 'Note sul razzismo coloniale fascista', *Storia Contemporanea*, 6 (1988: 1223–66); D. Fossa, 'Sane norme economiche e morali per i cittadini dell'Impero', *AC*, 26 Jan. 1939.

were thought to conflict with Italian interest. With a handful of exceptions, mission properties were summarily seized. No foreigner of any religion was allowed to set up schools of any class in IEA. The Italian clergy replaced the non-Italian Catholic missionaries. Men of talent and long experience, like the French prelate, Monsignor Jarousseau, a saintly old man with fifty years' residence in Ethiopia, beloved by all who came into contact with him, was replaced by an Italian who was undoubtedly a political bishop of the most ordinary kind. The largest beneficiaries were particularly the Consolata Mission, who played a crucial role in the Italian penetration and occupation of the country, and the Capuchin Order, and, to a smaller extent, the Verona Fathers. These missionaries ran most of the newly organized schools, almost all of them bearing the names of the Italian 'heroes' who fell in Italo-Ethiopian wars.[24] Like its early precursors, the Italian Catholic Church, whose missionary scramble for the Ethiopian land and soul had proved a disastrous failure in the past, had welcomed the Italian occupation in the mistaken belief that it would ease its proselytization activities in a country that had so far consistently frustrated its missionary endeavour. Indeed, Italian occupation made Ethiopia the exclusive field of operation to the Italian Catholic clergy. Missionary stations were expanded and Church administration restructured but no significant conversion was made. Nor did the missionaries, despite their resounding pronouncements on assisting the Ethiopians in the civilizing process, make any combined, and very few individual, attempts either to defend Ethiopian rights or condemn the excessive abuses by the Italians. On the contrary, crude racial policies were extended to liturgical spheres which had the effect of alienating the population and causing serious friction between the followers of the Latin rite, who operated under the auspices of the Sacred Congregation of Propaganda Fide, and the indigenous clergy of the EOC rite, subject to the Sacred Congregation for the Oriental Churchs.[25]

The ruthless imposition of the discriminatory rules was partly responsible for a protracted guerrilla war that sapped the settlers' moral and material resources and led to constant reshuffling of the administrative

[24] FO371/22020/J369/40/1, Lord Perth to FO, Rm 24 Jan. 1938.

[25] FO371/20209/J8291/4321/1, Mr Mallett [Holy See] to Anthony Eden, Rm 20 Oct. 1936: FO371/22020/J369/40/1, Lord Perth to Anthony Eden, Rm 24 Jan. 1938; *GP*, 22 Oct. 1936. As a rule, the native clergy was not allowed to perform functions, particularly celebrate the Holy Mass, on the altar destined to the Latin rite (i.e. the Whites) and across the country the Ethiopian clergymen of the EOC rite told the author strange tales relating to cases of reconsecration by the Italian priests of altars where functions of the Ethiopian liturgy took place on the grounds that their action profaned the holy altar.

machine. The pacification programme demanded an extensive road network. The Italians, setting great store by the imperial Roman dictum that roads are the basis of a military control and prosperity, within twenty-four months of their conquest completed no less than 2,000 miles of fully metalled and tar-treated roads at an average cost of £12,000 a mile and employing an Italian workforce which at the peak period stood as high as 60,000 men. As a result, provinces were brought closer to each other than ever before, greatly undermining the power of the unruly provincial élites. But the roads ate up a disproportionate amount of the budget and led the Empire into near bankruptcy. Financial stringency demanded the slowing down of the work and the expensive and restive European labour force was gradually replaced by local Ethiopian labour. By May 1939 the former was reduced to 12,000, while the latter had reached 52,000.[26]

Nor were military objectives that justified the construction of these roads achieved. In the governorates of Härär and Galla and Sidamo, organized armed resistance came to an end with the capture and subsequent execution of Ras Dästa in February 1937. But passive resistance continued in a variety of forms, fuelled by the harsh and arbitrary treatment especially after the assassination attempt on Graziani and the almost gratuitous failure to take practical measures that might win the confidence of the population. In the traditional Abyssinian heartland, such as Amara and Shäwa, Italian control was confined to areas close to highways and major towns. In the rest of the countryside, including other minor urban centres, the resistance front, albeit disunited, disorganized, badly armed, and ignored by the rest of the world, had a firm foothold. Indeed, it is difficult even with massive military force to subdue people fighting on their own ground.[27]

This situation aggravated the difficulties caused by what was virtually a permanent war economy, the inevitable result of sanction and the immense cost of conquest, and led to high inflation, chronic problems in balance of payments, and a reduction of commercial contacts with the outside world. The rigid measures of autarky increased taxation and created a serious shortage of imported goods. The beleaguered economy only benefited the arms dealers who made fat profits, while business

[26] Cf. Quaranta, *Ethiopia*, 74–93.

[27] FO371/22021/J1321/40/1, Sir M. Lampson to PSSFA, Cairo 21 Mar. 1938; FO371/22021/J1221/40/1, Mr Starling to Mr Campbell, London 25 Mar. 1938; Masotti, *Ricordi*, 38, 40, 42. Additional information on resistance given in Biblio.: see entries under Dämsé Wäldä Amanuél, Gärima Täfärra, Käbbädä Täsämma, R. Pankhurst, 'Ethiopian Patriots', Salome Gabre Egziabher.

suffered severely from rigid trade restrictions.[28] Trade was also worsened by soaring transport costs and expulsion of well-established foreign firms. Non-Italian businesses of any importance were compelled either to liquidate or enlist Italian interests in partnership.[29] An acute shortage of foreign currency prevented efficient use of the Addis Ababa–Djibouti railway, and the diversion of trade to roads only spiralled the costs.[30]

As in Eritrea, in Ethiopia the exclusion of foreign competition and monopolistic or near monopolistic rights gave the new Italian businesses a protected market. By the early 1940s about 4,007 industrial and 4,785 commercial firms with a total capital of about L.2,700,000,000 and L.1,100,000,000 respectively were authorized to work in IEA and another 4,452 applications were under consideration. Out of these 1,225 industrial and 1,435 commercial companies, with a capital of L.458,598,000 and L.603,322,000 respectively, applied to work in the four Ethiopian governorates.[31] Indeed, taking into account the unattractive conditions of the country, these developments were staggering. But only 400 industrial and 650 commercial firms actually made some investment in the Empire; the remainder simply acted as distributing or purchasing agents for their mother firms in Italy. Most importantly, it is hard to speculate on how many of these were new for many consisted of confiscated firms whose existence predated Italian occupation.[32] Even though the officialdom claimed that by early 1939 the capital sunk in industry amounted to L.135,500,000, this needs qualification.[33] As the economy was corporatist in character, primarily run by large parastatal corporations, the State

[28] FO371/22020/J657/40/1, CG Bird to Mr Lambert, AA 14 Jan. 1938; Pankhurst, 'Economic Verdict', 76–7.

[29] FO371/22020/J641/40/1, CG Bird to FO, AA 4 Jan. 1938; FO371/23376/J574/41/1, CG Bird to FO, AA 2 Jan. 1939; FO371/24635/J466/18/1, ACG Gibbs to Cavendish-Bentinck, AA 30 Dec. 1939.

[30] FO371/23380/J1776/296/1, CG Bird to Visc. Halifax, AA 1 Apr. 1939.

[31] FO371/24635/J412/18/1, Sir P. Loraine to Visc. Halifax, Rm 29 Jan. 1940. Not only are the authoritative figures given by the Italian sources conflicting, but most significantly some authors consider these simple authorizations as actual investments, see Villa Santa *et al.*, *Amedeo*, 272–5, a source also used quite uncritically by Sbacchi, 'Italian Colonialism', 438–40.

[32] The most unfortunate victims of this policy of *Italianization* of business were the Indian house of G. M. Mohammedally and Co., a leading import–export company established in 1888 and with an extensive business network, including agriculture and banking, throughout the Empire; the Arabian Trading Company, trading in skins and hides; the Ethiopian Mechanical Transport Co. engaged in road construction, and the French firm A. Beese. Cf. FO371: 22020/J641/40/1, CG Bird to FO, AA 4 Jan. 1938; 22021/J1221/40/1, Starling to Campbell, London 25 Mar, 1938; 22021/J1804/40/1, CG Bird to FO, AA 2 Apr. 1938.

[33] FO371/23380/J1776/296/1, CG Bird to Visc. Halifax, Ethiopia: Annual Report Economic A, AA 1 Apr. 1939.

remained the major investor. The corporations exercised a near monopoly in their particular line and were subject to serious criticism for their inefficiency, abuse, and anti-entrepreneurial attitudes even by normally tightlipped Fascist press and hierarchy.[34] Close observation reveals that industry was dominated by construction and transport firms—building, road-making, and hydraulic works—, and commerce by import and export and trade in foodstuff. Investment in these areas, where capital requirement was minimum, risks small, turnovers immediate and large, State subsidies and privileges considerable, was a bonanza. Otherwise, the Italian capital maintained its traditional circumspection and timidity and even those few willing private investors and enterprising merchants complained of the serious difficulties they had to contend with in their attempt to find their way through a complicated network of permits, red tapes, and restrictions of all types or of the large percentage of their profits being taken by state corporations in a variety of exactions. Outside Eritrea, the trade companies were largely concentrated around Addis Ababa.

Owing to the traders' severe lack of real information of and contact with the country, their inability to converse with all the intricacies of the Ethiopian business methods, and unfamiliarity with their psychology, with a few exceptions most trade dried up or was reduced to a fraction of its former level.[35] Contributing partly to this decline was also Italy's mounting desperation to earn foreign exchange which dictated the government to concentrate its economic policy not on stimulating exports to Italy but rather forcing goods on to the world market. Export to Italy of most of the principal products of Ethiopia—coffee, hides and skins, leather, oilseeds and beeswax, and others—was either forbidden or only allowed in small and controlled quantities. The restrictions rebounded on the economy by discouraging the bona-fide traders and, as exports fell considerably, the government was to confess its failure repeatedly by making frequent resort to raise the export quotas to Italy.[36] Trade was

[34] FO371: 22021/J2376/40/1, Lord Perth to FO, Rm 11 June 1938; 22021/J2926/40/1, CG Gibbs to FO, AA 5 July 1938; 23380/J296/296/1, CG Bird to FO, AA 30, Dec. 1938; 24635/J412/18/1, Sir P. Loraine to Visc. Halifax, Rm 29 Jan. 1940; ACS SPDR 44/242/R/39, Farinacci to Presidente, Cremona, 24 Apr. 1938 and 25 Dec. 1938.

[35] One such exception was the export of hides and skins (8,600 tons in 1937/8) which, despite some restrictions, seemed to compare favourably with the average annual exports of the pre-occupation years (6,400 tons during the years 1931–4). Cf. FO371: 22021/J2363/40/1, CG Bird to FO, AA 24 May 1938; 23380/J1776/296/1, CG Bird to Visc. Halifax, AA 1 Apr. 1939.

[36] Import to Italy of coffee e.g. from an initial 20% moved up to 30% and then to 40% of the total export from AI. Then a 1939 decree allowed the unrestricted import of the cheap Jimma coffee, while maintaining the ban on the import of the superior Härär coffee.

further disrupted by forced replacement of MTD with unstable Italian paper currency[37] and the requisition of agricultural produce from the Ethiopians below market price. Except in flourishing black-market export of lira, the policy of economic progress proved a dismal failure.[38]

## DEBATE OVER STRATEGIES

The series of pre-occupation expeditions and studies had equipped the colonial administration with considerable knowledge of Ethiopia's agricultural environment and productive potential. In the highlands of Eritrea, the system and use of land tenure was almost identical to that of Ethiopia. Yet surprisingly, despite their long contact with the area, the Italians claimed that, to their great dismay, their knowledge of Ethiopia at the time of the occupation was 'fragmentary and unreliable'.[39]

Technical and agricultural experts followed in the footsteps of military conquest. Data were collected concerning climatic conditions, natural environment, ploughing methods, terracing and cultivation systems, seeds and implements, as well as market conditions, aided by photographic materials, for the Agricultural Department (Direzione dei Servizi Agrari) in the MC. Their reports urged caution against the prevailing enthusiasm for mass settlement.[40]

See FO371/23380: J1776/296/1, CG Bird to Visc. Halifax, AA 1 Apr. 1939; J2209/296/1, ACG Gibbs to FO, AA 12 May 1939; J4814/296/1, ACG Gibbs to FO, AA 18 Sept. 1939; 'Notiziario agricolo commerciale: AOI', *AC* 32/7 (1938), 380–1.

[37] The Ethiopians accustomed to silver MTD, the currency of Ethiopia for a long time, viewed the Italian paper L. with suspicion and the government's inconsistency in its attempt fictitiously to fix the exchange rate did not help. Initially, the exchange rate was put at 5 MTD/L., then at 8.5. Later on 19 Feb. 1937 this was raised to 10.5 and again on 18 June to 13.5, only to be reduced to 10.5 on 25 Sept. All this took place against the background of the unofficial market rate where throughout the year MTD stood considerably higher than the L. In midsummer e.g. it rose to L.19 in AA and over L.20 elsewhere, with an average of between L.15 and L.16 per MTD. (Cf. ACS GRA 46/41/9, Lessona to Graziani, Rm 14 Apr. 1937; FO371/22020/J641/40/1, CG Bird to FO, AA 4 Jan. 1938; Villa Santa *et al.*, *Amedeo*, 280–2.)

[38] In 1938 the official rate of exchange for the lira was L.89 to the pound. However, in Italy the unofficial rate was as high as L.130, and in AA between L.170 and L.180 to the pound. Cf. FO371/23380/J1776/296/1, CG Bird to Visc. Halifax, AA 1 Apr. 1939; FO371/24635/J5/5/1, CG Gibbs to Visc. Halifax, AA 27 Nov. 1939; R. Pankhurst, 'A Chapter in Ethiopia's Commercial History: Developments during the Fascist occupation 1936–1941', *EO* 14/1 (1971), 47–67; id., 'Economic Verdict', 74–5.

[39] MAI, *La costruzione dell'Impero, AAI* 3/1 (Milan, 1940), 947.

[40] CD IAO AOI 1923, Promemoria per il Sign. Direttore Generale dell'AOI circa lo studio del problema fondiario ai fini della formazione di terreni per uso agricolo, Rm 1 July

Despite this rather muted judicious assessment by a few individuals, the effect of the facile victory combined with the fascist mentality lent weight to unite the press, the public, and the pulpit[41] behind Mussolini's extravagant plans. Those who captured the popular imagination were particularly the hand-picked journalists, uniting the nation—North and South, Catholics and Protestants (Waldensians), liberals and republicans, socialists and nationalists—in the cause of conquest.[42] The poet Vittorio Emanuele Bravetta eulogized:

Our People!
[to alien land] emigrate no longer
in order to suffer!
The colonists' fecund labour and drive
Will soon make entire Ethiopia blossom and thrive.

The rumbling *nägarit* sounds
are dead silent for long.
Long live [our] Emperor King!
War we'll wage at all costs
on whomever dares
to usurp from us
what is already a Roman land![43]

1936. In a footnote to the document A. Maugini, after having read the agricultural technicians' report, urges the Minister of Colonies (Lessona) 'to pour a bit of cold water on the ardour of agricultural colonization'. A. Di Crollalanza, 'La valorizzazione agricola dell'Impero', *REAI* 25/4 (1937), 491.

[41] Perhaps the speech made by the papal envoy, Mgr. Castellani, archbishop of Rhodes, on the occasion of his pontifical mass in AA was the most representative view of the Italian Catholic Church: 'I had come here to greet the officials, the soldiers, the blackshirts and the [Italian?] people in Abyssinia both as a brother for I too am an Italian, and as a father, for I had been sent to you by the Supreme Teacher of Christianity. I salute all the heroic soldiers of the army which the world marvels at, but which Heaven has no need to marvel at since it was their ally . . . Italy is the country destined by God to make civilization and the glory of the Church ever greater in the world. Thanks to the magnificent work of that great figure, the Duce, who combines the calmest balance with the daring of a hero, it will be the Roman Empire which will carry the cross of Christ into the world. It is the Duce who, in the opinion of the people, will always uphold the conception of just and lofty peace.' (FO371/20209/J8291/4321/1, Mr Mallet to Anthony Eden, Rm 28 Oct. 1936). Enclosed with this document is also Mr Mallet's despatch no. 140 which is the translation of Cardinal Tisserand's interview to the *GP*, 22 Oct. 1936 on the planned restructuring of the Catholic Church in Ethiopia.

[42] Del Boca, *Caduta*, 8–9; ACS MCP B18bis/261, Servizio radiofonico speciale per l'Impero, Rm 24 Nov. 1938.

[43] 'La nostra gente / or non emigra più / per sofrir! / Il fecondo lavor / dei coloni / tutta l'Etiopia farà / fiorir. / Il *negarit* / non rimbomba più. / Viva il Re Imperator! / Guerra a chiunque vuol / usurparci il suol / che romano è già!' V. Savona and M. Straniero, *Canti dell'Italia Fascista* (Milan, 1979, 261–2; Del Boca, *Caduta*, 9–10).

Once again Ethiopia was portrayed as the key solution to Italy's long, agonizing, and humiliating emigration problem, and as a fruitful field for Italian labour.

Such an optimistic view was shared and vigorously argued by the regime's intellectuals who maintained that the agricultural possibilities of the Empire were almost inexhaustible, contrasting it with the conditions of the other colonies now described as no more than 'a collection of deserts'.[44] Ethiopia was depicted as an El Dorado, an earthly paradise waiting to be exploited by Italian labour and technology, in order to make this wealth available to the civilized world.[45]

This propaganda momentarily succeeded in making large-scale settlement extremely popular in Italy, where unemployment and poor living conditions made the prospect of a better life very grim indeed. Foreign newspapers, sceptical about the optimistic assertions that became part of Fascist orthodoxy, were discredited and labelled malicious propaganda; Italians who failed to subscribe to the El Dorado myth were regarded as unpatriotic. Once the war was over, economic, political, and strategic factors made large-scale settlement seem even more plausible. Apart from providing the settlers with employment opportunities, the presence of a large Italian population in the Empire was thought to enable the colony to become self-sufficient economically and offer the motherland a protected market.[46]

The urgency of settlement was further dictated by the geographical distance separating the Empire from the motherland and the hostile colonial powers surrounding it. The settlers were thought to fulfil a function like that of the Siberian Cossacks in Tsarist Russia, both as a ready army to defend the Empire and, in case of need, the motherland, as well as a 'centre of radiation of Fascist and Italian civilization in Africa'.[47] The military factor remained an overriding argument of the pro-settlement lobby.[48] At a later stage, D. Fossa, the Fascist Party Inspector in IEA, described how: 'If it is the plough which traces the furrow it will be the sword which must defend it. The workers are soldiers and the soldiers

[44] L. Cipriani, *Un assurdo etnico: L'Impero etiopico* (Fl, 1935), 324; B. Pace, *L'Impero e la collaborazione internazionale in Africa* (Rm, 1938), 37.

[45] Lessona, *Memorie*, 353–5; Cipriani, *Assurdo etnico*, 323.

[46] R. Trevisani, 'Originalità nelle direttive e nei metodi di colonizzazione fascista', *REAI* 26 (1938), 538; C. Giglio, *La colonizzazione demografica dell'Impero* (Rm, 1939), 12; L. Diel, '*Behold our new Empire*'—*Mussolini* (London, 1939), 68–9.

[47] Giglio, *Colonizzazione*, 12; Trevisani, '*Originalità*', 583; J. Kulmer, 'The Return of Haile Sellassie', *Contemporary Review*, 159 (1941): 291.

[48] Giglio, *Colonizzazione*, 12; Terruzzi, 'L'economia dell'AI nel secondo anno dell'Impero', *REAI* 26 (1938), 366; G. De Michelis, 'La valorizzazione agricola dell'Impero', *REAI* 26/1 (1938), 6.

workers. Legionnaries are those who conquered the Empire, legionnaries will be those who with their toil will render it fertile.'[49]

Peasant settlement was the most archaic and expensive form of imperialism and yet the Fascist authors were at pains to contrast it with earlier types of colonization attempted by other powers, particularly the British, and to emphasize its uniqueness, based on the supposed fecundity of the Italian race. Their arguments, based on a few platitudes and generalizations, were remarkable for their pomposity rather than factual accuracy. Thus Nicolo Castellino claimed that the Italians had in their favour,

> all the conditions which other colonizing peoples lacked or lack: the numbers which were missing to the enterprise of the Portuguese, the youth which cannot fortify British imperialism; the dignity of race so often forgotten by the Spanish conquistadors and now abandoned and disowned by the French. . . . The fecundity of race with the eternal renovating spring of its children, will avoid for us the angry and constant complaint expressed by people who were once rulers but have now declined into an old age: *twenty-two English women live in Kenya—a million authentically Italian families are thriving in Italian East Africa.*[50]

Italian colonialists boasted that with their unique colonization techniques, they eschewed the plunder and exploitation that was so characteristic of the traditional 'industrial capitalist colonization' whereby a relatively small number of Europeans mercilessly exploited an indigenous population and squandered the colony's resources. A respectable academic stated, 'We cannot allow ourselves the luxury of "doing it the English way", that is to live and get rich at the expense of others. We must take to our Empire our own peasants, our own workers, our own artisans, which up to now wander around and are dispersed throughout the world.'[51] Another

[49] ACS MCP B236, Guerriero, Capitale e lavoro nell'Impero italiano in Etiopia; D. Fossa, *Lavoro italiano nell'Impero* (Milan: 1938), 484–6; IFPLAOI, *Tre anni di attività* (AA, 1940), 164–6.

[50] 'Il problema del meticciato', *Nuova Antologia*, 399 (1938), 394–5; (author's italics); cf. C. Poggio, *Politica economica imperiale con particolare riguardo all'AOI* (Hr, 1939), 35–8; A. da Marsanich, 'Per l'autonomia economica dell'Impero', *REAI* 26 (1938): 849–50; M. Pomilio, 'I problemi attuali dell'Impero nel pensiero di Attilio Terruzzi', *Autarchia Alimentare* (Aug. 1938): 7, also in *Alimentazione Italiana*, 13–31 Aug. 1938. Castellino's arguments, like those of most of the firebrand and militant Fascists, are obviously unfounded and, though effective as propaganda, do not stand up to close scrutiny. Even according to the old 1914 census, in Kenya the bulk of the female population belonging to the settler farming community alone numbered 1,794. At its peak in 1939—which is also the time Castellino wrote his article—the Italian population in IEA totalled about 130,000 and that of the entire empire, including Libya, about 305,000. Cf. D. A. Low and A. Smith (eds.), *History of East Africa*, V. iii (Oxford, 1976), 434; Segrè, *Fourth Shore*, 4.

[51] G. Gennari, 'La colonizzazione agraria di popolamento nell'economia corporativa dell'Impero: Osservazioni di un legionaro', *Georgofili*, 6/2 (1936), 511; id., 'L'agricoltura nell'AOI', *REAI* 25/12 (1937), 1916.

commented: 'Our imperialism differs profoundly from the English in an essential point: we have the demographic possibilities which the English do not have.'[52]

The Italians worried, however, that an influx of lower-class Italians might damage their prestige and dominance, as Italians and as Europeans, over subject Ethiopians.[53] The industrial-capitalist colonization tried to avoid this by employing a relatively small number of Europeans in managerial positions while the indigenous population did the heavy work. The most universally held view was that in the case of Italian colonialism, the 'prestige of the White race' would be maintained by educating the workers in an imperialist mentality and employing a vigorous policy of racial segregation. The point of reference were native reserves in Kenya and South Africa.[54] Such a device, combined with the 'prodigies of civilization that the White man introduces through the construction of roads, aqueducts, schools, buildings, and all forms of gigantic works', would keep the prestige of the White man undiminished.[55]

Settlement was conceived in the most grandiose terms. Soon after his victorious entry into Addis Ababa, Marshal Badoglio declared that the figure of one million settlers within a year 'was not an exaggeration'.[56] Another Fascist official talked glibly of 6,250,000 souls, consisting of 1,250,000 heads of families, who would be settled on 50,000 plots, each plot consisting of 1,000 ha. This was all to take place within the early part of the occupation.[57] Although by the end of 1939 it was clear these claims were unrealistic, Party officials and some journalists continued to perpetuate the illusion.[58]

Opinions were divided over the policy to be followed in transforming Ethiopia into a settlement colony. The MC, led by Lessona, argued for State-sponsored colonization. Lessona was a man with rich colonial experience. Once an enthusiastic student of colonial problems, before he became Minister he served as Under-Secretary to the Colonies (1929–36). During this period he was credited with spearheading a campaign that

[52] Giglio, *Colonizzazione*, 9.

[53] The Italians were aware that in the pre-occupation period the Ethiopians had a very low opinion of the Italians. Cf. R. Pankhurst, 'Italian and "Native" Labour during the Italian Fascist Occupation of Ethiopia, 1935–1941', *Ghana Social Service Journal*, 2/2 (Nov. 1972), 56.

[54] Giglio, *Colonizzazione*, 10; Gennari, 'Colonizzazione', 508; Lessona, *Memorie*, 298.

[55] Gennari, 'Agricultura', 1916; C. Poggiali, *Albori dell'Impero: L'Etiopia come è e come sarà* (Milan, 1938), 517–20.

[56] *GP*, 30 June 1936; Del Boca, *Caduta*, 199.

[57] CD IAO AOI 1778, C. Poggio, Rilievi e proposte per la pronta valorizzazione dell'Impero (Hr. Oct. 1936), 13–14. Later he had reduced this figure to 1,562,500 agricultural settlers. See C. Poggio, 'Colonizzazione capitalistica e demografica nell'AOI', *Illustrazione Coloniale* (Dec. 1939).

[58] *Dottrina Fàscista* (Jan. 1939), 97.

accelerated colonization under State direction at the expense of individual enterprise. His policy in Ethiopia owed much to his Libyan experience. Yet his main objective was to make of the Italian East African Empire a model colony where the shortcomings of Libyan colonization could be eliminated and experiments pursued on a wider scale.

Lessona envisaged four distinctive forms of colonization corresponding to four categories of lands.

1. National demographic colonization implemented by State-run colonization agencies. The peasants would settle in legions of militia and function as military garrisons. After a given period, the land under cultivation would become the property of the peasant.
2. Small settlements where small- and medium-sized estates would be worked by Italian farmers who had modest initial capital and know-how.
3. Industrial colonization where estates, impractical for demographic settlement, would be given to competent Fascist confederations which would work under the supervision of the MC and local government.
4. Native settlements would include estates under Ethiopian ownership. In view of both improving and increasing local production and elevating the lifestyle so Italian goods would be purchased, the Ethiopians would cultivate these lands while the government would provide them with technical expertise, including high-yield seeds and advanced agricultural tools.[59]

Lessona stressed demographic colonization, pointing out its importance and uniqueness. Demographic settlement had its precedents in American colonization, and aspects of it had also been practised by the French and the British in their colonies.[60] But Lessona rejected any comparison to the Italian scheme, including that of the French in Algeria which was traditionally seen by many Italians as a model.[61] Indeed, Lessona's programme had certain distinctive aspects: settlements were regional in character, whereby people of the same province in Italy populated an area which reproduced the atmosphere of the region from which they sprang. According to Lessona, this was devised with two functions

[59] ACS GRA 45/4/4, Lessona to GG AOI, p. 8; Lessona, *Memorie*, 303–4; R. Pankhurst, 'A Page of Ethiopian History: Italian Settlement Plans during the Fascist Occupation of 1936–1941', *EO* 13/2 (1970), 145.

[60] A typical ex. was the Tennessee Valley Authority, the French settlement scheme in Algeria. Cf. A. F. Robertson, *People and the State: An Anthropology of Planned Development* (Cambridge, 1984), 14–15, 107–8, 130–1.

[61] According to Lessona the Fascist corporative state was not comparable to others: 'L'agricoltura tripolitana', in *Scritti e discorsi coloniali* (Milan, 1935), 130.

in mind: first, it would rouse the ego of the provinces to compete with each other to provide initial capital expenses for agricultural development; secondly, it was hoped that it would facilitate group solidarity.[62]

However, there was nothing new in Lessona's advocacy for State intervention. Many before him had favoured State-run colonization agencies as the best means of providing the financial and technical resources that private entrepreneurs usually lacked. Nevertheless, Lessona was blamed for overlooking the fact that Italians had shown almost no desire in the past to emigrate to colonial territory. In fact, the sceptics pointed out that forty years previously Franchetti had pursued unsuccessfully an identical scheme in Eritrea.[63] Lessona retorted that Franchetti's plans had only one flaw: they had not been carried out energetically.[64] Like Franchetti, Lessona was unsympathetic to private initiative, knowing from his Libyan experience that it would lead to land speculation and the formation of *latifundia*. But Lessona was inconsistent. Even though he continued to maintain that the best lands be kept for demographic settlement, it was capitalist initiative that gained paramountcy, nor were the settlements he encouraged exclusively regional.[65]

State-supervised colonization was the brainchild of Armando Maugini, a man with much agricultural experience and knowledge. As an agricultural officer in Cyrenaica during World War I and as organizer of the colony's technical services, he had gained a rare insight into the problems of settlement schemes, later becoming the major architect of Libyan colonization. He was a convert to the cause of State-funded and -directed projects of land settlement, which he had opposed until 1932 when he was called upon to develop plans for the mass colonization of Cyrenaica. Initially, he offered qualified support; but when he realized that it was a politically motivated enterprise, he upheld it wholeheartedly, arguing strongly for the creation of a special corporation that would promote demographic goals rather than a reliance on private colonization.[66] During the Ethiopian conquest, Maugini became chief technical adviser to the MC as well as Director of the Istituto Agronomico Coloniale (Institute of Colonial Agriculture) in Florence,[67] a leading institution for the study of tropical agriculture and the training centre for agricultural technicians.

[62] Id., *Memorie*, 283–4.
[63] See Ch. 1, above.
[64] 'Lo stato Fascista e l'economia coloniale', in *Scritti*, 49–50.
[65] CD IAO AOI 1936: Ministro to GG, Rm 2 June 1937; IA to GG, Rm 23 June 1937.
[66] Segrè, *Fourth Shore*, 96.
[67] It is now known as the Istituto Agronomico per l'Oltremare (Overseas Agricultural Institute).

Maugini believed that the Italian presence would make itself felt in Ethiopia within a very brief period provided that there was goodwill and appropriate technology. But he maintained that colonization had to be centrally planned and preceded by a proper study. In his view demographic goals could be economically implemented by para-statal organizations with financial and technical resources—thus not exacting undue sacrifice from central government.[68]

Maugini distinguished between areas where work should start immediately and those requiring long-term development. But in this he was neither clear nor consistent. On the one hand, he argued that the programmes be started immediately, although with a limited range and clear objectives. On the other, he called for common sense and caution. He stressed that demographic colonization, even in its most economically ideal conditions, was a very costly affair, requiring solid financial, administrative, and technical organization.[69] He pinpointed the existence of large estates with excellent conditions for demographic colonization in many of the areas he visited; yet he claimed that, despite their healthy climate and enormous agricultural potential, their location did not qualify them as areas of immediate activity. Ethiopian competition, absence of adequate infrastructure, and poor markets made other areas unattractive. Maugini believed that, unless there was a guarantee of prosperity to offset the sacrifices of colonization, it would not be easy to persuade any prospective peasant settlers to move to Ethiopia.[70]

In dealing with Ethiopian peasant farming, Maugini maintained that this needed progressive improvement. According to him, improved standard of living of the Ethiopian masses would increase their purchasing power of Italian manufactured goods, whilst increased cereal, particularly wheat, production would contribute to the Empire's agricultural self-sufficiency. In his view, the improvement could be carried out in two stages; the first would involve the expansion of existing agricultural produce, using intensive methods and the construction of silos for storage; the second would be to introduce cash-crop agriculture or the expansion of existing crops, so that the colony could supply the raw materials needed by the motherland and earn itself foreign currency. He urged the government to use intensive propaganda and create the necessary incentives, especially by purchasing the produce.[71]

[68] ATdR 24/100, Relazione Prof. Maugini sul viaggio compiuto al seguito di S.E. Tassinari, gennaio–febbraio 1937, Rm 26 Mar. 1937, p. 16.

[69] Ibid., p. 15. [70] Ibid., p. 14. [71] Ibid., pp. 18–19.

Like Lessona, Maugini was unsympathetic to private agricultural enterprise.[72] He stressed, with the exception of a few experimental cases, that this be channelled into areas economically ready for capital investment (close to urban centres), as any attempt to direct it into remote zones, no matter how great their economic potential, would encounter insurmountable difficulties.[73]

Although not as loud as those advocates for State-sponsored mass settlements, the dissenting voices were considerably strong and were primarily expressed by Tassinari, the Under-Secretary at the Ministry of Agriculture. Tassinari had travelled extensively in Ethiopia in the company of Maugini and two other officials. As declared in his report, Tassinari went under instructions from the Duce to 'take note of all and everything: agriculture, economy, society, military and ethnology; report all of this to me in absolute truth'.[74]

Like Maugini, Tassinari stressed that colonization should be a gradual process, but unlike him, advanced the idea that the major task of implementation should be left to 'private initiative' which 'must be guided, directed and controlled by the State in line with the Fascist norms of life', while the government should provide the basic economic infrastructure. His view was that capitalistic colonization could create a number of conditions that would facilitate progressive Italian settlement, including a demographic one.[75]

The main reason determining Tassinari's position was economic. He estimated that the initial cost of settling a family, even in ideal economic conditions, would be enormous: one thousand families would need an expenditure of L.50,000,000, ten thousand families L.500,000,000, and so on. In addition to this, demographic settlement would demand roads, hospitals, schools, and churches, so that the lifestyle of the settler would not become 'debased' to the level of that of the Ethiopians.[76]

Tassinari's decision to publish a section of his data in the country's leading newspaper, *Corriere della Sera*, contradicting the MC's view and claiming that it corresponded to Mussolini's thought, worsened the already delicate relationship between himself and Lessona.[77] Yet Tassinari's conclusion was sustained by a commission of members of the CFA, despatched by Lessona himself with the 'task of reporting in detail the

[72] Ibid., pp. 17–18; Maugini, 'L'agricoltura nelle colonie: Esperienze e nuovi doveri', *AC* 30/11 (1936), 417. [73] ATdR 24/100, Relazione Maugini, pp. 16–17.

[74] CD IAO AOI 1990, G. Tassinari to Duce, Relazione del viaggio compiuto attraverso i territori dell'Impero, gennaio–febbraio 1937, Rm 23 Mar. 1937, 1.

[75] Ibid., 7–9, 15. [76] Ibid., pp. 7–8. [77] Lessona *Memorie*, 285.

impressions and considerations gained during their journey of an economic, financial, technical, and social character that could have a bearing on agricultural colonization as well as native farming'.[78] The Commission seemed to have a deep distrust of demographic colonization. Their report was strongly in favour of capitalist agriculture on the one hand, and of Ethiopian peasant farming on the other. After having thoroughly discussed the crucial problem of availability of colonizable land and the enormous economic cost necessary for demographic colonization, the Commission concluded that the 'tides of colonization and the activities of the nationals should be limited to those forms and entities that can develop by themselves, let us say, by virtue of their own energy, by their own direct initiative, whether encouraged or given all possible assistance or not, but stimulated by a healthy concept of economic profit.'[79]

They were even more emphatic in suggesting that the best means to achieve the goal of agrarian development of the Empire more rapidly and economically was by employing the local population. Dealing with demographic colonization, they stressed the fact that the country was not suited to a massive influx of colonists, as the basic economic infrastructures favourable to the purpose were virtually non-existent. They warned the colonial administration to move cautiously. According to the report, in this manner valuable experience could be gained and mistakes corrected as they arose.[80]

The result of these conflicting interests and opinions was a stalemate, with the MC uncertain how to proceed. Its early enthusiasm for demographic colonization was soon dampened, and attempts were made to defer its execution. When this was voiced, the advocates of an immediate transfer of masses of peasants felt outraged and accused the authorities of betrayal. Giuseppe Pironti di Campagna's statement was not an isolated feeling:

> Any policy that tends to limit and hinder the emigration of nationals towards the colonial territories frustrates the basic reasons of the foundation [of the Empire], betrays its defence, and is against 'the interests of the Empire'. Indeed, the very fact of delaying the flow of migration waves towards the colonies, after so many sacrifices have been made, is an absurdity contrary to our national cause and aversion to our interest that was established with the proclamation of the Empire.[81]

[78] ATdR 24/103, Relazione di una missione di agricoltori in AOI, 5 marzo–9 aprile 1937, Fl, 1937, p. 5.
[79] Ibid., p. 45. [80] Ibid. [81] *La difesa dell'Impero* (Rm, 1937), 66.

Towards the beginning of 1937 the controversy between the two factions became so acute that Lessona, blamed by the supporters of both capitalist as well as demographic colonization for trying to thwart every initiative, was forced to defend his position in parliament:

I would like to assert solemnly in front of you, honourable comrades, that the demographic colonization of IEA is and remains one of the fundamental objectives of the Fascist government. Certain groups have given themselves to doubting, stating that the implementation of the programme is neither easy nor practical. The government does not hide the difficulties as it did not fail to curb the enthusiasms. This is a part that, though unpleasant, we have assumed willingly knowing that problems of such magnitude do not admit of magical solutions but it is essential that they should be seriously studied and worked out. If, on the one hand, I feel compelled to quell, after all, the justified impatience of many comrades and important entrepreneurial classes, on the other hand, I have also to declare firmly that, should we abandon the decision of settling a large mass of Italian workers, we will frustrate one of the central pillars of the Ethiopian enterprise, so solemnly and repeatedly proclaimed by the Duce in his honourable speeches to the Italian people.[82]

Attempting to reconcile the political goals of colonization with the realities of the Ethiopian environment and Italy's financial resources, towards the end of March 1937 Lessona enlisted the co-operation of the Agricultural Council. This was one of the supreme bodies created by Lessona himself to co-ordinate economic activities in the colonies with those of metropolitan Italy. Its role can be described as the Minister's own mini-cabinet.[83]

The Council, the Consulta di Agricoltura, included men with considerable experience in colonial agriculture, almost all of them advocates of colonial expansion. Their conclusions would hold considerable weight with the policy-makers. A special commission was set up to study Lessona's programme in detail, in which most of the interests already operating or intending to operate in Ethiopia were represented. The two major Fascist farming associations CFLA and CFA, were represented by Dr Dallari and Ing. Romadoro. There was Nallo Mazzocchi-Alemanni, Inspector of the ONC; Conte Livio Gaetani, National Secretary of the Sindacato Nazionale Fascista Tecnici Agricoli (National Union of Agricultural Technicians'), an organization involved in establishing the Empire's agricultural services; Rienzi, Chief Secretary of the Consiglio Tecnico-

[82] *AI nel 1° anno dell'Impero* (Rm, 1937), 11. [83] Id., *Memorie*, 286.

Corporativo (Technical Corporative Council) of the MC. Maugini was the Commission's chairman and Dr Pirró its secretary.[84]

The Commission rejected the views held by advocates of immediate mass peasant transfer into Africa as simplistic. A strategy of gradual colonization was also dismissed as cowardly. But the Commission failed to give clear guidelines on how to proceed and its conflicting views are recorded in the minutes of meetings held between April and July 1937. The outcome of these deliberations was an endorsement of Lessona's ideas, which continued to mould Italian colonial land policy even after he was demoted from his ministerial position.[85]

Despite differences amongst the members in particular policies to be adopted, especially relating to the degree of State intervention, the Commission reached consensus on the main objective of Italian agricultural activity. It summed up the objective in three points:

1. To seek urgently and in the most complete manner self-sufficiency in foodstuffs, whereby the Empire would become independent from external supply.
2. To create centres of settlement and employment opportunities for Italian rural populations.
3. To secure low-cost raw materials for the Italian economy and, in the long run, create a new economy in the Empire, bent on capturing neighbouring foreign markets.[86]

Yet the question remained as to how these goals would be achieved. In this regard, the Commission only repeated the well-worn options of demographic colonization, capitalistic colonization, and native cultivations. A number of intermediary forms were excluded, and the mutual relationship and influence between these forms were not considered. The Commission itself made these limitations abundantly clear.[87]

[84] CD IAO AOI 1775, Processo verbale della Commissione di Competenza per L'Agricoltura, Seduta 7 Apr. 1937, Rm 12 Apr. 1937, p. 1. Most of the material on the background of the personalities was provided by Prof. Gian Carlo Stella under the title 'Qualche note su organizzazioni agricole Fasciste', Ravenna 3 Aug. 1985; see also 'Rassegna agraria coloniale', *AC* 31/9 (1937), 391.

[85] CD IAO AOI 1776, Relazione della Consulta per l'Agricoltura al MAI sui problemi relativi all'avvaloramento agrario dell'Impero, Rm 23 Sept. 1938, pp. 3–4. The report is published as Consulta, 'L'avvaloramento agrario dell'Impero', in *REAI* 25/10 (Oct. 1937), 1561–75.

[86] CD IAO AOI 1776, Relazione, p. 1, and Consulta, 'Avvaloramento', 1561. This same view would be retained by Lessona's successor, Terruzzi. See FO371/24635/J412/18/1, Sir P. Loraine to Visc. Halifax, Rm 29 Jan. 1940.

[87] CD IAO AOI 1776, Relazione, 2, and Consulta, 'Avvaloramento', 1561–2.

The Commission regarded the transplantation of Italian peasants and their way of life as the best means of guaranteeing Italian sovereignty, civilizing the indigenous population, and developing the colonial economy. It was of the conviction that, if earlier mistakes were corrected, the scheme had a good chance of success. The Commission urged the government to create a few trustworthy and efficient para-statal organizations, *enti di colonizzazione* (colonization agencies), whose chief purpose would be to transform into small independent landowners the landless and unemployed families who migrated from Italy. The agencies were envisaged as effective institutional tools, useful in eradicating many of the problems encountered by earlier planned settlements and in offseting the financial burden that weighed heavily on government. According to the plan, the *enti di colonizzazione*, directed and controlled by the State, were to provide the settlers with technical and financial assistance.[88]

Another central theme in the Commission's recommendations was that, unlike in Libya where environmental and other exigencies demanded a special approach, in IEA the idea of profit must be the guiding principle. It stressed the important role that private commercial farming had to play, particularly in the production of valuable raw materials. The Commission urged the government to give this form of business interest maximum freedom of action and not to saddle the prospective concessionaries with obligations to settle or employ immigrant families as had been the case in Libya.[89]

While the Commission carefully considered settler farming, only vague and fleeting interest was paid to indigenous agriculture. Of eleven meetings held, four were dedicated to demographic and capitalistic colonization, one for the examination of the relationship between these two, and only one dealt with native farming. The Commission maintained that indigenous farming should be subordinate to the interest of settler agriculture and opposed its modernization, particularly where it could compete with settler agriculture or where there was need of cheap labour.[90]

The Commission based its investigations on the premiss that agricultural transformation was an outcome of partnership between the State and enterprising individuals or agencies. Justifying limited State intervention on the grounds of the political and social goals of colonization, it called upon the government to cater for the basic infrastructures necessary for a civilized life: roads, law and order, economic and civic

[88] Consulta, 'Avvaloramento', 1563–7. [89] Ibid. 1568–71.

[90] Ibid. 1575; CD IAO AOI 1775, Seduta, 15 July 1937; CD IAO 1776, Relazione, pp. 2, 42.

organizations, market discipline, long-term credits. The government was also made responsible for the selection of settlers, the allocation of land, and African labour. The Commission stressed that colonization would remain purely academic unless the government took immediate action to create vast crown or domanial[91] lands and carry out a land survey of the country.[92]

## LIMITS AND NATURE OF LAND POLICY

In his report to the Ministry of Colonies on his fact-finding mission immediately after the conquest into the main areas controlled by the Italian army, Maugini confirmed the great agricultural potential of the Empire: 'The optimistic impressions of my first journey to the new territory are reinforced more than ever. At last, we have colonies that will guarantee positive results in the economic sector as well.'[93] However, he considered lack of sufficient vacant land as the single major handicap for Italy's ambitious programme of colonization. Maugini's solution to the land problem was *indemaniamento*—transforming into domanial lands the estates belonging to the ex-Emperor or rebels, or those confiscated from the Ethiopians, and the Church.[94] This was in line with the general practice followed by the Italians in other colonies. But Maugini vacillated, for in his view any attempt to establish domanial lands was fraught with serious difficulties: 'Domanial lands, in order to be disposable for colonization, should be freed from their life-long sharecroppers, which raises a delicate problem that is not only economic, but above all political and social. Thus domanial lands or land that has been confiscated is not necessarily to be equated with land disposable for colonization.'[95]

Maugini was alluding to the expropriation in Eritrea by Barattieri, the main factor behind Däjach Bahta Hagos's 1893 uprising that had brought to an abrupt halt Italy's early colonization attempts. Maugini alerted Lessona to base his land policies on sound knowledge of indigenous agrarian activities and customs.[96]

Indeed, the creation of domanial lands proved long and arduous. The task was given to Caroselli, an expert in colonial issues and Governor of

[91] The author uses the terms 'domanial land', 'crown land', 'state land', or 'public domain' interchangeably to indicate land belonging to the state.

[92] CD IAO AOI 1776, Relazione, 4, 6, and Consulta, 'Avvaloramento', 1562–3.

[93] ATdR 24/100, Relazione Maugini, p. 2.

[94] Ibid., p. 12.

[95] Ibid.

[96] Ibid., p. 13.

Somalia (1938–41). He was to explore the best method 'that will rapidly and with the Fascist spirit, but in consonance with the current political situation, allow the creation of domanial land'.[97]

To give the study an authoritative voice, Caroselli worked out a plan for a Five-Man Land Survey Commission established by Royal Decree and working directly under the MAI. The Commission included eminent personalities with wide colonial experience such as Massimo Colucci, Vallilio Erennio, Giangastone Bolla, and Arnaldo Bertola. All of them were well-established jurists. Colucci and Erennio were Supreme Court magistrates. Bolla and Bertola were university professors. Colucci, who became the president of the Commission, worked out a meticulous programme, involving at least six months' fieldwork in various parts of the colony, to collect the necessary economic and legal data on land institutions as well as the extent of Ethiopian ownership.[98] The Royal Decree instructed the Commission to study landownership and devise strategies necessary for the organization and operation of agriculture-related services.[99]

Before it even began, the programme was threatened by one crisis after the other. The Ministry of Justice was not prepared to release its officials. For the Ministry of Finance, the plan was too ambitious and costly, and ought to be stopped. The behaviour of these ministries exacerbated already tense relations with the MAI whose officials were stunned by 'the incomprehension and obstructionism of all sorts' coming from 'the ignorant and short-sighted bureaucrats'.[100]

As some members of the Commission were reluctant to be posted as permanent employees in the colony, the MAI felt that it was left with no alternative but to fight. With the decree of 3 June 1938 it looked as though the programme was rescued from oblivion and heading for success. Yet the expectations of many were irreparably frustrated by its lugubrious proceedings. Although to maintain a façade of success, the Commission published its findings in 1940, the truth was the mission was a total disaster. Owing to another turbulent year, the Commission was able to go to the colony only in early August 1939. Each member of the

[97] ASMAI PB5: Missione studi AOI, Ministro to Colucci, Rm, n.d. As its name suggests, 'Pack B' consists of a bundle of documents which at the time of research were in the process of reorganization.

[98] ASMAI PB5, Massimo Colucci to De Rubeis, Schema di studi sul regime fondiario in AOI, Fl 28 Sept. 1937.

[99] RD 3–6–1938, no. 965, Istituzione di una commissione per studi fondiari nell'AOI, in *GU* 160, 16 July 1938; 'Land Tenure in AI', *The Times*, 19 July 1938.

[100] ASMAI PB5, De Rubeis to Colucci, Rm 30 Sept. 1938.

Commission was assigned a particular field of research. Within a week of its arrival, the Commission was shaken by the abrupt departure of Bolla whose task was 'The Comparative Study of Systems of Agricultural Concessions in IEA'. Bolla claimed that the state of emergency declared by Italy on 15 August had made the position of the Commission untenable. But for Colucci, who dismissed Bolla's behaviour as 'cowardly and selfish' and his work as a useless medley of 'legalistic data' lacking originality, such a conclusion was reckless. The clash of personalities between the two men served only to frustrate the work.[101]

Further setbacks arose, provoked by the unstable military situation. Aware that no significant headway in demographic colonization and economic development could be made as long as the territory was not occupied effectively, organized, and pacified, the Italian government had concentrated most of its energies and resources on stamping out all traces of resistance with a relentless military campaign. But the authorities utterly failed to prove themselves masters of the Empire with the result that the situation badly strained their capacities, while their control of the country remained tenuous until the end.

One of the commission's plans was to rely for most of its activities on the preliminary works that were to be carried out by the local agricultural offices. This proved largely unfounded. The land office, known as Ispettorato dell'Agricoltura, set up on 19 January 1937, experienced chronic shortage of skilled manpower and crisis of leadership. It became operational only at the end of 1938, staffed with inadequate, overworked, demoralized, and apathetic personnel, concerned largely with its own personal economic advancement and the promotions that residence in the Empire entailed.[102] Most importantly, the Ispettorato lacked precise definition of its role *vis-à-vis* political institutions and its own regional and local branches. The latter often functioned as if they were autonomous agencies. Further complication came with the creation in the same year of the Land Reclamation and Agricultural Colonization Unit whose chief task was to allocate land for settlement. Lack of clear remit between the Ispettorato and the latter was the source of constant friction over their respective fields of action and, as a result, their activities suffered considerably.[103]

[101] ASMAI PB5, Colucci to Bolla, AA Oct. 1939; Colucci to DG CL MAI, Hr 15 Nov. 1939; Massimo Colucci to GG AA, n.d.

[102] CD IAO AOI 1832: IGA, Promemoria per S.E. il Prof. Maugini, AA 25 Sept. 1938, p. 2; IA Hr, Promemoria al Prof. Maugini, 12 Sept. 1938; IA AA, Promemoria per il Prof. Maugini, 2 Oct. 1938; IA As, 'Promemoria per il Prof. Maugini', 15 Aug. 1938.

[103] CD IAO AOI 1802, Giovanni Piani to De Benedictis, Hr 12 July 1939 where he refers to the strange relationship his Hr Agricultural Inspectorate office has with both the Land

Severely limited by these developments, the task of the Commission was further worsened by the difficulties in finding interpreters and the local authorities' lack co-operation. As a result, it was forced to cut short its programme and return home after a troubled four-month stay.[104]

As an attempt to provide a coherent land policy for the colonial administration, the Commission's findings were open to many criticisms. The survey of Ethiopian land tenure was based on inconclusive data from extremely restricted areas under Italian control. Ideally, study of land tenure demands a long and painstakingly careful examination of legal institutions, familiarity with the languages and cultures of each region, the presence of a sizeable personnel knowledgeable on the area, and substantial financial resources. The Commission had very few of these requirements.

Even though the restoration of the Abyssinian monarchy made them useless, the findings were illuminating. The Commission, rather than launching a new policy, codified practices that were already in place, providing the State with a legal framework to pursue its policy of land alienation from the Ethiopians.[105]

Since the occupation, such a policy was based on political expediency and much less on a clear and authoritative land regulation. Military tactics were combined with economic pressure, and the lands of rebel and exiled leaders were confiscated along with those of the royal family. These same measures had formed the pillar of Italian agricultural policy in Libya, Eritrea, and Somalia, where they were conceived as a means for asserting Italy's sovereignty as successor to the vanquished State.[106] It was thought that there was no other more conspicuous and efficient way of manifesting such sovereignty than punishing those implicated in challenging Italy's authority either with armed resistance or by means of propaganda abroad.

As a general rule, this category of land was added to the public domain. But a somewhat different interpretation was, as least initially, given to

Office as well as the Colonization Directorate because of allocating lands independently; Sbacchi, '*Italian Colonialism*', 422.

[104] ASMAI PB5, *Il libro fondiario nell'ordinamento della proprietà mobiliare dell'AOI*, AA Feb. 1940; MAE, *Avvaloramento*, 351; MAI, '*Valorizzazione*', 197–8; Sbacchi, 'Italian Colonialism', 406. An almost identical problem was faced by the local officials who attempted to complete the work started by the Commission. Cf. ASMAI PB5, Brotto to Colucci, Gm 18 Feb. 1940 and Dalle 26 Mar. 1940.

[105] ASMAI PB5, Colucci to be Rubeis, Schema, pp. 1–2, 5; MAI, 'Valorizzazione', 197.

[106] Cf. for Er see above, ch. 1; Segrè, *Fourth Shore*, 49–54; Mack Smith, *Mussolini's Empire*, 38–9; Hess *Italian Colonialism*, 112.

sequestrated lands. Part of these lands belonged either to people under arrest or executed, to active rebels or exiles against whom no measures of confiscation were declared, or to people assumed to be absent, i.e. away from their normal abode and whose whereabouts could not be traced.[107] The official view was that this group of land was not part of the public domain and the State, who administered them, was considered only a caretaker. The revenue from them was set aside, subtracting only what was needed for the upkeep of dependants and administrative costs.[108] But such a practice was rather an exception, and sequestration almost always led to confiscation, or compensation in kind or cash, or both, and involved no restitution of the original land.

Eager to amass land for colonization, the government also claimed as domanial any land which could be made so without political and social complications. This included land on which tax was owed, where Ethiopians were least able to prove their right, uncultivated land, or land the government thought had not been cultivated efficiently or was in excess of the need of the local population.[109] This view failed to take into account of grazing communage, land under fallow, or lands used for fuel or building material. It was assumed that the Ethiopians had no claims for such lands. The practices were sanctioned, claiming they were more humanitarian than the brutal spoliation allegedly adopted by the British in their colonies or the French in Algeria.[110]

Confiscation began in November 1935 when the lands of some leading northern nobility were declared state domain. Soon lands belonging to the royal family, the exiled leaders as well as resistance fighters in Härär, Galla and Sidamo, and Shäwa were also confiscated.[111] As the case of Achäfär stands to show, sometimes the entire region was subjected to

[107] MAI, 'Valorizzazione', 189–96; CD IAO Hr 1847, G Hr DCL, Verbale della III riunione del Comitato di Colonizzazione, n.d., pp. 3–4.

[108] MAI, 'Valorizzazione', 189; CD IAO 1847, Proprietà terriera, p. 3; Brotto, *Regime*, 113–14.

[109] ASMAI PB5, Mussolini to GG AOI, Rm 19 May 1936 [1938]; CD AOI 1926, Terruzzi to Duce, [Rm] 27 Apr. 1938.

[110] *Colonie*, 9 Apr. 1940; MAI. 'Valorizzazione', 185–96; Mack Smith, *Mussolini's Empire*, 109; G. Nasi, 'L'opera dell'Italia in Etiopia', in *Italia e Africa* (Rm, n.d.), 18; Massi, 'Economia', 448.

[111] DGG 30-7-1936, no. 135, Confisca beni del suddito coloniale blattenghietà Herui, in *GU* 2/8 (suppl.), 26 Apr. 1938; DGG 15-10-1937: no. 738, Confisca beni dei sudditi indigeni', in *GU* 2/22, 16 Nov. 1937 where the land of 61 people was confiscated; no. 751, Confisca beni di sudditi indigeni, in *GU* 2/23, 1 Dec. 1937 which affected the land of 47 people; no. 752, Confisca beni di sudditi indigeni, *GU* 2/23, 1 Dec. 1937, which implicated 70 people; Brotto, *Regime*, 107–12.

these measures whenever loyalty was questionable.[112] Few confiscations were preceded by a proper study of the status of the land.[113]

Even though we have no accurate information on their quantity and quality, it appears these were prime lands. In some areas, they amounted to thousands of hectares. In Lake Tana it was reportedly to be 1,000,000 ha. In Achäfär and Wagerat, lands confiscated from the population for their alleged sympathy with the patriotic leader Babile totalled 180,000 ha.[114] In Härär by the end of 1937 80,000 ha, and in a Galla and Sidamo subdistrict—Däräsa alone—49,655 ha, were made domanial.[115]

Although there was plenty of land, as Maugini also rightly pointed out, it was not vacant but densely populated and intensely cultivated. A bulk of it could not be profitably commercially farmed: although fertile, it consisted of fragmented small plots often intersected by other privately owned lands which the Italians pledged to honour. But protection of Ethiopian interests and development of settler agriculture remained difficult to reconcile, and the Italians had to make recourse to a number of devices. The succeeding chapters will tackle the nature and limitations of these and their repercussions on Italy's overall strategy of empire-building.

[112] DGov 9-2-1938, no. 73412, Indemaniamento del territorio dell'Uogherà (zona di Dabat), in *BUGA* 3/2–3 (Feb.–Mar. 1938); DGov 17-3-1938, no. 53303, Indemaniamento dell'Uogherà (zona Amba Ghiorghis), in *BUGA* 3/1/3 (Feb.–Mar. 1938); DGov 5-6-1937, no. 46144, Indemaniamento del territorio dell'Acefer, in *BUGA* 3/4–16 (Apr. 1938); DGov 8-6-1937, no. 46204, Indemaniamento dei territori del Ginfrancherà, Uorchemder, Gianorà, Tseghedè, Gusquam, in *BUGA* 3/4–16 (Apr. 1938).

[113] Brotto, *Regime*, 46.

[114] 'Notiziario agricolo commerciale: AOI', *AC* 31/4 (1937), 158–9; UA Gn, 'L'Acefer: Notizie di indole generale', *AC* 32/2 (1938): 56.

[115] Sbacchi, 'Italian Colonialism', 396–9; Brotto, *Regime*, 112–13; ACS GRA 44/36/1, A. Ricagno to GG DS AE, Assegnazione di terreni demaniali, AA 18 May 1937, where the list of confiscated lands in AA area are given with the name of their respective owners and the areas involved; *IC*, 15 Sept. 1938; Guido Guidi, Nel territorio dei GS. Rilievi ed aspetti politici, sociali, economici. La Residenza dei Darasa, Rm, 1938, quoted by Del Boca, *Conquista*, 198.

# 4
# *Military Settlement: The ONC*

## POLITICAL AND ECONOMIC FACTORS

Military settlement was part of the demographic colonization scheme. While later regional settlement schemes focused on a particular province of Italy, military settlement was nationwide, aiming to settle landless peasants from all over Italy. The experiment began while the military operation of the Ethiopian campaign was still underway and before the Agricultural Council began its operation. Much of the Council's opinion was influenced by this initiative. Contrary to later claims, the scheme did not come about as part of a broad and conscious vision, but was born of economic and strategic crisis.

Contrary to optimistic forecasts that the Empire would become a granary for Italy, it swallowed most of the metropolitan's food exports. In the second half of 1936 Italy supplied the Empire with 75,000 tons of wheat at a cost of L.43,000,000. The prospects for 1937 were no better. Even though in 1936 the Empire imported 45 per cent of all Italian exports, the situation was by no means beneficial to Italy as it did not bring in foreign currency and meant the loss of foreign markets. While the Empire absorbed the metropolitan's product, it gave little in return.[1]

It was against this critical situation that on 21 October 1936 Lessona, while visiting Addis Ababa, cabled to the president of the ONC, Di Crollalanza, to reactivate two agricultural stations, Holäta and Bishoftu:

> I have examined the question of agricultural colonization of the governorship of Addis Ababa territory. A number of domanial lands have been identified with a total area amounting to several thousands of hectares and others will be identified soon. . . . I count on the intervention of the ONC to translate into action the programme of the Capo. In the two centres of Holäta and Bishoftu, I will give you vast plots of land to immediately start developmental work using presently

[1] Sbacchi, 'Italian Colonialism', 360.

demobilized peasant militia in Africa. I request you to send immediately personnel with authorization to start work at once in agreement with the local government. Indeed, as this is a matter of extreme urgency I, therefore, appeal to your immediate personal attention.[2]

The choice of the ONC as the first colonization agency, i.e. the use of soldiers as settler farmers, seemed well calculated. The ONC was founded towards the end of World War I with the aim of providing ex-soldiers with a stable economic occupation. From the early days it had concentrated on land settlement projects which would become a familiar feature throughout the Depression in Europe and the United States.[3] The ONC had carried out a large number of settlement schemes throughout metropolitan Italy, often in combination with land reclamation. Its most publicized accomplishments were in the Pontine Marshes, south of Rome, where, working in partnership with the State, by the end of 1937 it had established 2,574 small farms, and distributed over 48,380 ha of land.[4] So, as an established agency—the ONC had capital and technical expertise—it would pose little financial burden on the government's already strained budget and could start the colonization programme immediately.[5]

The unsettled political situation made reliance on the ONC for the initial phase of demographic colonization even more important, while also dictating use of the military. The decision by the political authorities to confine settlements within gendarmerie posts, or *praesidia*, established by the military authorities, indicates the gravity of the situation.[6] As appears from Ciano's notes in December 1937, even someone like Italo Balbo, one of the founders of Fascism and Governor of Libya, whose name was linked with demographic colonization of Libya because of the mass migration of the Ventimila—the twenty thousand colonists that he had transported in a single mass convoy in October 1939[7]—used to advise

[2] ACS ONC 1/1, Lessona to MC, AA 20 Oct. 1938; Di Crollalanza, 'Relazione sui programmi di colonizzazione demografica nell 'Impero da parte dell'ONC', *REAI* 26/5 (May 1938), 739.

[3] See in USA the attempt in 1935 to settle colonists from Michigan and Minnesota in part of Alaska; cf. A. Stringer, 'The Red-Push Pioneers', *Saturday Evening Post*, 208 (28 Dec. 1938), 8. A somewhat similar project but with emphasis on regional development was the Tennesse Valley Authority; see A. F. Robertson, *People and the State*, 14–15.

[4] FO371/22021/J2512/40/1, Lord Perth to Visc. Halifax, Rm 24 June 1938.

[5] ATdR 24/100, Relazione Maugini, p. 15.

[6] CD IAO AOI 1936, Graziani to GG AOI AA 8 Apr. 1937; J. Kulmer, 'The Return of Haile Sellassie', *Contemporary Review*, 159 (1949), 291. Some of the key resistance leaders operating at Ambo region were Gäräsu Duké, Zäudé Asfaw, Mäsfen Seläshi, Blatta Takkälä, and Bulcho Folé, while Iticchia and Fitawurari Dästa Guno were active at Addis Aläm, between Awash—River Borga and Lake Holäta (ACS ONC, Rapporti di S.E. Graziani, AA 7 and 13 Nov. 1937).

[7] Segrè, *Fourth Shore*, 102–11.

'certain people not to go to Abyssinia because of conditions of grave insecurity in the country'.[8] In the light of this, the settlement of the ex-servicemen in the proximity of Addis Ababa was seen as a useful means to prevent popular insurgency in the region that displayed the 'most resistance to Italian rule'.[9] The area would form a formidable military outpost, providing a fence for Addis Ababa. The settler community, in case of emergency, could become soldiers. In the long term, it was through these groups that the cohesion of the Shäwan Christian Amharas would be broken and their strength weakened.

The scheme had its economic advantages too: with the settlement of demobilized soldiers, already present in Ethiopia, a considerable part of the enormous economic cost of bringing agricultural families from Italy could be spared. It was Mussolini's belief that out of almost 500,000 Italian soldiers in Ethiopia, at least 400,000 of them were farmers.[10] It was said that these could be lured by land concessions and economic incentives to permanently settle in the Empire. The decree of 19 December 1936 tried to do just that by conferring priority to the soldiers of the Ethiopian campaign in any future land allotment.[11]

Established towards the end of the 1920s, the two farms at Bishoftu and Holäta were model farms owned by Emperor Haile Sellassie, producing cereals, milk and its by-products, chicken, tobacco, and grapes in amounts sufficient to meet most of the capital's needs. Of the two, Bishoftu was better developed, equipped with several stocks of modern farm tools, considerable variety of agricultural machinery, complete with cattle-breeding and veterinary departments, a dairy, and oil-pressing plants.[12]

[8] *Ciano Diary, 1937–1938* (London, 1952), 44.

[9] E. L. P. Newman, *The New Abyssinia* (London, 1938), 103.

[10] Acs ONC 3/A/2, Verbale della riunione tenuta il 26.11.1937, AA 26 Nov. 1937; Sbacchi, 'Italian Colonialism', 353.

[11] Legge 10-6-1937, no. 1029, Conversione in legge del RD 19-12-1936, no. 2467, che conferisce un diritto di preferenza nella concessione dell terre dell'AOI a coloro che hanno ivi partecipato alle operazioni militari in qualità di combattenti, in *GU* 2/15, 1 Aug. 1937, and *GURI*, 10 July 1937; 'Rassegna Agraria Coloniale', *AC* 31/5 (1937), 202.

[12] A. Zervos, *L'Empire d'Éthiopie* (Alexandria, 1936), 130–1; Pankhurst, *Economic History*, 208; Poggiali, *Albori*, 198–200. When the Italian government handed over to the ONC, the inventory of Bishoftu Farm included, in addition to those already mentioned, 2 tractors in good working order, 1 threshing machine, 2 sowing-machines of which 1 for cotton, 3 trucks, and electric-generator engines; pig-processing machinery; water-boiler, ploughshares, a number of residential premises, engine shed, and cattle-shelter. Despite its poor state, the livestock was sizeable consisting of 331 bovine, 80 ovine, 69 equine, 28 swine, and 4 poultry. Among the farm produce stocked in the warehouse there were 424 sacks of grain of which 135 contained peas, 113 chick-peas, 104 barley, 55 *ṭef*, 9 beans, 7 wheat, and 1 lentils. Two hectares of its land are occupied by vineyard. By comparison Holäta was more modest but, as it was a military training centre and royal holiday resort, it had extensive and more

In an interview towards the end of 1936, Lessona made it clear that he wanted to make the two stations part of a large-scale plan of White settlement in which the best parts of Shäwa Province, especially those around Addis Ababa, would be included. The scheme would be based on the employment of Ethiopian labour. As a result, the Empire's capital

> will be surrounded from all sides by an area which will be intensively cultivated and densely populated by Italians. In fact, what we aim at in this particularly favoured zone is the formation of small farmowning Italians. The Duce has said that the Fascist Empire will be the empire for population and indeed, the Capo's order will be put into effect through this form of colonization.[13]

The two localities, even though not necessarily endowed with the agricultural potential that the Fascist propaganda machinery attributed to them, appeared to be admirably suited for European settlement. The climate is temperate and the land exceptionally endowed with volcanic and argillaceous soil, clearly capable of producing a wide variety of crops twice a year.[14] Surrounded by five picturesque lakes, Bishoftu Farm lies amidst sparsely wooded rolling countryside that is interrupted now and then by mounds, hills, and elevations of all sorts, and distantly by Qalitti massif and mountain chains facing three lakes known as Bishoftu, Hora-Arsodi, and Bishoftu-Guda. It stands at medium altitude of between 1,800 and 1,900 m at 50 km South-East of Addis Ababa, close to the Djibuti railway, the capital's main outlet to the sea, and was flanked by the imperial highway to Moggio—an important town lying at about 25 km southwards.[15] The neighbouring district of Ada, which would experience an unprecedented agricultural boom after independence, is renowned for its high-quality *ťéf* and burgeoning commercial agriculture and in the past had served as Emperor Menilek's *madbét*, provisioning his court and Addis Ababa's churches.[16] Holäta, located at between 2,200 and 2,400 m above sea level, was 40 km from Addis Ababa on the main road running westward to Näqämté and, during Haile Sellassie's regime, was

complex buildings. It had a water-mill belonging to Empress Mänän, and a tractor. With the exception of forage and a 1,000 sq m thriving vineyard with 2,000 plants, the cultivation of cereal crop was negligible. The livestock population comprised 90 ovine, 62 oxen, 35 equine (ACS ONC 1/3, Taticchi to SC ONC, AA 1 Mar. 1937, and 27 Feb. 1937).

[13] *IOM*, 20 Dec. 1936, p. 19.

[14] CD IAO AOI 1338, Vincenzo Manusia, Relazione sui risultati ottenuti nel campo di orientamento agrario di Biscioftù nel 1938; Di Crollalanza, 'Relazione', 739.

[15] ACS ONC 3/A/1, Benigno Fagotti, Esperienze culturali eseguite nell'AAB dell'ONC (1937–8), AA n.d., pp. 3–4.

[16] J. McCann, *A Great Agrarian Cycle? A History of Agricultural Productivity and Demographic Change in Highland Ethiopia 1900–1987* (Boston, Mass., 1988), 16.

a military training centre (see Fig. 2). The location of the two farms close to Addis Ababa assured their products easy access to the largest and the most lucrative market of the Empire as well as to luxury goods, spare parts for machinery, and rapid military assistance.[17] In addition, the presence at Holäta of a mixed population of Amhara and the largely tenant—*ťisäña*—Oromo was thought to be an advantage, as it would prevent concerted local opposition. All this, combined with the favourable topography and attractive climate for European settlement, made the two places appear admirably suited to start the first experiment in demographic colonization 'in a complete form, with relative speed, minimum expenditure, with minor risks and greater advantages'.[18]

Di Crollalanza and Mazzocchi Alemanni worked out plans of settlement that won the consensus of both the Governor General and the Ministry of Colonies, and met with Mussolini's approval and encouragement. The scheme envisaged an initial settlement of 100 families in each of the two farms, on 10,000 ha each—later upgraded to 12,000 ha. These were to be provided by the Governor-General before June 1937, the beginning of the big rains. In this way, the agency would be in a position to make full use of the land. At the end of the big rains, towards mid-September, the ONC would begin to build the first 200 colonists' houses, 100 on each farm. The plan was to settle 1,000 families on each farm within a period of four to five years in 50,000–60,000 ha.[19] But the scheme soon run into a welter of physical, technical, and administrative difficulties and failed ignominiously to come anywhere near achieving the goals set out by its planners.

## LAND PROBLEMS

On 2 November 1936 Di Crollalanza despatched the first batch of eleven employees composed, in accordance with Lessona's recommendations, of three directors, an engineer, an administrator, four foremen, and two land surveyors. The two farms established their own separate budget and administration, headed by their respective directors[20] and staffed by a

[17] MAI, 'Valorizzazione', 268–9; *CI*, 22 Feb. 1939; Quaranta, *Ethiopia*, p. vii; Newman, 'New Abyssinia', 103.

[18] Di Crollalanza, 'Valorizzazione', 492.

[19] ACS ONC 1/1: Graziani to MC, AA 1 Mar. 1937; Lessona to GG, Rm 10 Mar. 1937.

[20] All the heads of the farms were educated to graduate-degree level and had considerable working experience within Italy. Of the three managers, with the exception of Benigno Fagotti, the head of Bishoftu Farm, two were already directors at one of several of the ONC's farms: Giuseppe Taticchi at Sabaudia Farm and Angelo Ponzetti, the head of Holäta

skeleton personnel, handpicked by Rome to whom the farms were directly answerable for the minutest detail of their activities. The office at Addis Ababa provided overall co-ordination and technical assistance and until the end of occupation, its head, the agronomist Giuseppe Taticchi, acted as the general manager of all the ONC's activities in IEA and as a link between the agency and the central government authorities.[21]

The men selected for the task can hardly be described as the best-seasoned human material. Even though their number grew gradually with later addition of supporting administrative and supervisory personnel, it remained absurdly inadequate in relation to their task.[22] They were in their mid-thirties on average and engaged on varying terms of employment: only a handful were permanent, the rest were either on a fixed or indefinite period of contract, but tied to the agency for a minimum period of three years before any eventual change of employment or re-entry to Italy—a period considered to be sufficient for the agency to recover its investment in their transportation. Yet many drifted away whenever a better opportunity opened up and the political authorities turned a deaf ear to the agency's pressure to repatriate them on the ground of their breach of contract.[23]

Most of the employees had some sort of technical and professional qualifications, but no experience of work outside their own home environment and the need for their special training was recognized only towards the end. For these men the colonial life was a mixed blessing. Its hardships and many deprivations were depressingly frustrating and there were also aspects that were reasonably attractive: unlike home, supervision from superiors was minimal and the salary considerably higher. Those with even few means could afford to employ Ethiopian servants to take care of their domestic chores. In Ethiopia, an unskilled Italian labourer, for example, earned a wage of L.30 a day. In Italy such a labourer would have been quite lucky if he was able to make even L.15 a day. The salary of senior personnel in Ethiopia was even better.[24] Obviously these earnings

Farm, at Altura, while the chief technician, Angelo Balconi, was employed at the central headquarters. Cf. ACS ONC 5/2, Di Crollalanza to presidenza del consiglio dei ministri, Rm 7 Nov. 1936.

[21] ACS ONC: 3/A/1, ONC adunanza tenuta da S.E. il presidente verbale, n.d. [AA 2 Feb. 1937]; 3/A/2, Verbale della riunione tenuta il 26.11.1937 in AA, AA 26 Nov. 1937.

[22] ACS ONC 3/B/1, Azienda AOI, Relazione mese di dicembre 1941 e gennaio 1941.

[23] ACS ONC 1/2: MAI to ONC, Rm 26 Oct. 1937, and 29 Dec. 1937; Di Crollalanza to MAI, Rm 13 Jan. 1937 [8].

[24] ACS ONC 3/B, Pasquale Pistone to SC ONC, AA 18 Mar. 1939; FO371/23380/J575/296/1, CG Bird to PSSFA, AA 20 Jan. 1939; *CI*, 15 Nov. 1939.

may not mean much in a situation where the constantly shrinking value of the lira and the high cost of living threatened to wipe out most of the takings.[25] Yet the various fringe benefits attached to service in the colonies cushioned them from the deleterious effects of the prevailing rampant inflation.[26] Of course, these salaries and benefits were pretty low compared with those paid by private-sector or other parastatal agencies. But the ONC, unlike other organizations, was a non-profit-making public body engaged in the service of a nation's supreme objective and the assumption was that those who worked with the agency in order to make this possible gained spiritually; but in practice the agency paid periodically performance-linked allowances that it was thought would fairly compensate for the comparatively low salaries.[27] There was, of course, boundless opportunity to supplement salary with earnings from part-time pursuits—both lawful and unlawful, including petty trade with the Ethiopians, or pinching from the latter's meagre wages, embezzlements of the agency's money, currency speculation.[28] As the inventory of personal possessions reveals, even a wretched employee who had come with a handful of belongings had amassed considerable wealth within a short time of his arrival from Italy.[29] Of course, the proportion of men involved in this kind of activity was relatively small even though those responsible were the cream of the agency's personnel once described as the best by the management. Indisputably, the agency dealt severely with these people by dismissing or repatriating them immediately. However, the official view that the agency's personnel were, on the whole, honest, courageous, and industrious people

[25] It is stated that in 1936 life at ONC farms in Ethiopia was four times more costly than in Agro-Pontino, Italy. (Cf. ACS ONC 3/A/1, Taticchi to ONC Rm, AA 4 Dec. 1936).

[26] The allowances tended to be greater than the salary itself and included a wide range of benefits: some were paid to all, such as bed and board allowance, active service allowance, family supplement, colonial service and residential hardship allowance; others were related to positions of responsibility and duty (office allowance), or performance (performance award), or skill (foreign-language-knowledge allowance). From the total was deducted a small %age for party membership (not all of them were members, including Taticchi himself), income tax, surtax, personal income tax. In addition the employer had to fulfil all the legal obligations to which the employees in metropolitan Italy were subject, such as contribution to pension funds and medical insurance. Cf. ACS ONC 2, Trattamento del personale destinato in colonia; ACS ONC 5, Di Crollalanza, Trattamento del personale dell'ONC destinato in AOI, n.d.

[27] ACS ONC 5, Specchio degli emolumenti assegnati al personale di enti parastatali dislocati in AOI, n.d.; ACS ONC 3/A/1, ONC adunanza tenuta da S.E. il presidente: Verbale, n.d. [AA 11 Feb. 1937].

[28] ACS ONC 7: Taticchi to SC, AA 29 Mar. 1939; Fagotti to SC Rm 7 Apr. 1939.

[29] The inventory of the personal belongings is given in ACS ONC 5, AH and AB Personale.

who did a superb job needs qualification.[30] One of the management's constant complaints was that a large segment of its workforce was unruly, disobedient, and apathetic.[31]

The administrative set-up itself remained one of the ONC's disturbing problems which threatened the smooth functioning and efficiency of its farms. ONC documents starkly reveal tales of bitter quarrels, managerial disputes, charges of financial mismanagement and administrative incompetence and interference particularly between the directors of the two farms and the head of the nominally autonomous technical department, on one hand, and Taticchi, on the other. Irascible by temperament and an administrator cast in a typical Fascist mould, Taticchi was averse to criticism and showed little tolerance of individualists and enthusiasts with initiative. He passed severe strictures on those who did not agree with the style of his administration or neatly fit his description of acceptable Fascist behaviour. These undignified wrangles, which occasionally burst into open confrontation,[32] were largely attributable not to policy differences, as the disputants often alleged in the heat of controversy, but to personal antipathy and to win their President's favour in Rome. Di Crollalanza had considerably overcome the difficult problem of maintaining control at a distance by asserting his right through constant pressure to be kept informed of everything that went on in the farm. Yet keeping peace

[30] Even the Duke of Aosta is reported to have said that the ONC's personnel are 'men who need to be restrained from work rather than spurred'. Cf. ACS ONC 3/B/3, Todaro to Presidente ONC, AA 7 Apr. 1938.

[31] ACS ONC 3/B, Pistone to ONC, AA 18 May 1939.

[32] A case in point is an incident involving Taticchi with Pasquale Pistone and Mazzucato. As a newly appointed director of Bishoftu Farm, Pasquale had sent his appraisal of the farm to Di Crollalanza where he reported serious operational shortcomings that implicated also Taticchi. The latter, having taken the matter as a personal attack, reports how 'his repressed anger boiled up' over an argument during a visit to the farm in the course of which he told Pistone: 'I reminded him that we were talking not as two officials but as two men. I accused him of being a liar and coward, and challenged him to a duel' (ACS ONC 7, Taticchi to SC, AA 3 July 1938). Equally illuminating is the ex. of Mazzucato. Taticchi had strongly supported Mazzucato's candidacy for chief technical officer against Di Crollalanza's hesitation to employ the latter because he had no military background. However, it was not long before Taticchi came to be disenchanted with his one time protégé when he discovered that he was making independent decisions without his authorization. To Di Crollalanza's own anger, this is what he says about the person whom he had earlier praised as a well-qualified and hard-working man: 'As a technician, he is incompetent; as an administrator, he is a fool: lax with his subordinates and inferiors, and deceitful and stupidly obstinate with his superiors. Morally, one can hardly say he is irreproachable.' Crollalanza's response was to remind Taticchi to refrain from harshness and to confine his evaluation to competence and not character assassination (ACS ONC 3/A/2: Taticchi to SC ONC, AA 14 July 1940; Di Crollalanza to Taticchi, Rm 7 Aug. 1940; for other similar exx. see ACS ONC 2, Benigno Fagotti to Di Crollalanza, Bishoftu 29 Aug. 1937).

between his warring officials remained a daunting prospect. Appeals to what he described as the uniquely Fascist ethos of deference to hierarachy and discipline, fused by sense of comradeship and collaboration, had only momentary success as did his personal interventions and the strategy of alternating paternalist exhortations with staunch authoritarian admonition.[33] Unless voluntary resignation was forthcoming on grounds of ill-health or a better job offer, repatriation of the undesirable official remained the ultimate resort. Within their brief lifetime, the farms experienced rapid changeover of officials which, combined with the chronic staffing shortage and lackadaisical workforce, left the farms in a parlous state, administratively and operationally.

Lessona did not specify the area to be granted to the ONC and little was known about its legal status except that it was State land. Although Di Crollalanza later claimed he had been promised 50,000 ha, what was sure was that Lessona agreed to grant immediately 1,000 ha at Holäta and 500 ha at Bishoftu; no details were given about where and when additional lands promised 'within a short period' would be made available.[34]

It was generally believed that in the two areas a total of between 25,000 and 27,000 ha were available for immediate colonization.[35] Bishoftu was believed to have 12,000–15,000 ha, much irrigated by the rivers of Wädächa, Bäläla, or those of neighbouring lakes. Holäta was assumed to have an unspecified number of several thousands of hectares of fairly fertile undulating plains, well watered by two rainy seasons, the big rains between July and August and the small rains in April–May. Considerable use could also be made of the nearby Holätta River which, despite its meagre flow, offered multiple possibilities for irrigation.[36]

Of the several thousand hectares promised, only a few hundred were made available at Holäta where a hurriedly selected 120 demobilized militia of Sabaudia Division were accommodated by the Governor General as salaried workers at L.20 a day. The ONC officialdom had reservations about the merits of the scheme, yet accepted it reluctantly and with few modifications.[37] Contrary to the agreement, the Governor General

[33] ACS ONC 3/A/2: Verbale della riunione tenuta il 26.11. 1937 in AA, AA 26 Nov. 1937; Di Crollalanza to Taticchi, Rm 7 Aug. 1940.

[34] Del Boca, *Caduta*, 202; CD IAO AOI 1936, Di Crollalanza to MAI, 13 May 1938.

[35] Pankhurst, 'Italian Settlement', 148–9.

[36] Di Crollalanza, 'Relazione', 739–40.

[37] L.20 settlement was reached after the discontent of the settlers with the early mobility-allowance payment of L.5 per day. Ibid. 739; id., 'L'avvaloramento agricolo dell'Impero: Esperienze e realizzazioni dell'ONC', *REAI* 27/11 (Nov. 1939), 1198; CD IAO AOI 1839, Promemoria per S.E. il Capo del governo sull'attività colonizzatrice dell'ONC in AOI, 7 July 1938, p. 2; Poggiali, *Albori*, 219.

continued to procrastinate in the handing over of Bishoftu Farm. But most frustrating of all was the fact that necessary equipment for initial working of the land was lacking and twenty tractors were lying idle for three months at Djibuti port. As a result, the expertise of the agency's staff already in place and the militia workforce was underutilized. The situation so heavily strained the agency's budget that at the end of the contract, it claimed that it ended up paying half a million lira in salary alone.[38]

Di Crollalanza appealed to Lessona for his strong and authoritative intervention without which, he stated, the enterprise would face a 'guaranteed collapse in its social purpose and a very certain economic disaster'.[39] The situation also worried Mussolini who, for his own propaganda purpose, wanted to transform the land confiscated from fugitive Emperor Haile Sellassie into model farms. Di Crollalanza left for Addis Ababa to speed things up, arriving in January 1937 with the inspector of the Pontine Marshes, Nallo Mazzocchi Alemanni.[40] Even though he succeeded in setting up part of the Bishoftu farm, his views on the whole were not encouraging. As appears from his report to Mussolini, the two areas, as most parts of the Empire, far from being 'the earthly paradise that the unwarranted optimism of the amateurs, the colourful descriptions of the impressionist journalism, the oversight of some pseudo-technicians, and the uncritical information of a few returners portrayed or hinted at',[41] were of limited productivity. In fact, the farms were occupied by a considerable number of Ethiopian farmers, mostly *ťisäña* or *tekläña*—who cultivated the land to which they had usufructuary right. The State land itself was composed of small plots, scattered amidst non-State lands. Holäta Farm, made up mainly of two separate blocks, had about 480 ha, with around 250 ha of these situated at Qagliu, a location fairly distant from the centre.[42] Most of the unoccupied part was grassland suitable only for cattle-grazing and requiring substantial investment to cultivate it productively. The fertile section—in all about 150 ha—was thickly inhabited by about 200 Ethiopian families—a mixture of *ťisäña* and *tekläña*

[38] CD IAO AOI 1839, Di Crollalanza, Promemoria, 7 July 1938, p. 2; the military authorities provided board. Cf. Di Crollalanza, 'Avvaloramento' (1939), 1198.

[39] ACS ONC 5, Di Crollalanza to MC, Rm 17/Dec. 1936.

[40] Del Boca, *Caduta*, 201. With him there was also Ing. Ugo Tedaro, the Head of Reclamation Service (cf. ATdR 24/124, Di Crollalanza, 2° Promemoria per S.E. il Capo del Governo', Jan. 1938, p. 1).

[41] Di Crollalanza, 'Valorizzazione', 489, id., 'Relazione', 740.

[42] Near *gebbi* there were 200 ha, part occupied by vineyard; another two blocks of about 2,061 sq m each within *gebbi*, occupied by the Italian servicemen. Away from *gebbi*, there was a mixture of farmland and grassland of about 45 ha located around the ex-Empress Mänän's country residence.

—who offered their services at the ex-imperial court as servants, guards, storekeepers, and forestry-men in exchange for the land they exploited. The bulk of Bishoftu Farm was situated at two localities, Foqa (624 ha) and Qalitti (310.47 ha). The rest of the domanial lands, made up of several small plots sandwiched between private ownerships, were largely grasslands and forests and, like Foqa, which is flooded by the nearby Wädächa River, they would need extensive work before they became cultivable.[43]

Although the ONC took over the two farms within a relatively short time of the fall of the Addis Ababa regime, it claimed that most of the equipment was lying waste and in a state of neglect. Dwellings and shelters at Holäta made of *čeqa* (mud), which formed the imperial *gebbi*, were taken over by Italian military authorities. Bishoftu farms were destroyed by the resistance fighters except for the storage, millstones, and a number of small rectangular huts used for animals. Even though the farms had been productive in the past, Di Crollalanza claimed no land improvement had been carried out.[44]

Early enthusiasm for utilizing these lands was soon replaced by cautious realism and uncertainty. Di Crollalanza, having established the complexity of the local conditions, had come to a more sober appraisal of the scheme and joined the ranks of those who came to be known as procrastinators.[45] Following Tassinari, he urged the colonial administration 'to call a halt on several undesirable projects' and advised it to take a 'cautious and piecemeal approach to each developmental initiative', including demographic colonization.[46]

This shift in outlook also tempered Di Crollalanza's attitude towards the ONC's settlement programme. Like any other demographic colonization, he wanted it to proceed 'gradually and with a realistic sense of timing'.[47] In his view, the limited agricultural knowledge of the country and the numerous practical problems, combined with the agency's own restricted technical and financial resources, provided no other option. Yet reflecting on Mussolini's plan, he wanted to ensure that the ONC's activities

[43] ACS ONC 3/A/1, Taticchi, Relazione settimanale: AH, n.d. [23 Nov. 1936]; ACS ONC 5/1, Taticchi, Relazione sui terreni della valle dell'Auasc [AA], 28 Apr. 1938; CD IAO 794, Ciocca, Elementi, p. 46.

[44] CD IAO: 739; ACS ONC 3/A/1, Taticchi to SC ONC, AA 29 Dec. 1936.

[45] CD IAO AOI 1839, Di Crollalanza, Promemoria, 7 July 1938, p. 3; Di Crollalanza, 'Avvaloramento' (1938), 713.

[46] CD IAO AOI 1839, Di Crollalanza, Promemoria, 7 July 1938, p. 2.

[47] ATdR 24/124, Di Crollalanza, ONC; Promemoria, Feb. 1937, p. 3.

should not serve merely as a laboratory test case, but it should be an essential part of these pioneer farms to supply the government, by their scale, organization, and economy, not with theoretical or even scientific models, but with elements of concrete study and judgement that would serve to regulate the conduct of subsequent demographic colonization.[48]

Land remained one of the chief constraining factors and the government, despite the agreement, failed to maintain its word. Of the first 12,000 ha promised before the beginning of the big rains of 1937, only 3,000 ha were made available. About three-quarters of these were uncultivable; part was rocky, and part marshy, needing intensive reclamation work. A year later the whole concession amounted to 3,216 ha, 1,516 at the Holäta Farm; of this, only 495 ha were granted under formal agreement. Of the remaining 1,800 ha at Bishoftu Farm, 900 ha were uncultivable.[49]

As the ONC's frustration increased and its financial resources became stretched, Di Crollalanza made a strong protest to the MAI urging the handing-over of the remaining land. He also made a passionate personal appeal to Mussolini for his direct intervention.[50] But as a result, only an additional 5,000 ha were secured at Holäta and 700 ha at Bishoftu, registering a total of 6,600 ha—7,400 ha short of that promised.[51] Later on, 500 ha were added to the Holätta Farm. In 1940 the total holding of this farm had the composition given in Table 6.

Bishoftu Farm was even more modest, consisting of only 2,000 ha. By 1939 reclamation work was underway diverting part of the Wädächa River, whose seasonal floods had serious damaging effects on the Bishoftu Farm, to a new bed made of reinforced concrete. The government had allocated L.6,000,000. But the work had only a limited success as by 1940 only some 1,800 ha, or just over a tenth of the 15,000 ha available, were recovered, with a cost amounting to L.3,500,000 approximately. Thus, although 24,000 ha was initially to be handed over by the end of May 1937, by 1940 only about 7,500 ha had been secured as a result of Di Crollalanza's strenuous struggle with the colonial authorities and petitions to Mussolini.[52]

The expansion of the farms followed a scheme developed by Mazzocchi

[48] CD IAO AOI 1839, Di Crollalanza, Promemoria, 7 July 1938, p. 3.

[49] Ibid., pp. 3, 8. ACS ONC 3/A: Direttore del servizio to GG, AA 15 Aug. 1937; Di Crollalanza to MAI, p. 3. ATdR 24/124, Di Crollalanza, 2° Promemoria, Jan. 1938, pp. 2–3; Di Crollalanza, 'Relazione', 747.

[50] ACS ONC 1/1, Di Crollalanza to MAI, p. 4; ATdR 24/124, Di Crollalanza, 2° Promemoria, Jan. 1938, p. 13; CD IAO AOI 1839, Di Crollalanza, Promemoria, 7 July 1938, p. 7; Villa Santa *et al.*, *Amedeo*, 232.

[51] CD IAO AOI 1839, p. 8.

[52] Ibid., p. 11; ATdR 24/124, Di Crollalanza, 2° Promemoria, Jan. 1938, p. 13.

TABLE 6. Land Allocation at Holäta Farm (ha)

| Type of land | No. of ha |
|---|---|
| Arable | 2,060 |
| Pasture | 943 |
| Garden | 16 |
| Fallow | 1,646 |
| Unproductive | 301 |
| Scrubs and woodland | 130 |
| Vineyard | 4 |
| Water | 335 |
| Buildings and road | 65 |
| TOTAL | 5,500 |

*Source:* MAI; 'Valorizzazione', 265; 'L'ONC per l'avvaloramento agricolo dell'Etiopia', *Colonie*, 6 July 1940.

Alemanni. In his attempt to make each farm an unbroken unit and free large areas of land needed for settlement, Mazzocchi Alemanni had worked out an ingenuous method believed to meet the needs of both settlers and Ethiopians. It was known as *accorpamento per permute*—annexation through a system of exchange. The central concern of the scheme was to avoid any interpenetration, or Italian settlements in the gaps between Ethiopian cultivations. This view gradually became of paramount importance as the settlers failed to live up to the official doctrine of racial prestige. Ethiopians were 'persuaded' to move off the land pinpointed for colonization and offered other land in exchange.[53]

It was thought that the annexation system would enable the formation of large unbroken domanial lands on which settlement could begin. Whenever possible, exchange would involve respect for the principle of fair compensation. Differences in value were to be offset by the concession of larger plots, or the award of special prizes, or by carrying out some valuable works such as digging wells, making drinking-troughs, or assisting in the building of dwellings. Only the landlords were to be displaced; the tenants, as a rule, would remain unaffected and continue to farm

[53] ATdR 24/124, Di Crollalanza, 2° Promemoria, Jan. 1938, p. 3; Di Crollalanza, 'Relazione', 740.

within the settler's agricultural enterprise, forming an integral part of each farm.

With this programme the settlers' farms could be an organic whole. Fragmentation into small units scattered amid Ethiopian farms was thought not only to be uneconomic but also to have serious repercussions for the prestige of the White man.[54] Under the scheme, the Ethiopian would assist the colonist, his new landlord who, although alien, was taken for granted as being more beneficial to the Ethiopian because of his assumed racial superiority. Ethiopian assistance involved both working on the settler's farm and looking after his livestock.[55] As a result, the Ethiopian farmer would remain attached to the land and become a stable element in the population. The political advantages of such stability for peaceful Italian penetration into the country were obvious.[56] Mazzocchi Alemanni explained:

> The most important element in the execution of exchanges lies in the fact *that the Ethiopian population is not removed from the area*; while the owner is transferred on to the new land given to him in exchange, the peasant, instead, remains on the original land. This responds to the already-mentioned economic contingencies of associating in a number of ways, but essentially more or less through customary forms of sharecropping, the Ethiopian labour (farmers and shepherds) to the metropolitan farm. In fact, the political significance of such a policy, whereby, for example, 50 owners are transferred while 2,000 peasants remain in place, are obvious, for the Ethiopian population only in this manner, i.e. by remaining in their abode and in their day-to-day activities, can contribute to the metropolitan's work of developing the country and, while ensuring themselves a better life, become the most peaceful auxiliaries for the penetration of our civilizing activity.[57]

In the mean time on lands deemed subject to exchange, the ONC was authorized to carry out construction work and, provided that there was no Ethiopian opposition, to farm unused lands on a provisional basis. Once the survey was over, and the Ethiopian claims were met, the agency was issued with a permanent title to the land.

As a whole the scheme looked more optimistic than rational and its implementation seemed deceptively easy. But the planners gave little or

[54] Di Crollalanza, 'Relazione', 741; M. Montefoschi, 'I centri agricoli di Olettà e Biscioftù: Successo di un esperimento', *IC* (May 1939).

[55] FO371/22021/J2512/40/1, Lord Perth to PSSFA, Rm 24 June 1938; ATdR 24/124, Di Crollalanza, 2° Promemoria, Jan. 1938, pp. 9–10.

[56] FO371/22021/J2512/40/1, Perth to PSSFA, Rm 24 June 1938; Di Crollalanza, 'Relazione', 742.

[57] Mazzocchi Alemanni, 'Orientamenti nella valorizzazione demografica dell'Impero: Prime realizzazioni dell'ONC', *AC* 32/4 (Apr. 1938), 166.

no importance to the social and political vagaries which became a source of considerable frustration. In fact, exchange of land, though easy in theory and simple in formulation, proved to be an arduous task. From a technical point of view, it involved complicated surveying work ranging from identifying and mapping the colonizable lands, finding comparable exchange, and assessing the dwellings to be abolished, to winning the confidence of the Ethiopians who were often reluctant to disclose their possessions. And, as a large part of the existing land records had been destroyed prior to the occupation of Addis Ababa, most of this work had to be done afresh. Qualified staff were extremely few and, no matter how determined these were, the practical difficulties on the ground made their task a difficult one. They worked in an environment where communication and transport were poor, to say the least, and political conditions extremely volatile.[58]

Despite a number of incentives attached to it, several hundred Ethiopian farmers were unwilling to relinquish their land or work for new masters. Technical experts, apprehensive of the political repercussions of land alienation, proceeded in a scrupulously logical fashion. At a preliminary meeting in the presence of political authorities, the local Land Survey Committee, established in June 1938, explained the government's guidelines to the Ethiopians. The Committee included a representative of either the agricultural department or the government office, with the district chief and the village chief. Afterwards, the Ethiopian was shown the land put at his disposal, allowing him certain choice among other available land; effort was made to ensure that the new land was agriculturally equal to, if not better than, the one to be exchanged. When this was impossible, the authorities increased proportionally the size of the land. After a formal acceptance by the Ethiopian, a Committee of Elders defined the boundaries of the new land. The transfer normally took place after the harvest period in order to allow the Ethiopian to prepare the new land for sowing before the big rains. The Ethiopian was allowed to take with him whatever mobile property he had, such as fences or *tukul*. The government normally paid L.30–100 towards rehousing, a prima-facie value established by a special committee after an on-the-spot-inspection. The trees were also paid for with a proviso that they should not be removed.[59]

The transferee was settled in an area considered to be of no relevance

[58] ACS ONC 3/A/1, Taticchi to Presidente ONC, AA. 27 Mar. 1937.

[59] T. Moreschini, 'Il problema della disponibilità delle terre per la colonizzazione nello Sc: Prime esperienze', *Georgofili*, 7/7 (1941): 108–9.

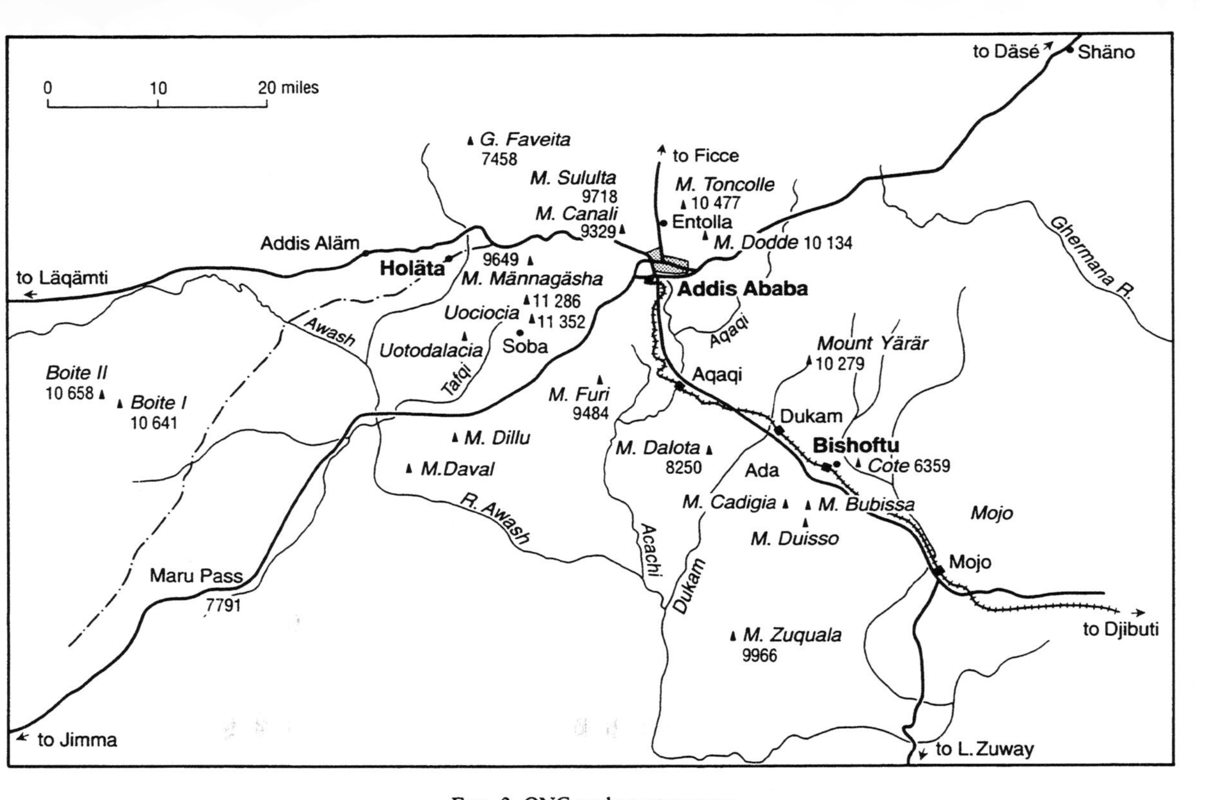


FIG. 2. ONC settlement centres
(*Source*: Quaranta, *Ethiopia*, 41.)

for future metropolitan colonization, far removed from the motorway or important communication networks. Cases where these arrangements met with the transferee's approval were not uncommon, but this was reportedly for fear of possible further displacement, a not unusual occurrence. However, many others refused such arrangements, often under the pretext of the government's inability to control the area and guarantee safety. The transferee occupied the new land under identical rights, ownership or otherwise, that he had held over the previous land. With such a plan, the authorities avoided impairing future decisions by the Land Survey Commission which was expected to make a definite pronouncement on the legal status of the Ethiopian's claims to the land.[60]

Resistance to relocation took a variety of forms. Sometimes delay tactics were used to frustrate the actual transfer; the Ethiopians either went ahead with sowing the land to be evacuated so the authorities were faced with a *fait accompli* or they protested to the central authority whose intervention often forced the ONC to suspend all its construction and farming activities.[61] There were cases when the Ethiopian repeatedly failed to turn up for an appointment on the date fixed and, later on with a great show of sincerity, apologized on the grounds of his inability to remember it. No practical remedy could be found except, as one technical officer noted, 'a remarkable dose of patience'.[62]

At other times it was the chief, who had previously accepted with seemingly great enthusiasm, who secretly connived with the transferee. This was often the case when the transferee went to settle in an area that fell outside the jurisdiction of the chief. But with a generous pecuniary reward hanging upon each successful operation, many chiefs co-operated.[63]

Nor was the situation easier when dealing with the settlement of the tenants, the *ṭisäña*, who claimed no ownership or usufructuary rights. When they were given the choice of either following their former landlord, provided that he consented, or accepting new arrangements made by the Italian authorities, most of them opted to follow their former landlords; nevertheless, the agency attempted to ensure that a consider-

[60] T. Moreschini, 'Il problema della disponibità delle terre la colonizzione nello Sc: Prime esperienze', *Georgofili*, 7/7 (1941): 108–9.

[61] A typical case was that at Holäta Farm. When seven prominent Ethiopian landowners —Buruk Badeg, Fäyesa H. Mikaél, Mämherä Estifanos, Betwäddäd Häylä Giyorgis, Däbbalä Qukare, Ras Nadäw, and Korcho Damo—owning a substantial tract of land (363.83 ha) between them failed to turn up to sign the contracts, in spite of protests, the agency was told to stop all the ongoing farming and construction work (ACS ONC 5/A/1, Ten. Dr Vincenzo Terranova [R. Residenza of Olettà], Holäta 7 Feb. 1938).

[62] Moreschini, 'Problema', 109. [63] Ibid.

able number of them remained in place. This was dictated by political considerations and the economic advantages to the settlement. There was fear that massive displacement would alienate the peasantry to the point of open warfare, and the agency's success depended on the availability of sufficient manpower.[64] In the face of such political and economic imperatives, the agency maintained that its interests would be better served by employing most of the tenants on their original land. The general tendency was to tie them, through a variety of home-grown *metayer* systems, to its farms as sharecroppers, with one obvious difference: the White colonist supplanted the Ethiopian landlord. So, at the boundary of each colonist's farm, the agency settled two or more Ethiopian families to supply him with labour.

According to this plan, the relationship with the settler's farm took one of the three forms.[65] Whenever surplus land was available, the Ethiopian family was provided with a plot estimated to be sufficient enough to supply him with basic foodstuffs and grazing-land. As the new landowner, the colonist received one-half or one-third of the produce from this land and as employer, he engaged his labour by paying him a salary at the market rate.[66] In the second form, instead of giving a certain percentage of his product, the Ethiopian supplied the colonist *x* days of labour during various periods of the year for a salary considerably lower than that of the market rate. Both the dates and the salary were to be fixed in advance. Under the third form, the colonist engaged, under a contract of an annual salary, payable partly in cash and partly in kind, the labour of the Ethiopian family who was normally allowed to make use of a small piece of land, usually located at the outskirts of the farm. Under this arrangement, the Ethiopian normally worked at least 200 days a year, 8 hours a day, for a monthly salary of L.150. Of this L.100 was paid in kind, assessed at the local market rate, and L.50 in cash. This same arrangement applied to other members of the Ethiopian's family in case the colonist decided to employ them. The only exception was that children under 16 and women were paid at half the rate.[67]

The imposition of these contracts proved frustratingly arduous. The Ethiopians were happy to conclude informal agreements with the individual

[64] Ibid. 109–10; ACS ONC: 4/33, Di Crollalanza to Vice-GG, AA 17 June 1938; 3/A/1, Taticchi, Concorso degli indigeni alla colonizzazione demografica, n.d. [15 Oct. 1938].

[65] ACS ONC 15, Di Crollalanza to Vice-GG, AA 17 May 1938.

[66] CD IAO AOI 1839, Di Crollalanza, Promemoria, July 1938, p. 17.

[67] Ibid.; Di Crollalanza, 'Avvaloramento' (1939), 1206–7. As regards salary payment and labour problem see Ch. 6.

colonists while continuing to reject any attempt by the agency to formalize it. Only towards the end of 1938 were some practical results achieved when the government ended its indecision to come to the aid of the agency by openly supporting the third form, which remained the most widespread, particularly at Holäta Farm where it was claimed that labour was relatively abundant.[68] But the first contract, which had its foundation on the customary practices of *erbo* and *sisso*, had to undergo a complex and drastic modification before it became a predominant form largely at Bishoftu Farm. In fact, none of the parties were favourably disposed to this form. The agency saw it as an ideal means for an economical management of the farms for it provided the settler with an income, reduced the cost of production to a minimum, and put into plough hitherto uncultivated land. But despite such benefits, the enterprise entailed loss of considerable land. Therefore, even the agency welcomed it only as a temporary expedient, tolerable until the settler stood on his own feet. But it was disapproved of by the government in Addis Ababa, and unwanted by the Ethiopians. Addis Ababa's government was unsympathetic partly because it had the potential to aid racial intercourse with the possibility of putting Italian prestige into disrepute and partly because its experiences with this form of contract elsewhere had been unsavoury.[69] The Ethiopians, for their part, pointed out the lands allocated for them were uncultivated and would demand substantial time and investment before they became fully productive.[70]

Under the new agreement, a compromise was struck. Land was divided into three categories depending on whether (1) it had been under cultivation for at least the last three consecutive years, or (2) uncultivated for the same length of time, or (3) part cultivated and part uncultivated within the last three years. The settler was entitled to one-third of the produce from any previously cultivated land. The Ethiopian kept for himself the entire produce from the uncultivated land. On the other hand, all the three contracts imposed on him the obligation to supply his labour at a specified time, ox and seeds at his own expense, and to transport the share of the colonist.[71]

The second form was reportedly based on the experiences of the

[68] The text of the contract is given in App. 1 and 2, below.

[69] ACS ONC 3/A/1, Taticchi, Concorso.

[70] ACS ONC 4/33, Taticchi, Relazione, [AA] 13 Apr. 1938.

[71] ACS ONC 3/B/1, Taticchi to SC, AA 24 July 1938; ACS ONC 3/A/1, Taticchi, Concorso.

Villabruzzi[72] in Somalia and that prevailing among the Kenyan settlers,[73] and it also seemed to have precedents in the Babitchev Farm within Ethiopia. But it did not last. As ONC officials were to realize, the contract was founded on ill-construed comparisons, with emphasis on superficial similarities and without taking into account the remarkably different agrarian environment between the countries. Unlike most of the Kenyan and Somali ethnic groups, who, with few exceptions like the Kikuyu, were originally largely mixed farmers combining crop cultivation with cattle-herding, the Ethiopians at the ONC's farms were pre-eminently settled agriculturalists who used quite sophisticated farming techniques and many advanced tools. The squatters in Kenya, who were normally given unworked land, and the Africans at Villabruzzi Village, might engage themselves in rich crop farming and had holdings relatively larger than the plots granted to the Ethiopians by the ONC. Most importantly, unlike the ONC's distinctively subsistence-based demographic colonization, Kenyan settlers were engaged in large-scale industrial farming. By the same contrast, the holdings of the tenants at Babitchev Farm were either irrigated or irrigable and their Ethiopian cultivators had no restrictions on the type of crops they should grow.

Perhaps one of the most important factors contributing to the Ethiopian resistance was the mental attitude of the subject people towards their colonial masters. Ethiopians had traditionally looked down upon the Italians with contempt, and the poor quality of the settlers and their engagement in manual labour, a traditionally despised occupation, served only to reinforce this view. This perhaps may not be the case in the two neighbouring countries—Kenya, whose settlers were largely genteel, belonging not to the dregs but to the most important section of the White society,

[72] The abbreviation stood for Villagio del Duca d'Abruzzio (Duke of Abruzzi Village).

[73] Taticchi toured Kenya twice from 13 Dec. 1936 to 10 Jan. 1937 and from 2 to 30 June 1937 staying each time twenty-nine days, bringing back with him substantial material relating to settler farming and labour organization. He established profitable business contacts with a farm owned by an Italian settler, Vincenzini, specializing in cross-breeding of local sheep with Romney Marsh, and ox with Freisian and Shorthorn, at Sarangetti near Lake Victoria. Experiments in Ethiopia with both these breeds, and others imported from Italy, were only at their initial stage but the results were poor (ACS ONC 3/A/B, Taticchi to SC ONC, AA 29 Jan. 1937; ACS ONC 4: Presidente to MAI, 10 July 1937; and MAI to ONC, Rm 20 Sept. 1937; ACS ONC 5, Taticchi to SC ONC, AA 9 Aug. 1937). Taticchi's initial enthusiastic proposal to apply the Kenyan labour model, stipulated between the European settlers and the Africans, was turned down as unrealistic by Mazzocchi because of huge differences in the customs, conditions, and populations of the two countries (ACS ONC 4/33, Taticchi, Relazione popolazione indigena, AA 13 Apr. 1938; ibid., Di Crollalanza to DONC, Rm 18 May 1938).

and Somalia. In the 1930s Kenya was recovering from the height of its depression, and employment prospects in Somalia, after a brief period of unprecedented prosperity brought about by the Ethiopian war, had become grim. As for the Ethiopians, the employment boom in road construction, deforestation, and housing industries, especially within the environs of ONC farms, had put them in a better position to resist the agency's badly paid work in search for greater rewards and opportunities offered elsewhere.[74]

Despite the obvious advantages it enjoyed, the economy of ONC farms was built on precarious foundations. In order to fend off such fragility, regulations were placed on the new Ethiopian tenant cultivator so that he could easily be controlled and directed towards securing the eventual economic domination of the Italian colonist. The plots at the disposal of the Ethiopians had a maximum size of ½ ha and normally were of modest productivity. Emphasis was laid on the condition that the Ethiopian should cultivate it using only traditional methods. Such a device, combined with the limited fertility of the Ethiopian's land, guaranteed a profit margin to the colonist. Otherwise, it was argued, the latter, new to the environment, unfamiliar with local farming, and using costly modern machinery, could not survive competition from the Ethiopian. Underlying this assumption was that the Ethiopian, unlike the settler peasant, could produce cheaply and had a lifestyle of limited requirements.[75] The Ethiopian could not own at any one time more than four cattle and eight sheep and any numerical changes either by way of birth, purchase, death, or sale were to be reported for approval by the ONC. These were not simply preventive quarantine measures against the multitude of cattle diseases but a clear attempt by the agency to artificially insulate the settler's economic interest against challenges from the Ethiopian production.[76]

The process of land exchange, impossibly slow and complex under the best of conditions, was further compounded by conflicts between political directives and technical imperatives of colonization. In the early years the agency officials and the government authorities were at daggers drawn. The former claimed that it felt marginalized at a time of traumatic experience in an alien land, accusing the government authorities of 'waging a clandestine war, multiplying obstacles particularly at a time the agency

[74] ACS ONC 4/33, Benigno Fagotti, Relazione, Bishoftù 12 Mar. 1938; Pankhurst, *Economic History*, 61–4.

[75] [C. Giglio], 'Rapporti della colonizzazione demografica con la colonizzazione capitalistica e l'agricoltura indigena', *REAI* 26/10 (Oct. 1938): 1564.

[76] Di Crollalanza, 'Avvaloramento' (1939), 1207; Moreschini, 'Problema', 110.

desperately needed co-operation to overcome its early and inevitable difficulties, . . . demeaning the officials as lunatic fringe . . . treating the agency on a par with a private company,' or simply as 'Lessona's stooge'.[77] The authorities saw the conduct of the agency's officials as liable to involve the government in conflict with the Ethiopians and talked of even repatriating the head of the ONC for his 'incomprehension of colonial matters'.[78] However, whatever differences existed between the agency, anxious to obtain land for settlement, and the colonial administration in Addis Ababa, desirous to avoid local conflict, they were not fundamentally over the policy of alienating Ethiopian land to the settler. The disagreement was only on the timing, scale, and method of such alienation. Obsessed with their narrow interest, ONC officials were oblivious to the awkward predicament of the Addis Ababa government who, despite its agreement on the paramountcy of demographic colonization, favoured phased implementation of the programme. Massive alienation was regarded as a politically sensitive issue not only because it would mostly hurt government's alleged allies—the Oromos, an ethnic group assumed to be largely sympathetic to the Italian conquest—but also for its possible international complications. These important facts that shaped government's attitude are summed up by Graziani:

> The creation of 50,000 ha of unbroken farms will have vast and serious political repercussions, for it demands massive transfer of the original inhabitants to settle them elsewhere . . . and cause inevitable discomfort and large-scale discontent. The victims will be not only the Amharas but the Gallas too who have been oppressed by the previous regime and welcomed the Italian occupation with exultation . . . Such discontent among the population will be exploited by the superpowers and elements hostile to the Italian occupation.[79]

The squabbling continued under the Duke of Aosta who also criticized the way the existing land-alienation system itself was carried out and urged the local authorities to proceed cautiously and gradually, using not the 'rough and reckless measures that are always unwelcome by the Ethiopians, but the slow work of persuasion'.[80] However, despite such sound arguments the ducal regime's land policy was not alien to the ambiguities and inconsistencies that had characterized its predecessors. Although it was opposed to alienation of land until the land regulations

[77] ACS ONC 3/A/1, Fagotti to Mazzocchi Alemanni, AA 10 Dec. 1936.
[78] ACS ONC 5, Taticchi to [Mazzocchi-Alemanni], AA 9 Mar. 1937.
[79] ACS ONC 1/1, Graziani to MC, AA 1 Mar. 1937.
[80] Villa Santa *et al.*, *Amedeo*, 238.

were put on a more satisfactory ground and the feasibility of colonization itself was studied,[81] it maintained that until such time the policy of amalgamation be 'limited to lands that are for one reason or another uncultivated, rather than depriving the Ethiopians of the land they use no matter how inefficiently this may be'.[82] In fact it had become clear to all but the most prejudiced Fascist firebrands that it made little sense to the notion of civilizing mission to deprive the Ethiopians of their often well-cultivated lands in order to grant them to the settler practically as a gift, while making the unfortunate Ethiopians start afresh on clearing a jungle. 'If so,' ONC was asked, 'where is the contribution [of the national] whose very intervention is supposed to aid expanded farming? As it appears from the [ONC] programme the role of the national is exclusively confined to take over as manager *vis-à-vis* the native. Otherwise, there is hardly any other reason to justify his presence'.[83]

Aosta's government policy reflected the recommendations of two economic experts, G. Cusmano and Roberto Corvo, authorized by the Viceroy himself. In their secret report of April 1938, they made clear that the country was not suited for a massive influx of Italian peasants. In their view, at the early phase of colonization, private enterprise and capital should be encouraged as the dominant mode of development. In order to avoid the opening of the Empire's resources to land speculators, the report urged that colonization be restricted to undisputed public domanial lands and be strictly controlled by the State through a central co-ordinating body. The experts equally rejected the existing system of land transactions. Their report pointed out that expropriation, even with extensive compensation, would transform the best peasantry, mostly the Oromo, into a multitude of have-nots, who would subsequently swell the resistance movement.[84]

The Duke of Aosta's regime's unsympathetic stand towards the existing policy of land alienation seemed to offer adequate guarantee that the lands occupied by the Ethiopians would not be despoiled for the benefit

[81] 'Land grants presume ascertaining of land rights. If we do not create a public domain, we cannot weigh up either the true political implications or the relative importance of concessions; we have to give the Ethiopians the certainty that the lands acknowledged as their properties will be strictly cultivated by themselves.' (Ibid. 212; CD IAO AOI 1936, Amedeo D'Aosta to MAI, 5 Jan. 1938.)

[82] ACS ONC 4: Taticchi to Di Crollalanza, AA 8 Feb. 1938; ibid., Di Crollalanza to Taticchi, Rm 6 Feb. 1938.

[83] ACS ONC 5/A, GG to ONC, AA [date illeg.] Nov. 1939.

[84] ATdR 24/128, Cusmano G. and Roberto Corvo, [Relazione sulla situazione economica AOI] AA 1 Apr. 1938, pp. 8–9.

of the ONC's settlers. Yet within a few months of his statement, 1,406 Holäta families with 5,115 members were removed and 1,685 houses destroyed in order to settle only 46 Italian families. In case of plot no. 37, for example, where an Italian family of 7 individuals was accommodated, 760 Ethiopian houses were disposed of, and 634 families with their 2,303 members displaced.[85] Again in early 1939 the decree of the *residente* of Holäta ordered several Ethiopians to move off to other land received in exchange. Among those badly affected by this decree were the settlers on plot no. 79, allocated to the colonist Lazzerini Benvenuto and his 3 family members. This entailed forced removal of 40 skilled agriculturalists of Benishangul families (along with their numerous cattle), settled by Emperor Menilek, who practised irrigated vegetable-garden farming at Guntuta.[86] Though at a reduced scale, alienation of land continued until the end of 1940,[87] the year in which the policy began to experience some serious reversals as a result of intensified nationalist activities in places like Qalitti.[88] Only then did the government strongly insist on its previous pusillanimous demand that the ONC should either revise its existing programme by contracting the size of the individual plots,[89] or cultivating valuable industrial crops, or looking for domanial lands elsewhere.[90] Earlier the agency's suggestion for government to purchase the Ethiopian lands had been rejected as an excessively burdensome alternative, and 'by no means less complicated than the system so far pursued'.[91]

The ONC strongly resisted any reduction in the size of the farms but grudgingly accepted the government's demand to divert part of its work

[85] ACS ONC 33, AAH: censimento indigeni che abitano sui poderi dell'azienda, Holäta 11 Apr. 1938.

[86] ACS ONC 15, Di Crollalanza, 'Trasferimento famiglie coloniche nelle AH e AB dell'ONC', Rm 16 Feb. 1939; ACS ONC 3/B/1, Taticchi to SC: Relazione, AA 11 Mar 1939.

[87] In Dec. 1940 ONC was allowed a limited expansion to Dukam to an area 2 km wide and 1 km long (ACS ONC 5/A, Nasi to Com. G. Ada, AA 22 Dec. 1940).

[88] As its inability to provide protection became difficult, the government forced the agency to restitute 847 ha of land located part at Airport zone between Ada-Moggio, and part at Qalitti where the local clergy had managed consistently to frustrate the agency's efforts at sharecropping (cf. ACS ONC 5/A/1, Taticchi to SC, AA 3 Dec. 1939).

[89] On the issue of reduction of the size of the settler's plot, the Duke of Aosta's position was ambiguous. Contrary to earlier reports, lately he seemed to distance himself from the almost unanimously shared view of his government functionaries. Cf. ibid; ACS ONC 3/A/2, Taticchi to SC ONC: Udienza concessa al direttore dell'azienda da S.A.R. il Vice Re, AA 16 Dec. 1939.

[90] ACS ONC 3/A/2, Taticchi to SC ONC, 16 Dec. 1939; ACS ONC 5/A: GG to ONC, AA 15 Oct. 1939; Sergio Fornari [Acting ONC head] to SC ONC, AA 10 Nov. 1939; Nasi to Com. di G Ada, AA 3 Dec. 1940, 22 Dec. 1940.

[91] ACS ONC S/A, GG to ONC, AA 15 Oct. 1939.

to outside Shäwa, to which the agency had so far paid scant interest until 1940 when it lost the take-over of a number of well-established farms to other private and para-statal agencies.[92] Addis Ababa government also prevailed in its demand that the ONC, who had so far evaded its authority and acted as an autonomous organ responsible only to Rome, should submit, like the rest of the demographic colonization agencies, its farm organization and development strategies to the Governor's office.[93]

Having shown an incredible dose of incompetence in handling the issue of diverting the ONC's activities to other parts, in February 1940 the agency's officials decided to set up farms at Gawani in Dänkäl, at Wänqi, in the Upper Awash River basin, at Märäquo in Guraghé region, and at Goré in Illubabor.[94] The four areas were selected on the basis that they possessed most of the prerequisites for demographic colonization in terms of climate, communication network, and topography, and were reported to have extensive 'promising lands'.[95] But the decision came too late, and the plan remained on the drawing board. Only at Wänqi did the ONC succeed in sowing 2,300 ha out of 7,500 ha of land obtained towards the end of 1938 while Gawani, though ideal for large-scale irrigated farming, was soon discovered to be unsuitable for settlement, at least in the short term.[96]

As the exchange-of-land scheme proved unworkable, the ONC, like most Italian enterprises, capitalist as well as demographic, made ample

[92] Several promising farms were put at the disposal of ONC, two in Shäwa and three in Härärgé: in Shäwa, there was the 300–400 ha Mullu Farm, belonging to an English gentleman, Dan Sandford, and Babitchev at Bishoftu; in Häräghé, Fadis, located a short distance from Mäheso train station, and Baka farms (both belonging to Emperor Haile Sellassie), and Täklä Häwaryat Farm at Čärčär. Reasons for the rejection of Mullu Farm were understandable, because it was under the control of the resistance fighters—whom Taticchi describes as 'roving bandits whose vexatious demand for tributes and killings had led the natives to abandon their farming activities'. But the excuses for the farms in Härärgé were as flimsy as those for the Babitchev Farm. The latter was granted to Ente Romagna whose director Taticchi, having discovered his blunder, unsuccessfully strived to cunningly dissuade him from taking it over. Cf. ACS ONC 3/A/1, Taticchi to SC, AA 30 Apr. 1937; ACS ONC 5/A/1, Taticchi to SC, AA 12 Apr. 1938; ACS ONC 5/A/Mullo, Taticchi to SC, AA 10 Aug. 1938; CD IAO AOI 1936, Sottosegretario di Stato Terruzzi to GG AOI, Rm 26 Apr. 1938.

[93] ACS ONC 5/A, GG to ONC, AA 10 Nov. 1939, and Nasi to Com. di G. Ada, AA 22 Dec. 1940; ACS ONC 5/A/1, Taticchi to SC, AA 3 Dec. 1939.

[94] The concessions were formalized at a 14 Feb. 1940 meeting held by Minister Terruzzi, Duke of Aosta, and Di Crollalanza. An extensive grant was also planned at Limmu, Jimma. Cf. ACS ONC 3/A/2, Di Crollalanza to ONC Direzione AOI: Piano regolatore delle attività dell Opera in AOI, [Rm] 10 Apr. 1940; Villa Santa *et al.*, *Amedeo*, 232; Quaranta, *Ethiopia*, 48.

[95] Quaranta, *Ethiopia*, 48.

[96] ACS ONC 3/A/2, Di Crollalanza, Piano regolatore.

use of the *metayer* system to farm lands that their Ethiopian claimants were unable to cultivate. These lands fell under what came to be known as zones of influence which entitled the Italian titular of the adjoining land to cultivate them jointly with Ethiopian claimants.[97]

Like the exchange system, the objective of the zone of influence system was to concentrate the best land in the hands of Italian settlers at the expense of the Ethiopian cultivator and guarantee the settler cheap Ethiopian labour. Theoretically, all colonial policies were intended to foster these goals, and the Italians were not an exception. But they differed in their pride in their mission to civilize the Ethiopians and their persistent claims for their system's unique concern for humanitarian values. But owing to discontent caused by the abuse of the zone of influence by settlers, the system was suppressed. Many Ethiopians questioned the Italians' true intentions and viewed the scheme as a stratagem for robbing them of their land under the pretext of cultivating it temporarily. Some were reluctant even to consider the deal. Even those who had initially accepted had second thoughts. Some refused to sow and others, instead of sowing, ate the seed. In other instances, the chiefs in charge of seed distribution failed to deliver. In this, Chief Mammo Tädächa of Bishoftu excelled. The co-operation that was gained rested on bribery and other forms of allurement.[98]

## FARM ORGANIZATION AND SETTLEMENT PROGRAMMES

The ONC's settlement programme aimed to avoid the fragmentation and dispersion of the farms, secure the conditions of safety, make little demands on technical organization, and safeguard racial prestige.[99] Scattered homesteads as at the Pontine Marshes were rejected since those in a country 'so new and so far from the motherland' were considered a safety hazard. Concentrated villages were also discarded as this would necessitate a relatively long journey to and from the farm.[100] Instead homesteads of as many as eight adjoining holdings were grouped together as spokes

[97] CD IAO AOI 1930: Graziani to DS ACP *et al.*, Incremento culture alimentari, AA 11 May 1937; Promemoria Per il Vice Re, n.d. CD IAO AOI 1837, P. Bono, Memoranda sulla situazione degli agricoltori italiani in Etiopia, 19 Feb. 1941.

[98] Sbacchi, 'Italian Colonialism', 368. [99] Montefoschi, 'Centri agricoli'.

[100] FO371/22021/J2512/40/1, Perth to PSSFA, Rm 24 June 1938; ATdR 24/124, Di Crollalanza, 2° Promemoria, Jan. 1938, p. 10; Di Crollalanza, 'Avvaloramento' (1939); 1198; MAI, 'Valorizzazione', 263–4.

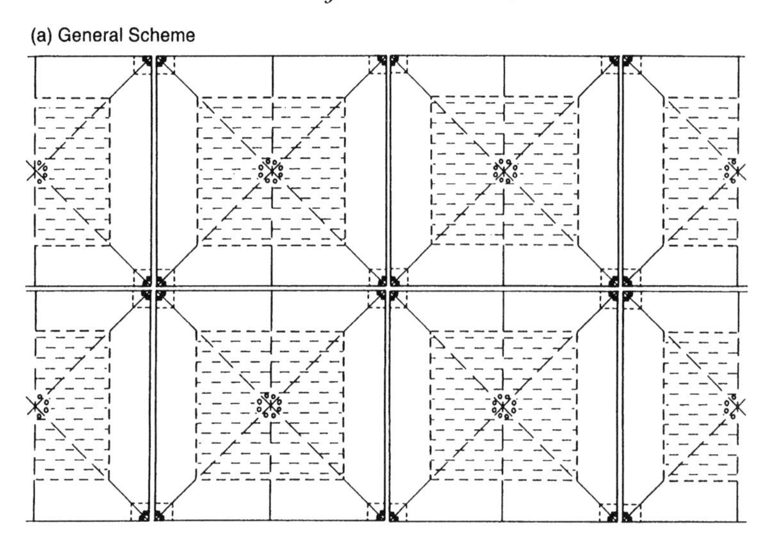


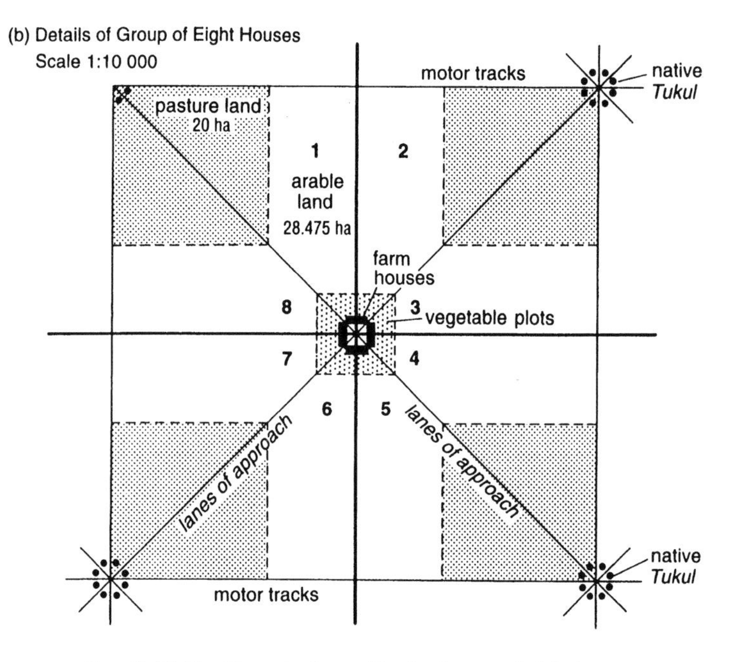


FIG. 3. ONC settlement scheme: Details of group of eight houses (*Source*: Mazzocchi Alemanni, 'Orientamenti', 21; MAE, 'Valorizzazione', 265.)

of a wheel, triangular in shape (see Fig. 3). Near the apex of each eight triangles were placed the individual farmhouses. This arrangement, reminiscent of the Roman *agri vectigales*,[101] enabled the sharing of a bakery and wells located at the centre and reconciled economy and security: within a radius of a couple of hundred yards lived about forty armed and able-bodied farmers, a large enough number to offer mutual assistance and protection.[102] The houses, situated inside their plots, were grouped around a vast central courtyard, easily transformable into a fort in case of emergency.[103] This system also removed the need to build an extensive interfarm communication network.[104]

Each farm measured 50–60 ha, an area larger than the average holding in Italy. A number of reasons determined this size: whereas in Italy cultivation was intensive in character, it was expected that, initially, a more extensive system would show the quickest returns in Africa. In Italy the peasant, helped by members of his family, normally did all the work on his holding; in Ethiopia, the employment of Ethiopian labour would enable a larger area to be farmed. Furthermore, the colonist had to save sufficient money to transfer the land into his ownership. Other considerations included the increase of the settler's own family, which in the course of time would demand the splitting of the farm between the heirs of original settlers; limited productivity of the land; and the tendency of tropical land to rapidly decline in fertility. Thus, ample precautions were taken for future exigencies.[105] It was on the grounds of these complex considerations that Di Crollalanza had consistently refused the Addis Ababa government's repeated demand to reduce the size to 30–5 ha.[106]

Each farm was divided into three parts (see Fig. 3). In the immediate vicinity of each farmhouse a small patch was reserved for vegetables and fruits intended for the settler's daily consumption; here he could plant fruit trees and vines. At the periphery of the farm there were *c.*20 ha destined for forages and cattle-grazing. The arable land, amounting to 35–40 ha was located in the centre, where diverse crops could be cultivated with the assistance of Ethiopian labour. The Ethiopians' residences were situated on the outskirts of the farm, at 1.5–2 km from the house of

[101] FO371/22021/J2512/40/1, Perth to PSSFA, Rm 24 June 1938; Quaranta, *Ethiopia*, 44; MAE, 'Valorizzazione', 264; Di Crollalanza, 'Relazione', 743.

[102] Di Crollalanza, 'Avvaloramento', 1200.

[103] MAI, 'Valorizzazione', 263–4.

[104] Di Crollalanza, 'Relazione', 741–2.

[105] Montefoschi, 'Centri agricoli'; MAI, 'Valorizzazione', 264.

[106] ACS ONC 5/A/1, Taticchi to SC, AA 3 Dec. 1939.

the settler—distant enough to keep the two races apart, yet close enough to control the Ethiopian and promote economic interaction.[107]

The farmhouse itself was planned and built in such a way as to promote the prestige and decorum of the White man in Africa. Even though most of the hitherto-discussed schemes were within the framework laid down by the Agricultural Council, in its housing programme the ONC, like some of the other agencies, followed slightly different criteria. The Agricultural Council had debated whether to build temporary or permanent dwellings. Its conclusions were mixed. On the one hand, it emphasized the advantages of low-cost housing in that it saved resources which could be allocated to transplanting a large number of unemployed families. But it equally made clear that if their quality were inferior to that which the peasants had at home, they might fail in luring them to Ethiopia. On the other hand, it argued that 'the elevated cost of rural constructions could, by itself alone, be seriously detrimental to demographic colonization, compromising from the start its happy development'.

The Agricultural Council left it to the policy-makers whether to opt for a few families with comfortable and expensive houses or the maximum possible number of peasants in more modest dwellings. Yet its own views were quite clear: it recommended that 'complete, perfect solutions should be avoided and transitory measures adopted'.[108] The ONC followed the middle ground: it built permanent houses capable of extension. The underlying assumption was that such houses decreased the total cost of reclamation, thus enabling the peasant to pay off his debts and save for extension.[109]

Concern for economy dictated maximum use of local materials, the rational utilization of space, and the provision of basic furniture. Yet in line with the Agricultural Council's advice, racial prestige demanded that the dwellings excelled the Ethiopians' homes—usually made of wooden walls with mud and thatched roofs—in comfort, design, and architectural beauty. The settler's house was a bungalow 13 m long and 3.4 m high. At Holäta, where basalt was abundant, stone was used; at Bishoftu, bricks and mud. It had a plain tiled roof of Italian design so that the farmer would feel at home. One colonial writer claimed that the sight of each roof gave 'the visitor an impression of a bit of Italy transported as if by

[107] Quaranta, *Ethiopia*, 44; Di Crollalanza, 'Relazione', 743.

[108] Consulta, 'Avvaloramento', 1565; CD IAO AOI 1776, Relazione, 14. For the actual debate see CD IAO 1775 Seduta, 30 Apr. 1937, Rm 5 May 1937, pp. 1–2.

[109] [C. Giglio], 'Di alcuni problemi della colonizzazione demografica', *REAI* 26/12 (Dec. 1938), 1875–6; Di Crollalanza, 'Avvaloramento' (1939), 1200–1.

magic to this distant part of the world'. The interior consisted of two bedrooms

equipped with a few simple pieces of furniture, and of a large, comfortable lounge–kitchen in the centre. Access is gained through a kind of veranda formed by an overhanging roof and a niche sunk into the façade, which is coated with white rough-cast . . . A wooden gate and a fence of pleasing appearance accentuate the rustic note.[110]

A foreign journalist admiringly, but mistakenly, described it 'as a solid-stone built bungalow' with 'large windows looking out on a really beautiful countryside with mountains and lakes like those in the Alps'.[111]

Each of the houses was intended to accommodate a family of up to six and to offer all the comfort and the simple amenities to which the Italian peasants were accustomed.[112] The family shared, with the rest of the farm members, water from wells, normally 30–60 m deep, pumped up by a motor; a grain mill; three or four communal ovens; and a store house.[113] All the houses were provided with a *zäriba*, a shelter built to local design to keep farm animals, agricultural tools, and machinery. The cost of each house at Bishoftu was L.38,000 and the estimate for Holäta was L.31,000–35,000 or L.10,000 a room. The agency personnel's houses had an additional three rooms and an extra toilet with water being supplied from a tank inside the house itself. There were also some differences in finishing touches and use of materials on the understanding that, because of their managerial status, they had greater needs than those of the average settler. It was meant to provide them with the minimum comfort that their position entailed and to offset the sacrifices imposed upon them as a result of their residence in a still largely unknown land. The total cost of each of these houses was L.109,000 at Holäta, and L.115,000 at the Bishoftu Farm.[114]

According to the plan, groups of farms were to be joined together into *aziende*, or co-operatives, built with a State budget, where institutions that catered for the larger cultural and social needs of the farm were set up. These rural centres, or *villaggio nazionale*, were to include facilities such as shops, the agency's administrative headquarters, government offices, Fascist headquarters, church, school, cinema, clinic, post-office and co-operative granary, supply store, and dispensary.[115]

[110] Quaranta, *Ethiopia*, 47. [111] Diel, '*Behold*', 184.

[112] [Giglio], 'Alcuni problemi', 47; Quaranta, *Ethiopia*, 47.

[113] Pankhurst, 'Italian Settlement', 149; Quaranta, *Ethiopia*, 43.

[114] Di Crollalanza, 'Relazione', 747–8, 750–2, 757–9; [Giglio], 'Alcuni problemi', 1876.

[115] Di Crollalanza, 'Relazione', 745, 749; id., 'Avvaloramento', 57; Montefoschi, 'Centri agricoli'; Quaranta, *Ethiopia*, 43; Berretta, '*Amedeo*', 258–9.

The settlement programme was a joint undertaking between the State, the ONC, and the colonists and its purpose was to bring the farms to economic autarky so that the colonist's plot could be productive in the shortest possible time. The ONC was to provide expertise and financial help so the settler could, to use the ONC's own much-cherished phrase, 'pass into the responsibility and dignity of a small landowner'.[116] The details of such a transition were regulated by a contract between the State, the agency, and the settler, as well as the Ethiopian peasants.

Under the contract, the State provided the land tax-free, developed the basic infrastructures of communication network, public amenities, and rudimentary social services. It was also the State's duty to select the prospective settlers in co-operation with the ONC and look after the transportation of families.[117]

The ONC drew up the development plans and implemented them; reclaimed the land and allotted it to the families; built houses and provided agricultural implements, seeds, livestock, as well as direction and supervision, technical advice, and the necessary capital; catered for the material and moral needs of the colonists; and recruited Ethiopian labour.[118] The ONC planned to regain all or a substantial part of its investment in five to seven years. It was hoped that, thanks to the amount of land in his possession, the agency's support, and the unfailing assistance of the Ethiopian with his cheap and experienced labour, the settler would repay his debts within the targeted period.[119]

The contracts that regulated the relationship of the colonist and the agency, as approved by the government authorities, consisted of four parts:

1. An initial transformation period in which the agency, assisted by the settler, did the preliminary work to develop the farm. During this year-long period, the agency would clear, drain, and plough the land; enclose the fields; provide the plots with communication networks; and immediately before the rains begin sowing operations. Contrary to other

[116] Montefoschi, 'Centri agricoli'; CD IAO AOI 1839, Di Crollalanza, Promemoria, July 1938, pp. 14–15.

[117] Di Crollalanza, 'Relazione', 745; CD IAO AOI 1839, Di Crollalanza, Promemoria, July 1938, p. 11.

[118] Di Crollalanza, 'Relazione', 749; id., 'Avvaloramento' (1939), 1205–6; Quaranta, *Ethiopia*, 42.

[119] CD IAO AOI 1839, Di Crollalanza, Promemoria, July 1938, p. 14; Di Crollalanza, 'Relazione' 749. ONC considered a period of between six and seven years as sufficient to transform a peasant into a landowner (cf. id., 'Avvaloramento' (1939), 1205–6; Quaranta, *Ethiopia*, 42).

demographic colonization agencies, the ONC, frustrated with the early settlers who, as we have seen, were waged, had a programme to give the farmer his estate upon his arrival, thus elevating him from the very start to the status of a colonizer rather than keeping him as a sharecropper. This, of course, was a device to attach the prospective settler to his farm, a fact that Di Crollalanze himself openly acknowledged.[120]

2. A phase in which the farm would begin to produce a yield. At this stage, the settler, provided with the necessary equipment and an adequate loan, would take over the farm and Ethiopian labour to help him reap the first harvest while the building work of his house was under way; his status would become that of manager and director of his farm, although under the direct control and with the assistance of the agency which fixed the cultivation methods, norms of husbandry, and care of animals, and marketed surplus products. The change in the settler's social status was not purely nominal. No matter how inferior his social and educational background was, from the start he was expected to be an employer of the Ethiopian whose economic life, as an employee, would remain tightly dependent upon that of the colonist farmowner.

3. With his house built and harvest time at an end, the settler would be joined by his family.

4. At the advanced stage, the farm's production would be regularized, while the family would live on the provision that it had secured from the preceding year's harvest.

The colonist would now enter the phase of redemption whereby, now working a productive farm, he made long-term arrangements with the ONC for a gradual repayment of the total cost of his farm.[121] In the mean time he continued to benefit from the agency's technical and financial assistance. The ONC registered his payments as credits in the account book, or *libretto colonico*. The booklet also included a cumulative annual balance which would establish the moment when he could become an independent landowner, and a list of his livestock.[122]

[120] CD IAO AOI 1936, ONC Azienda AO: Disciplinare provvisorio per i rapporti fra l'ONC ed i colonizzatori nazionali, AA 10 June 1938, p. 1; CD IAO 1839, Di Crollalanza, Promemoria, July 1938, p. 14; Di Crollalanza, 'Relazione', 745; id., 'Avvaloramento', 1205.

[121] CD IAO AOI 1936, Disciplinare, pp. 1, 3; CD IAO AOI 1839, Di Crollalanza Promemoria, July 1938, p. 15; Di Crollalanza, 'Relazione', 749; id., 'Avvaloramento' (1939), 1205–6.

[122] CD IAO AOI 1936, Disciplinare, p. 3; CD IAO AOI 1839, p. 15; Di Crollalanza, 'Avvaloramento' (1939), 1206; Mazzocchi Alemanni, 'Orientamenti', 24–5.

## PROBLEMS OF RECRUITMENT AND DISCIPLINE

The shortage of prospective settlers and the low quality of those remaining in the Empire was disturbing. As appears from a British consular report, of those who saw the country few were keen to remain. Out of the fifty Italian officers, prisoners, and others interviewed, only one was reported to have wished to settle. Of another fifty, not even one wished to stay. The same report suggested that a similar attitude had been reported by other foreigners.[123]

Yet there was no dearth of applications, particularly in the early years. By April 1938 the ONC had received 297 applications—29 of them for managerial positions—reaching 600 by the end of the year. Of these only a few were judged suitable and accepted. But what worried the ONC most were the long-term prospects of the scheme, for the responses of the Italian local authorities (*communi*) to its recruitment and co-operation campaign were disappointingly few.[124]

The selection of the colonists also proved arduous. In line with the Agricultural Council's recommendation, reflecting Fascist political thinking, families rather than individuals were to be recruited to ensure stability. The ideal family was that composed of, at least, four able-bodied and healthy individuals willing to make the Empire their permanent home. A good war record was also important, particularly in the Abyssinian Campaign, as was their attitude and morality, especially towards the Ethiopians whom they were going to 'civilize', and their occupational background and experience.[125] Vetting family members for their past political record became compulsory only after mid-1939. However, even after that, affiliation to the official Fascist Party or an active career in it remained of no paramount importance for initial selection, provided that the candidate had a sympathetic or not openly hostile attitude towards it.[126] Yet political

[123] FO371/20167/J5632/45/1, Gibbs to PSSFA, AA 21 June 1936.

[124] ACS ONC 15, Taticchi to SC ONC, AA 12 Apr. 1938, and 16 Dec. 1938.

[125] CD IAO: 1775, Seduta, 5 June 1937, p. 3; 1776 Relazione, p. 12. AOI 1936, Disciplinare, p. 2; Consulta, 'Avvaloramento', 1564; Di Crollalanza, 'Relazione', 744; id., 'Avvaloramento' (1939), 1205.

[126] The information about the colonists was not often candidly accurate. When the candidate's relation to the Party was not too clear the statement to this effect remained too general, and as a balance emphasis was laid upon other factors, such as religion and race, as in the following ex.: 'Taddia Antonio, son of the late Francesco and Corsini Adelaide, born on 17 May 1888 at Piave di Cento, and resident at N°. II Via Dante, Bologna . . . currently living in AA, appears to be a man of *good political conduct* and married [which was not the case] and *belongs to the Aryan race and professes Catholic religion*. He had the following criminal records . . .'. As Taddia, twice sentenced for bankruptcy, had a criminal history no

disloyalty, to the same degree as mischievous behaviour towards the Ethiopians, constituted sufficient ground for the dismissal of a colonist and, ultimately, the dissolution of the contract. Circumstances that affected the working efficiency of the colonist's family and compromised the prospects of land redemption had the same consequence, as did indiscipline towards the ONC's staff, quarrelsome and immoral behaviour, and as criminal conviction.[127]

The ONC's task was not only to transform the recruits into small independent landowners but also instil in them certain mental habits such as self-discipline, thrift, and independence, that were essential for the success of the social and political goals of colonization. Di Crollalanza stated:

In allotment of farms absolute priority will be given to families of colonists who have been fighters in East Africa and, among them, to those deserving it and having proper skills, a large family, enterprising spirit, and the will to win, with labour and adaptation to initial hardships, the ownership of land that will be granted to them, only in so far as they prove themselves to be worthy of it.[128]

Despite the ONC's claim to follow rigorous selection procedures, in reality most of its recruits did not live up to expectations. Mostly drawn from among the lowest elements of the Italian society, the colonist population not only remained small and largely unstable but also offered little to guarantee the success of the scheme. From the start of its operation up to the end of 1940 the ONC recruited a total of 300 colonists. Of these

description is given on his moral quality. Otherwise a general statement is made together with the individual's political background as in the following case of Girardi Lidio, one of the recruits of the Duke of Aosta: 'Girardi Lidio di Giovanni appears to be of *good moral and political conduct, with no criminal history nor pending moral* [*offences*] [*author's emphases throughout*]. He belongs to the Italian race and Catholic religion. Even though he has not subscribed to PNF, he is one of its sympathizers.' Cf. ACS ONC 15, Taticchi to SC, AA 9 Jan. 1939. ACS ONC 3/B/1: Sergio Nannini to ONC, Rm 4 Mar. 1939 and 28 Apr. 1939; Taticchi, Azienda AOI: Relazione mese di settembre 1940, n.d. ACS ONC 15, PNF, Foglio di disposizioni, 16 June 1939.

[127] CD IAO AOI 1936, Disciplinare, p. 5; Di Crollalanza, 'Avvaloramento' (1939), 1206. The Italians were unanimous that the secret for success of ONC's initiative ultimately lay in the choice of Italian colonist. Echoing this view, the head of the ONC in IEA, Giuseppe Taticchi, portrayed the ideal settler not to be 'the crafty speculator or the feeble amateur who surrenders at the sight of first hardship, or the conceited blockhead who, scorning experience and advice, comes to Africa sure to pluck lemons and oranges, but the strong, calm and collected, persevering, sober and thrifty Italian colonist who is fond of a large family and able to observe, learn, work, and be tolerant' (ACS ONC 3/A, Taticchi, Concorso degli indigeni alla colonizzazione demografica, n.d. [Oct. 1938]).

[128] Di Crollalanza, 'Valorizzazione', 494.

only 93 stayed.[129] Of between 120 and 150 (figures vary) initial recruits, only 60 persevered until the end. The ONC eschewed responsibility for such a modest performance, blaming the military authorities of the time for their poor method of selection. Yet of about another 150 settlers, whom the ONC claimed to have recruited 'with special care', less than half remained.

No matter how imaginative its policies and serious its initiative may be, the ONC was faced with the same problems which had affected earlier settlements in Eritrea and other Italian colonies. The background and the quality of the settlers were as disappointing as their numbers. The so-called authentic farmers were a tiny minority. Of eighty-one Holäta colonists, for example, only about four were credited as being excellent, the rest were described as mediocre or bad. Among these there were men of pensionable age; unmarried individuals, incapable of bringing up a family of their own; and married men reluctant to be joined by their families. Some had never farmed before coming to Ethiopia. Included among these were footmen, masons, bricklayers, bankrupt businessmen, and robbers.[130] Others showed adversion to fieldwork and attempted to enter the much more profitable field of commerce as soon as they realized that only meagre gains could be made from farming. Substantial numbers of these were adventurers who came to Ethiopia not with the intention to settle permanently, but to make a quick fortune before the inevitable return to the motherland. The 1940 field reports suggest that, despite the management's attempt to confine them to strictly agricultural activities, for those colonists who did not abandon their farms completely, farming formed a negligible part-time pursuit while their main interest lay elsewhere: in retail business, ranging from perfumeries to bakeries, from foodstores to cereal trade; others worked as absentee landlords while sharecropping with the Ethiopians.[131] Women and those with some sort of education, having found the physical demands of farming and the ennui of countryside quite unbearable, sought their fortunes in the city of Addis Ababa either as domestics or clerical workers, or ran away in search of a lover. Most of those who remained were the least educated people or elderly who had little social and economic alternatives. Cases of expulsion for dishonest or criminal conduct towards Ethiopians as well as fraud

[129] CD IAO AOI 1839, Di Crollalanza, Promemoria, July 1938, p. 14; Di Crollalanza, 'Relazione', 739, 744; id., 'Avvaloramento' (1939), 1198; Villa Santa *et al.*, *Amedeo*, 230.

[130] ACS ONC 3/B/1, Sergio Nannini [commissario] to ONC, Rm 4 Mar. 1939.

[131] ACS ONC 15, Curtone Vincenzo to Presidente ONC, Ada 3 June 1940; ACS ONC 3/B/1 Taticchi, Azienda AOI: Relazione mesi di dicembre 1940 e gennaio 1941, n.d.

were not rare, and some settlers were continuously in and out of prison.[132] After a probationary period of about four months, the ONC expelled those unable to transform themselves into colonizers. These included individuals lacking strong commitment, technical skill, and above all the tendency to save money and passion for fieldwork.[133]

Unless working on the White man's farm could be construed as such, in terms of 'civilizing' the Ethiopians, the colonists had often very little to offer. Substantial numbers of them had not the scantiest idea of what the civilizing mission meant. Management field reports talk of the families who 'have gone native' and whose 'innate carelessness and poor tendency to basic hygiene' have shown 'obstinate resistance' even to severe measures, such as physical punishment. In such embarrassing circumstances, the agency was left with no option but resignation and hope that 'such inevitable cases' would be eliminated in due course.[134]

The colonists' attitude towards the ONC's settlement progamme was even far more damaging. As the agency sadly realized, a substantial number of early recruits were attracted by the salary and did not intend to become lifelong farmers. Di Crollalanza's repeated attack on the salary system seems to illustrate how strong the wage mentality was among the settlers. Echoing the dominant thinking of most Italian colonial advocates, Di Crollalanza maintained that for Italy, poor in capital but rich in manpower, a salary system was costly. Its subsequent abolition caused many to leave.[135]

But most upsetting was the growing inability of the ONC to enforce its contracts with the severity it wished. This was especially so after family reunions which introduced new and insuperable difficulties into a situation that was already complex. Some families had constant arguments with their neighbours and maintaining peace between them became arduous. Families from different regions often squabbled. The head families,

[132] ACS ONC 3/B/1: Sergio Nannini to ONC, Rm 4 Mar. 1939; Daodiace [GSc] to ONC, AA, 11 Oct. 1940; Taticchi to GG AOI, AA 15 Oct. 1940; Sbacchi, 'Italian Colonialism', 422.

[133] ACS ONC 3/B/1, Taticchi to SC, AA 24 July 1938; CD IAO AOI 1839, p. 14; Villa Santa *et al.*, *Amedeo*, 230; Di Crollalanza, 'Relazione', 744.

[134] The difficulties of enforcing basic health measures were frustrating as the case of plot no. 3 serves to illustrate. Its colonist, Angelo Coniglio, of Campo Felice di Rocella (Palermo), has reduced himself to the status of the native and his wife Giuseppina Angelo 'has not the faintest idea of hygiene. She never combs her hair nor does she dress up her children nor wash them. She does not clean the house even though the ONC supplies the family with naphtha for the pavement. The assistant farm-manager visits the family three or four days per week. He insists that the house be kept clean and the children tidy but to no avail,' (ACS ONC 29, Dati e notizie riguardanti l'azienda di Olettà, n.d. [AA 15 Sept. 1939].)

[135] CD IAO AOI 1839, Di Crollalanza, Promemoria, July 1938, 14; Di Crollalanza, 'Relazione', 744; Mazzocchi Alemanni, 'Orientamenti', 24.

who had the duty to execute the agency's orders and keep the family labour intact, failed to provide discipline and strong leadership as expected. Those families with extensive kinship ties that included nephews and in-laws often quarrelled and, as the farms became more productive, the conflict intensified and their dispute often ended in the breakdown of the family unit.[136] Rather than tolerating any family disintegration or the risk of setting dangerous precedents, the ONC's policy was to settle the matter by repatriating the defaulting family. Repatriation was effective as a deterrent but faced many obstacles.[137] Although the elimination of troublesome and unproductive families was considered fair and easy to impose, it involved enormous loss of time and money especially when the litigants were hard workers and had established considerable credit with the agency. In such circumstances the ideal solution would have been either the repartition of the property between the disputants, liquidation, or appeal to the court of law. But partition and liquidation were thought to give an unfair advantage to the disputants over the agency's long-term interest and any court decision in favour of the litigants was believed to seriously damage the agency's credibility and set an unacceptable precedent to other families. In order to avoid such embarrassing setbacks the ONC resorted to informal lobbying with the Party officials and the government authorities, highlighting its main concern before bringing the case to court.[138]

But this method was not always successful largely because the two bodies seldom supported the agency's position. Indeed, political interference haunted the ONC the most and its repatriation policy remained one source of endless conflicts and disputes. The first test case came when ONC decided on the repatriation of four unwilling settlers. But frustration heightened when the difficulties of funding it became insurmountable and disappointed Taticchi finally exclaimed: 'We needed to repatriate four colonists and there nobody seems prepared to take responsibility for funding it. The army [General Staff] keeps refusing and the Employment Office shilly-shallying. Meanwhile the four colonists wanted to be repatriated. To the army it would cost nothing.'[139]

Major controversy clustered mainly around the repatriation or dismissal of three groups: drifters or runaways who declared themselves to

[136] ACS ONC 3/B/1, Cesare Curati, Holäta 3 Aug. 1939.

[137] ACS ONC 15, Di Crollalanza to ONC, Rm 22 June 1939; ACS 3/B/1, Taticchi to GG AOI, AA 11 Oct. 1939.

[138] ACS ONC 3/B/1, Taticchi, Azione AOI, Relazione mese di settembre 1940.

[139] ACS ONC 4/16, Taticchi to Mazzocchi, Alemanni, AA 13 Jan. 1939.

be farmers supposedly to get free travel and then left or disappeared from the farms claiming they had no farming background; one-person families who were unmarried or unwilling to be joined by their partners in Italy. (Much to the government's chagrin, this latter group were given the difficult choice between repatriation and reunion or setting up a family with members of the local settler community within a fixed period); members implicated in more or less serious crime, ranging from murder or attempted murder of Ethiopians whose cattle trespassed the settler's wheatfields or grazing land,[140] to petty misdemeanours or constant rows with the neighbouring families.[141] The government's reason for opposing repatriation was quite understandable: not only did it fear the adverse publicity which disgruntled and repatriated families might cause but also the considerable financial cost that such a repatriation entailed. But the government addressed its opposition on politically safe ground: it claimed the agency's policy was harsh and self-defeating because 'by taking away, for reasons of simple temporary alteration of the family unit, the farms that the colonists have developed so far through their hard labour, the agency simply ends up frustrating the very objectives of the demographic colonization whose main purpose is to install settlers definitely on land'.[142]

The ONC agreed that the settlement of the peasants was its paramount objective. But it stated that grant of land is not a right but a reward to the very best, and those who make themselves unworthy had to be punished

[140] One of the inevitable consequences of appropriation of grazing land for settlement purposes was restriction of movement of the Ethiopian stock and imposition of fines on slightest trespass of the settlers' lands. Considering that an average pay for an Ethiopian working for the agency was L.4 per day, the fines were severe, ranging between L.10 and L.40 per head. As it appears from the ONC's otherwise-haphazard book-keeping the revenue from this source was considerable. Cases of such trespasses leading to confrontation with the settlers and even murder of the Ethiopians are not uncommon. One such case that embroiled the ONC with the government was that of Savron Giuseppe, an ex-protégé of Marshal Badoglio, detained for an attempted murder of an Ethiopian whose cattle damaged his wheatfields. (ACS ONC 15, Taticchi to SC, AA 9 Jan. 1939. ACS ONC 3/B/1: Taticchi to SC, AA 11 Mar. 1939; Taticchi to GG AOI, AA 11 Oct. 1940, and 15 Oct. 1940.)

[141] An illuminating ex. in this instance is given by the case of 50-year-old Moscatelli Pellegrino, a 'man with violent temperament capable of committing extraordinary acts for trivial reason. On one occasion in Aug. [1940], he fired several gunshots in the course of an argument, causing unjustified alarm to the extent that local authorities demanded the immediate removal of his family which had in fact thinned down from what it was when it arrived from Italy. Of his two daughters one, Filomena [single], provides services in AA; the other, Atene, was forced to leave [the farm] with her husband and children because of continuous family dispute. He himself had ceased to be a colonist.' (ACS ONC 3/B/1: Sergio Fornari to SC ONC: Relazione mensile mese di settembre 1939, AA 18 Oct. 1939; Daodiace [GG Sc] to ONC: Famiglie coloniche dell'ONC, AA 11 and 15 Oct. 1940.)

[142] ACS ONC 3/B/1, Taticchi to GG AOI, AA 15 Oct. 1940.

by being denied such a gift.[143] These disputes were settled with the coming of Ethiopian Independence and the subsequent dissolution of the ONC but not before they had the effect of aiding the colonists to break the contracts with impunity against which neither the agency's authoritarian philosophy of harsh discipline nor close supervision offered effective remedy. The approaching of the Ethiopian War of Independence only heightened the agency's impotence. Some colonists refused to hand over their crops; others sold their cattle for substantial profits and appropriated the proceeds.[144] The growing complaints of the colonist families that had begun since their reunion and had become a 'constant and daily affair' intensified. Its pronouncements notwithstanding, the management failed in its bid 'to nip the settlers' pretentions in the bud'. With their mind increasingly set to milk extra benefits from the agency, the settlers often successfully pitted one authority against the other, further worsening the already tense relationship between the farm administration.[145]

The ONC's recruitment method and the slow pace of its work excluded many ex-soldiers and members of the workers' militia (*centurie lavoratori*). This situation forced the government to devise other ways to accommodate the growing number of applicants from this group.[146]

## ACHIEVEMENTS AND LIMITATIONS

The ONC claimed that the delay in land allocation had serious repercussions for the rest of its activities, most importantly for cultivation and

[143] ACS ONC 3/B/1, Taticchi to GG AOI, AA 11 Oct. 1940.

[144] ACS ONC 3/B/1, Taticchi to GG AOI, AA 15 Oct. 1940; Taticchi, Azienda AOI: relazione mesi di dicembre 1940 e gennaio 1941, n.d.

[145] The stage for settlers' self-assertion was set within a few days of their arrival with a quite successful strike action by refusing to send their children to school until they were picked up from individual homes. (ACS ONC 15, Taticchi, 'Colonizzatori', AA 12 Apr. 1939; ACS ONC 7: Taticchi to Direttore AH, AA 3 July 1939; Cesare Curti to Direttore ONC AOI, AAH 8 July 1939. ACS ONC 29, Angelo Tuttoilmondo [Segreatario Federale PNF SC] to Sergio Fornari, AA 31 Aug. 1939; ACS ONC 15, Sergio Fornari to SC ONC: Materiali prelevati per i componenti le famiglie coloniche trasferite in AOI, AA 28 Sept. 1939; ACS ONC 3/B/1, Taticchi, Azienda AOI: relazione mese di settembre 1940, n.d.

[146] One such initiative was the creation of La Prima Centuria Agricola di Precolonizzazione in Gudär and Mäkanisa, near AA, aimed at settling destitute peasant militia. The initiative, manned by the Agricultural Offices which provided it with inputs and management, gave mixed results. The plan was to develop by the end of 1940 43 farms (12–14 ha each) in 650 ha at Mäkanisa and 149 (10 ha each) in 1,800 ha at Gudär. Eventually, 24 farms were developed at Mäkanisa and 45 at Gudär, at an average cost of L.26,000 per farm (cf. *CI*, 10 Oct. 1939; *GP*, 19 June 1939; Villa Santa *et al.*, *Amedeo*, 233–5).

housing programmes. No matter how well founded these factors may be, the truth is the agency never managed to cultivate even those lands under its indisputable possession. The ONC experienced a shortage of skill and of agricultural machinery, and was faced with the relatively high cost of Ethiopian labour.

Both Holäta and Bishoftu Farms were dominantly vertisol and their management remained one of the ONC's key problems. Vertisol is rich in organic matter and has high water-holding capacity. Owing to the residual moisture, vegetation was luxuriant even during harsh summer seasons, giving the impression even to a sophisticated observer of extremely fertile soil. Such appearances are deceptive, however, for black soil requires advanced managerial skills to be productive. In winter it has the disadvantage of water-logging which suffocates crop growth and development as was the case at Holäta where in 1937 over 103 ha of wheat-crop died immediately after germination due to excessive humidity.[147] In summer it not only shrinks but also cracks which leads to root-pruning that eventually reduces the yield. Modern technology had overcome these shortcomings by skilful soil-management techniques such as proper drainage using cumber-beds, Broad-Bed Furrows (BBF), and fertilizer application. Over the centuries the Ethiopian farmers had intuitively adjusted sowing dates to rainfall patterns, the quality of the soil, and the altitude of the land and distinguished a variety of crops that seemed to fit these climatic combinations. One of the commonly used methods was late sowing and drainage. In the absence of comprehensive scientific studies of the local eco-system to guide the ONC's work, the Ethiopians' farming methods could have been an alternative option. But having scornfully dismissed them as primitive, the ONC initially followed a procedure of trial and error which in most cases proved disastrous.[148]

[147] ACS ONC 3/B/1, Taticchi to SC, AA 3 Mar. 1938.

[148] G. Mazzoni, 'Prima Impostazione del problema dei frumenti nelle terre alte dell'AOI', *AC* 35/4 (1941), 142. He describes some of the techniques used by the Ethiopians in their fight against the smut and other wheat diseases. CD IAO: Ex AOI Sc 1329, A. Ciccarone, Relazione sulla missione nel territorio di Ambó, 20 Nov. 1940; Ex AOI Sc 1337, Vasco Gatti, Attività del campo di orientamento di Olettà nel 1937, Mar. 1938; Sc 1338, Benigno Fagotti, ONC AAB, Relazione dell'anno agrario 1937. Ethiopian elders narrate an interesting case relating to ONC's farm at Bishoftu. When the first crop was sown during the big rains in July the harvest totally failed. Then ONC was forced to consult local notables. Qäñazmach Zärräfu, a chief renowned for his wit and jokes, was approached. Zärräfu was reluctant to air his view claiming that he was in no position to give any adequate advice to them as their knowledge far exceeds his own. But when the ONC insisted, he told them that in the future they should sow not in July but in June, i.e. when the big rains begin. Subsequently over 1,000 Ethiopians were requisitioned to farm parts of the land that was unsuitable for tractor use by their own oxen while the Italians ploughed the outer stretches by tractors. The crop

However, constant crop failure made late sowing an inevitable option. But, because of poor forecasting large areas had to be farmed within a short period, forcing the ONC to rely on drainage ditch and minimum tillage just to cover the seeds and plough in the grass. As water-logged soil made mechanical farming impractical, time-consuming, and costly, this method too was abandoned.[149]

The war-like situation withheld the release of tractors originally promised to the ONC by the army. Broken-down tractors were reportedly abundant, but lack of spare parts made their repair difficult. The Italian tractors were not strong enough for tropical agriculture and ploughing virgin land.[150]

Lack of foreign currency made purchase of foreign tractors unaffordable. Although by 1939 the ONC had 4 Italian tractors of modest capacity, 40 Hanomag, and 5 Caterpillar, because of scarcity of fuel and shortage of spare parts, they were not used efficiently.[151] In 1938 the price for agricultural fuel in Italy was L.0.81 per kg., while in Ethiopia it was L.2.77 per kg. The ONC threatened to increase the price of foodstuffs unless the government took immediate action to lower fuel prices. Di Crollalanza himself brought to the attention of Mussolini the implications of such action to his policy of autarky and requested the Duce's urgent intervention.[152] In July 1938 a formula acceptable to both the farmers and the oil companies was sanctioned by the government.[153] Yet a stringent budget

failed to grow and when the Italians reproached him for his incorrect information, Qäñazmach Zärräfu's answer was swift: 'You should have known better. Is it on my advice that you came all the way from Italy and occupied Ethiopia? Why do you use me as an excuse for the crop failure? If God has allowed too much rain what can I do about it? The best course to take would be to try it again next time but changing the date'. (Information gained from talking to one elder of the area.)

[149] Other settler farms also were confronted with similar problems. Cf. CD IAO 1921, Fuzzi, Relazione del Presidente, 11 Jan. 1940, pp. 8–10.

[150] ACS ONC 3/A/1, Balconi to Taticchi, AA 2 July 1937; ACS ONC 3/B, Pasquale Pistone to Presidente ONC, AA 18 Mar. 1939; ACS ONC 7, Cesare Curti to Presidente ONC, Holäta 19 Mar, 1939. ACS ONC 3/B/1, Fornari to SC: Relazione mensile, AA 13 July 1939; Relazione mensile mese di luglio 1939, AA 18 Aug. 1939; Relazione mensile mese di agosto, AA 25 Sept. 1939; Relazione mensile mese di settembre, AA 18 Oct. 1939. CD IAO AOI 1800, Promemoria per il Prof. Maugini: Carburanti agricoli in AOI, n.d. [31 Dec. 1939], pp. 2–3; M. Nastrucci, 'Lavorazioni del terreno nell'Impero,' *AC* 32/5 (1938), 226–7; G. Vitali, 'Il problema delle Lavorazioni nell'AI', *AC* 35/8 (1941), 324–5.

[151] ACS ONC 3/B, Pistone to Presidente ONC, Holätä, 19 Mar. 1939; CD IAO 1800, Promemoria Per Maugini, pp. 1–3.

[152] CD IAO AOI 1839, Di Crollalanza, Promemoria, July 1938, p. 19.

[153] DGG 22-8-1938, no. 1026, Approvazione per la esecuzione del DGG 23-7-1938, no. 828, relativo ai prezzi dei carburanti agricoli, in *GU* 3/17 (suppl.), 6 Sept. 1938, pp. 609–13.

led the ONC to use gas and coal instead of diesel fuel. This had its drawbacks, as only a few engines could be converted. The ONC's attempts at converting its German Hanomag tractors at Holäta Farm were only successful with 6 out of 24. The high altitude also made the tractors function at half-capacity.[154] More importantly, while their maintenance rendered them highly uneconomical, the yields from such mechanically ploughed farms were not significantly higher than those obtained using ox-plough methods.

As oxen seemed to offer better prospects, the ONC planned to replenish its farms with oxen stock, both for farming and breeding purposes. This was expected not only to minimize dependance on machinery and reduce the enormous fuel cost but also serve as a source of profitable cattle industry which the agency hoped to make the basis of its agrarian economy. By 1941 the ONC had increased its oxen population by more than sevenfold—from 115 in 1937 to 887 in 1941.[155] In the mean time the use of tractors was being reduced considerably. In 1937, for example, only ten oxen were used at Bishoftu but in 1940 all farming at both farms was carried out with almost exclusive use of oxpower.[156] But facts on the ground soon proved that oxpower was not an attractive proposition. The ONC had apparently purchased most of its cattle from among the animals seized by the army during the course of its numerous punitive expeditions and mopping-up operations. But the army most often practised a devastating scorched-earth policy, indiscriminately destroying property and razing villages to the ground which effectively made oxen a scarce commodity. In fact, despite the sevenfold upswing in the price of oxen

[154] ACS ONC 3/B/1, Sergio Fornari to SC, Relazione mensile mese di settembre, AA 18 Oct. 1939; ACS ONC 4/2, Taticchi to SC, AA 22 Aug. 1940; Sbacchi, 'Italian Colonialism', 375–6; Nastrucci, 'Lavorazioni', 226–7; Vitali, 'Il Problema', 324.

[155] This figure excludes the cattle for breeding. According to the 1941 inventory, the livestock totalled 2,672. Of these, 885 [115] were farm-oxen, 403 [211] cows, 279 [8] steers, 254 [67] calves, and 111 [63] pigs. Ovine and equine population numbered 2,180 with 788 [63] sheep, 1,078 [107] goats, and 314 [104] horses. (Figures inside brackets relate to the number of animals predating the takeover of the two farms by ONC.) The early results of experiments with wool-bearing sheep and cattle-breeding were disastrous, for most animals imported from Kenya and Europe died. Vague attempts made to cross-breed some of these animals with the local variety seemed to offer a better prospect. But it is doubtful whether the agency with its slender financial resources would have been able to make a success of this industry. Cf. ACS ONC 1/3, Taticchi to SC ONC, AA 27 Feb. 1937 (AH), and 1 Mar. 1937 (AB); ACS ONC 4: Di Crollalanza to MAI, Rm 10 July 1937; MAI to ONC, Rm 20 Sept. 1937; ACS ONC 5, Inventario materiale ONC AAH and Inventario generale dell'AAB ONC, AA 2 Apr. 1941; FO371/24635/J5/5/1, CG Gibbs to Visc. Halifax, AA 27 Nov. 1939.

[156] ACS ONC 3/B/1, Taticchi to SC, AA 29 Nov. 1938; ACS ONC 3/A/2, Taticchi to SC, Relazione febbraio 1940.

since the early days of the occupation and the constantly soaring market price, the Ethiopians were unwilling to sell their precious cattle.[157] On the other hand, rampant theft brought to the agency substantial loss of cattle and the measures pursued to fight it proved hopelessly ineffective.[158]

While these developments made adequate supply of oxen extremely difficult, ox-power, though advisable for normal ploughing of already-broken land, was shown to be a poor substitute for tractors when farming virgin land and demanded a large supply of Ethiopian labour. In fact, in addition to those attached to colonists' farms through the mechanisms of sharecropping, the ONC had engaged a sizeable number of Ethiopian day-labourers in a wide variety of fields. At a peak agricultural season their number stood between 200 and 300. Of these between 30 and 40 per cent was made up of women and between 11 and 15 per cent of children, both employed almost exclusively as farm-labourers for wages ranging from L.1 to L.4 per day. However, as the agency soon found out, this category of Ethiopian workforce did not prove either sufficiently cheap nor abundant, particularly at peak periods, nor reliable. In 1937, for example, the entire 80-ha crop at Qalitti had to rot in the fields as the Ethiopians failed to keep their harvesting agreement.[159]

Cultivation proved hard, particularly at Bishoftu Farm. Except for 200 ha ploughed in earlier years, Bishoftu land was covered with acacia stumps cut by local people without removing the roots. An abundance of couch grass often demanded two or three ploughings. With only a short time left before the rains, the work was carried out late into the night, using electric light. In this way, 2,700 ha were ploughed and of these 2,050 ha were harrowed; 1,179 ha were sown, over 940 ha by the settlers and 230 ha by Ethiopians on the *metayer* system. Sowing did not start until late July and, following the failure of seeds to arrive from Kenya and Italy, using local grains. As shown in Table 9 emphasis was on diversification of crops even though, as the Italians' main staple food, wheat remained

[157] ACS ONC 3/A/2, Taticchi to SC, Relazione febbraio 1940.

[158] CD IAO AOI 1839, Di Crollalanza, Promemoria, July 1938, p. 18; Di Crollalanza, 'Avvaloramento' (1939), 1203; ACS ONC 4/39, Crollalanza to Ugo Cavallero [Commando Generale delle forze armate], AA 14 June 1938. ACS ONC: 3/B/1, Sergio Fornari to SC, Relazione mensile di mese settembre, AA 18 Oct. 1939; Taticchi to SC, Relazione sull'attività dell'azienda nel mese di aprile 1940, AA 29 May 1940. ACS ONC 3/A/2, Taticchi to SC, Relazione febbraio 1940.

[159] ACS ONC 5: Taticchi to SC Roma, Relazione mensile, AA 3 Mar. 1937; Taticchi to SC, AA 3 Feb. 1938. ACS ONC 3/B/1, Taticchi to SC, Relazione mensile, AA 3 Mar. 1938; CD IAO 1839, Di Crollalanza, Promemoria, July 1938, p. 8; Di Crollalanza, 'Relazione', 743; id., 'Avvaloramento' (1939), 1199, 1207.

TABLE 7. Areas of Ethiopian Employment

| Types of employment | Bishoftu Farm | | Holäta Farm | |
|---|---|---|---|---|
| | No. of employees | Average salary (L.) | No. of employees | Average salary (L.) |
| Fishing | 5 | 6.00 | — | — |
| Guardsmen | 6 | 7.90 | 8 | 5.80 |
| Interpreters | 1 | 15.70 | 1 | 13.33 |
| Farm-labourers | 27 | 6.58 | 61 | 3.60 |
| Mill | 1 | 4.00 | 15 | 7.90 |
| Armed patrols | — | — | 24 | 5.05 |
| Shepherds | 2 | 6.00 | — | — |
| Tractor-drivers | 13 | 10.15 | 24 | 7.70 |
| Warehouse | 7 | 6.80 | — | — |
| Mechanical workshop | 4 | 6.80 | — | — |
| TOTAL | 66 | — | 133 | — |

*Source:* ONC 7: 'AAH: Note delle opere di quindicina dal 16–31.5.1939'; 'AAH: Nota delle opere della quindicina dal 1–15.10.1939'; 'AAB: Operai indigeni: quindicinale dal 16–31.6.1939'; 'AAB: Operai indigeni: Quindicinale dal 1–15.10.1939'.

predominant. Implicit in such a strategy was as much to produce locally for domestic consumption as to minimize risk and gain useful experience.[160]

Mechanical threshing took place in early December 1937 amid growing jubilation and publicity, with a sample being despatched to the Duce. In the following years the pattern of cultivation did not change, except for the introduction of experimental crops and the diminishing size of the area under cultivation. From 1938 low-yielding Ethiopian wheat was replaced with fast-growing and presumed rust-resistant Italian wheat, *mentana* and *quaderna*; and in 1939 experiments with oilseeds and castor oil were underway.[161] Yet despite its alleged superior qualities, Italian wheat results were poor. Soon it was realized that compared with Ethiopian

[160] ATdR 24/124, Di Crollalanza, Promemoria, Feb. 1937, p. 8; CD IAO AOI 1338, Manusio, Relazione Sui Resultati; Mazzocchi, 'Orientamenti', 22; id., 'DiAlcuni problemi', 1877; Di Crollalanza, 'Relazione', 750; id., 'Avvaloramento' (1939), 1202.

[161] Pankhurst, 'Italian Settlement', 149; Sbacchi, 'Italian Colonialism', 373.

wheat, it was vulnerable to *pucinia graminis*, locally known as *wag*—a microscopic parasite which lived and fed on the plant tissue, arresting the plant's growth and reducing the crop's productive potential. *Wag* brought disastrous damages in 1938–40. In 1940 the pathogen's damage to the Italian agricultural enterprise was so massive that the year was declared the 'year of wheat rust', whereas most of the luxuriant wheat-crop fields of the preceding year had been wiped out by locust.[162]

Despite encouraging reports of crop yields of 30.7 ql per ha in 1937, from the archival reports it appears that the crop was frustratingly below expectations. Depending on the area, the average ranged from 0.2 ql to over 5 ql per ha. At Bishoftu from more than 600 ha sown, only about 1,800 ql were harvested and from over 430 ha sown at Holäta a mere 386 ql were obtained.[163] The officialdom put the blame on the delay in sowing and the low quality of the grain provided by the military authorities.[164] The results of subsequent years were rather erratic. In 1938 2,500 ha were reclaimed, 1,935 ha ploughed, and only 450 ha sown with an average crop yield of 4 ql per ha. In 1939 the pattern of cultivation was as in previous years, except for the additional production of oilseeds and experiments in castor oil.[165] The production slightly improved, ostensibly with extensive use of Kenyan Wheat, which in a few cases proved to be both a high-yield and rust-resistant crop. Where successful, the average yield almost doubled in some areas, rising to 7 ql per ha.[166] However, as

[162] CD IAO: Ex AOI 606, A. Ciccarone, La coltivazione del frumento nell'AOI ed il problema delle ruggini, 1941; Ex AOI Sc 1329, A. Ciccarone, Relazione sulla missione nel territorio di Ambò; idem, 'Note sulla biologia della "Nebbia del Frumento" (*Erysiphe graminis D. C.*) nello Sc', *AC* 35/6 (1941), 233; A. Maugini, 'Appunti sulle prospettive agricole dell'Impero', *AC* 35/11 (1941), 406–7; CD IAO AOI 1221, E. Castellani: 'Ruggine e granicoltura nell'Africa tropicale montana', 8 Mar. 1942; ACS ONC 3/B/1, Pistone to DONC, Bishoftu 1 Aug. 1939; 'Problemi fitopatologici dell'Impero', *AC* 34/1 (1940), 10–11; G. Mazzoni, 'Prima impostazione', 42.

[163] ACS ONC 3/B/1, Taticchi to SC, AA 3 Mar. 1938.

[164] CD IAO: AOI 1338 Fagotti, AAB. Relazione dell'anno agrario 1937, n.d.; Sc 1338, Campo sperimentale di Bisciοftù presso ONC 1937–8 n.d.; AOI 1839, Di Crollalanza, Promemoria, July 1938, p. 5. ACS ONC 3/B/3, Promemoria to Vice Re, AA 6 Apr. 1938.

[165] MAI, 'Valorizzazione', 269; Pankhurst, 'Italian Settlement', 149.

[166] CD IAO Sc 1801, Attuale organizzazione dei servizi dell'agricoltura nel G Sc ed esigenze per il prossimo futuro, n.d., pp. 15–16. Kenyan Wheat was a homegrown variety known as Kenya NB1 and 500. Trial in 1938 at 625 ha of land gave a mixed result with an average yield of 13.04 ql per ha, with a maximum of 25.2 ql per ha and a minimum of 5.7 ql per ha. The difference in yield depended on a number of factors, among them the period of sowing, the altitude, and condition of the land. But the experiment was too limited to warrant it to be the best variety. Despite this fact, the ONC, tempted by its resistance to rust, decided to grow exclusively this variety at its Holäta farm in 1940. On the other hand, a motley variety of indigenous wheat was tried in very small plots. Some of them gave much higher results than the Kenya variety. In case of S463 Ethiopia variety the highest yield was

the rising transport costs had made modern fertilizers uneconomical, improved yield was not expected for a long time. As a result, the ONC was forced to use manure instead. By the end of 1939 it had about 2,000 livestock, mainly oxen and a few sheep, yet the manure they provided was far from adequate to meet the agency's needs.[167]

Things proceeded relatively smoothly at Holäta Farm, as the topography of the land simplified draining operations, while the presence of the Holäta River eased irrigation. As a result, Holäta Farm was more advanced, although by no means extensive (see Table 9). In 1939, for example, out of 5,400 ha provided by the colonial administration, some 2,157 ha were sown—over 1,105 ha by Italian settlers using 24 tractors, and 1,052 ha by Ethiopians operating on the *metayer* basis. Each settler was assisted by two or more Ethiopian families, and the pattern of cultivation in principle was identical to that at Bishoftu.

The results of the auxiliary industries were mixed. Between 1937 and 1938 the ONC was granted the exclusive rights to exploit a number of important concessions: fishery at three of Ada's five picturesque lakes, namely Bishoftu, Bishoftu Guda, and Hora-Asodi; Männagäsha Forest at Holäta; vegetable gardens at the two farms. The revenues from these sources, combined with those from the two existing mills at both farms and large distillery at Mäkanisa, were expected to insulate the finances of the agency's farms from rising costs at least by offsetting most of the administrative expenses.[168] In line with the agreement stipulated between the government and itself, the ONC supplied the Addis Ababa market weekly with fresh fish and garden vegetables and for some time these two remained the most lucrative sources.[169] Fishing, however, remained confined to Bishoftu Lake, manned by an Italian and four skilled Ethiopian fishermen. Most coveted for its uniqueness and high quality, the Bishoftu

42 ql per ha and 32 ql per ha for S106 Ib.454. And yet efforts made towards the promotion of these crops were modest.

[167] Di Crollalanza, 'Avvaloramento' (1939), 1202, 1204; according to Berretta there were 1,400 oxen and 400 sheep (*Amedeo*, 259); ACS ONC 4/20/1, Taticchi to GG, AA 15 Aug. 1937.

[168] ACS ONC 4/25, G Civile AA R. Presidenza di Ada to Commando Settore Occidentale Ferrovia, Relazione sui laghi di Ada, Ada 6 Nov. 1937; ACS ONC 22/G, Fagotti to ONC, Bishoftu 24 June 1937. ACS ONC 22: [Taticchi], ONC SC, Relazione settimanale 'Stralcio', AA 29 May 1937; GG AOI DS AE Commando Legione Milizia Nazionale Forestale [Ing. Guglielmo Giordano], Verbale di consegna del bosco demaniale di Mannagaccià [*sic*] assegnato in concessione all'ONC, AA n.d. [15 Feb. 1938]; GG AOI CSFA [Ugo Cavallero], Cessione all'ONC della distilleria di Macanise, AA 16 Aug. 1938.

[169] ACS ONC 3/A/1, Taticchi to ONC, AA 26 Sept. 1937. ACS ONC 22/G: Fagotti to ONC, Bishoftu 24 June 1937; Taticchi to SC, AA 22 Dec. 1938.

TABLE 8. Wheat production by ONC settlers in the key blocks of Bishoftu Farm, 1937 and 1938

| Place | Area sown (ha) | | Total produce (ql) | | Average yield per ha | |
|---|---|---|---|---|---|---|
| | 1937 | 1938 | 1937 | 1938 | 1937 | 1938 |
| Dämbi | 122.00 | * | 30.00 | * | 0.24 | * |
| Dukam | 90.00 | 73.00 | 537.30 | 128.18 | 5.97 | 1.75 |
| Foqa | 83.00 | 20.00 | 457.11 | 274.80 | 5.50 | 2.20 |
| Qagimma | 40.00 | 80.00 | 105.10 | 120.68 | 2.62 | 1.50 |
| Qalitti | 279.00 | 85.00 | 676.80 | 187.09 | 2.42 | 13.74 |
| TOTAL | 614.00 | 258.00 | 1,806.31 | 710.75 | 2.94 | 2.75 |

*Note:* * Indicates that no information is available.

*Source:* ONC: 3/B/1, Taticchi to SC, 'Relazione mensile', AA 3 Mar. 1938; 3/B, Pasquale Pistone to Presidente ONC, AAB 18 Mar. 1939.

fish, a rare cross-breed between fish introduced in early 1930s from Lake Arämaya in Härärgé by Emperor Haile Sellassie and that of the local lake, Lake Zuway, commanded a price twice higher than that prevailing in Addis Ababa market, L.25 and then L.16 per kg, against L.15 and L.8 per kg for its lesser but steadily growing competitors, the supply of which largely came from far more distant lake regions of Robbi, Sumba, and Zuway. The press made scathing criticism against higher pricing and monopolistic speculation but the ONC shrugged it off.[170]

However, the fish trade failed to withstand the constantly rising transport costs, and the ill-equipped management's propensity and greed for profit tended to shrug off the prognosis of market trend, and the strength of the competitors who had expanded their sources of supply. With the arrival of fish from as far afield as the ports of Djibuti, Asäb, Meŝwa, Lake Arämaya, and those lakes of neighbouring Awash River, the ONC's market share disastrously shrunk, and in early 1939 the price fell by half, further declining to L.5 per kg. However, with the increased demand for fish that followed the government's August 1939 restriction on meat consumption,

[170] 'Perchè il pesce di Biscioftù costa L.16?', *CI* 14 July 1938; ACS ONC 4/25, Di Crollalanza to Direttore del *CI*, Rm 26 June 1938.

TABLE 9. Types of crops sown at the ONC's farms[a]

| Crop type | Area sown (ha) | | | | | |
|---|---|---|---|---|---|---|
| | Bishoftu Farm | | Holäta Farm | | | |
| | 1937 | | 1937 | | 1939 | |
| | Direct | Metayer | Direct | Metayer | Direct | Metayer |
| Wheat | 624 | 76 | 233 | 200 | 549[b] | 112[c] |
| Chickpeas | 184 | 66 | — | 40 | — | — |
| Téf | 105 | 32 | 23 | 60 | 12 | 305 |
| Beans and peas | — | 39 | — | 70 | 354 | 396 |
| Barley | 10 | 16 | 12 | 30 | 111 | 153 |
| Vegetables | 12 | 1 | 18 | — | — | — |
| Oats | — | — | 42 | — | 33 | 7 |
| Linseed | — | — | — | 10 | 12 | 46 |
| Lentils | — | — | — | — | 10 | 23 |
| Medicinal plants | — | — | — | — | 9 | — |
| Nug | — | — | 2 | — | 6 | 3 |
| Maize | 4 | 5 | 10 | — | 1 | — |
| Mixed herbs | 2 | — | — | — | — | — |
| Experimental field | 1 | — | 4 | — | — | — |
| Durrah | — | — | — | — | — | 7 |
| Lupin | — | — | — | — | 4 | — |
| Colza oilseed | — | — | — | — | 4 | — |
| Vines | 4 | — | 3 | — | — | — |
| Castor | — | — | 4 | — | — | — |
| Potato | — | — | 8 | — | — | — |
| TOTAL | 946 | 235 | 359 | 410 | 1,105 | 1,052 |

[a] There are minor discrepancies in the statistical data between both the published and unpublished sources.

[b] 485 ha of Kenya and 64 Ethiopian.

[c] 78 ha of Kenya and 34 Ethiopian.

*Source:* CD IAO AOI 1839, Di Crollalanza, Promemoria, July 1938; p. 9; MAI, 'Valorizzazione', 266; Fossa, *Lavoro Italiano*, 472–3; Pankhurst, 'Italian Settlement', 148, 150; ONC 5, Taticchi to SC, AA 20 Oct. 1937.

the price rose again to L.8 per kg against L.3.5–4 for the fish sold by the rest of the fish-merchants. But the rise in demand coincided with substantial contraction in supply caused by the partial pauperization of Lake Bishoftu and the low temperature of the waters that drove the fish to the bottom of the lake. As the implementation of planned expansion to the other lakes became untenable, the equipment at the agency's disposal made deepwater fishing prohibitive, and the administrative and transport costs soared, by early 1941 fishing was totally abandoned.[171]

The agency's market garden also experienced a serious crisis, partly injured by stiff competition from cheaper Ethiopian vegetables arriving from distant Härärgé and the nearby regions of Addis Ababa, and by the end of 1938 Holäta garden alone was about L.100,000 in the red.[172] Nor were the prospects of Bishoftu market garden bright and in 1939, as its management became taxing, the ONC was forced to hand it over to Italian sharecroppers.[173] The mills lay idle periodically because of shortage of wheat or, until a dam was built, lack of water. Only forestry exploitation afforded promise. A saw-mill, set up at Holäta, provided both the settlement and the city with wood.[174] With Italy's entry to war the auxiliary industries were propped up with feverish intensity in order to make the agency economically self-sufficient. Even though they were able to furnish a substantial part of the administrative and general expenses with the forestry as their backbone, contrary to the official claim, their overall productivity did not improve.[175]

The overall achievements of the ONC were quite negligible, as the figures for houses built and the colonists settled show. The ONC, anxious to make economic housing a reality, resorted to a number of contractual schemes which met with limited success. Experiments with piecework contracts using the veterans, individual farmers, and professional con-

[171] ACS ONC 22/G: Di Crollalanza to Carlo Boidi, Rm 4 Aug. 1939; Taticchi to SC, AA 28 Nov. 1939; Di Crollalanza to ONC, Rm 12 Dec. 1939; Taticchi to SC, AA 1 Jan. 1940. ACS ONC 3/B/1: Pistone to DONC, Bishoftu 1 July 1939; Sergio Fornari to SC, AA 13 July 1939, 18 Aug. 1939, 25 Sept. 1939; Relazione di mensile del mese di settembre, AA 18 Oct. 1939.

[172] ACS ONC 7, Cesare Curti to Presidente, Holäta 19 Mar. 1939.

[173] ACS ONC 3/B, Pasquale Pistone to Presidente, AA 18 Mar. 1939.

[174] Di Crollalanza, 'Relazione', 748; Quaranta, *Ethiopia*, 43; Montefoschi 'Centri agricoli'.

[175] However, this was partly due to the fact that most of the able-bodied settlers were mobilized to war, which undoubtedly reduced the administrative cost substantially. Cf. ACS ONC 17, Taticchi to Commissario Straordinario dell'ONC, Relazione sull'attività svolta dal Dr Giuseppe Taticchi Direttore dell'Azienda AO in AA dal settembre 1939 al febbraio 1946, 21 June 1946.

tractors all foundered.[176] Private companies soon discovered that building materials were scarce, technical experts and supervisory personnel rare, and their financial resources inadequate. Appalled by the immensity of the task and the small and uncertain prospect of profits, the companies invested only a fraction of their capital.[177]

The Fagioli Company, which had successfully bid for a number of contracts, was a typical example. Out of L.1,000,000 it claimed to have invested in the contract, only L.200,000 was transferred to the construction. The effect was to hold up salary payments, alienating the workers and thus causing a shortage in the workforce. As a result, the dwellings were not completed by the date agreed, June 1938. The contract was taken over by Romanola Construction Company, which made no real progress and was forced to give up.[178]

The house-building project was not completed until 1940, two years behind schedule. Even then, out of the planned 600 houses, less than 100 were built. Of these, at the end of 1940 84 were at Holäta and 16 at Bishoftu, some still incomplete. Only 93 families were settled.[179] Many of the houses were of low quality and unhygienic, despite the ONC's promise to provide its colonists with economic, decent houses. After a heavy rain, complaints arose over the excessive use of the local material, *čeqa*, and the undue economy made on cement. A number of walls collapsed, serious cracks appeared, and houses without ventilation were blackened by smoke. At least one settler family had to take refuge next door when it found itself in the open after a storm swept away the ceiling of its house one night in the rainy season.[180] Falling houses spread fear among the settlers, shattering their confidence in their safety and durability while on the other hand undermining the prestige of the agency.[181] Lack of washing- and toilet-facilities caused considerable discomfort.[182]

[176] ACS ONC 4/20/20, Balconi to SC ONC, AA 27 July 1937; ibid. 1/3, Taticchi to Di Crollalanza, AA 8 Mar. 1938.

[177] ACS ONC 3/A, Balconi to ONC, AA 8 Aug. 1937; Sbacchi, 'Italian Colonialism', 425.

[178] ACS ONC 3/B/2: Di Crollalanza to Taticchi, Rm 20 Feb. 1938; Ing. Mazzuccato, Relazione lavori al 31.7.1938; Situazione lavori, AA 2 Oct. 1938, and 16 Oct. 1938. ACS ONC 4/20/2: Di Crollalanza to Ing. Lino Fagioli, Holäta 20 Mar. 1938; Balconi, Impresa Lino Fagioli, AA 15 Feb. 1938; Taticchi to SC, AA 25 Oct. 1938. Sbacchi, 'Italian Colonialism', 426.

[179] Villa Santa *et al.*, *Amedeo*, 230–1.

[180] ACS ONC 7: Taticchi to Direttore AH, AA 3 July 1939; Cesare Curti to Direttore Azienda AOI, AAH 8 July 1939.

[181] ACS ONC 3/B/2, Mazzuccato to SC ONC, n.d.

[182] ACS ONC 3/B/2, Ing. Mazzuccato, Relazione lavori al 31.7.1938; Del Boca, *Caduta*, 204–5; Sbacchi, 'Italian, Colonialism', 427.

Such a picture contrasted with the carnival atmosphere that had accompanied the departure of the colonists' families from Italy. The first group of fifty-six families[183] sailed from Brindisi 20 February 1939 in the steamship *Italia* and arrived in Addis Ababa on 4 March. An enthusiastic reception awaited them. Led by the ONC director, the column was led through the streets of Addis Ababa lined with cheering crowds, to the workers' militia camp, where they were met by the inspector of the PNF, the Supreme Commander of the Militia, government officials, and Church notables with the new Apostolic Delegate, Monsignor Castellani, welcoming them 10 km from the gate of Addis Ababa.[184] Soon they were accommodated in the main hall, duly decorated for the occasion, entertained by the Viceroy (who had earlier personally inspected their new homes), and fed on a lavish dinner served by members of the Fascist Women's Organization. Then the convoy proceeded to their respective farms where the family heads were waiting.[185]

The press was silent on the reception that awaited the second group, while giving a detailed account of their colourful and spectacular departure from Italy. Twenty-eight families, left on 1 May 1939 on the steamship *Urania*. This 95-strong group was made up almost entirely of women and children: of the latter there were 3 under the age of 3, 22 between 3 and 12, while the remaining 31 were 12 and above.[186] Their departure was also organized by *Commissariato per le Migrazioni e la Colonizzazione Interne e nell'Impero* (Commission for Internal Migration and Colonization) with the assistance of other government officials. For the send-off the Segretario Federale dei Fasci di Combattimento (The Secretary of local black Shirts militia), who personally saw to the assistance given to the departees, the head of the ONC's press office, the senior health inspector, and the head of the Department of Colonization from the Commission for Migration were present. These vied with officials of the Opera Maternità e Infanzia, (Maternity and Nursery Agency) with the administrators and workers of the Fascist Women Organization, and Rural Housewives, to lavish every care on the departees.[187]

When the steamship approached the port, the departees, many of them in the uniform of the Fascist organization to which they belonged, left the Casa del Mietitore (House of the Harvester), and walked through the streets of the city amidst chants and cheers of the well-wisher crowd. The

183 They consisted of 79 men, 86 women, and 72 children.
184 ACS ONC 3/A/2, Taticchi to SC ONC, AA 8 Mar. 1939.
185 *GI*, 8 Mar. 1939; *IL*, 15 Mar. 1939; *Messaggero*, 7 Mar. 1939.
186 *CI*, 16 May 1939. 187 *Messaggero*, 2 May 1939.

local *Fascio* were much in view, with representation from all the organizations directly or indirectly involved in the Empire. Standing on the quay, carrying banners and streamers, they chanted slogans to the Duce and praised the Fascist regime, as local clergy blessed the steamship on its voyage to Meŝwa.[188]

The festive atmosphere was believed to make less acute the anxieties the colonists might feel at leaving home. But most importantly, organized in the style of Balbo's Ventimila of Libya, it was a symbolic reminder that the days of humiliating exodus when emigrants left crying and cursing their homeland were over. It was a sign of the changed times. Everything was arranged to make the families feel important. But this official parade also masks a quite disquieting but unreported story. The journey, particularly from the port to inland, was rough and risky. Some colonists were sick and hospitalized, suffering from cold, bronchitis, and pneumonia. Military rations, though abundant and edible, were distasteful and when Taticchi pointed out this to his superiors in Rome, the remark was 'What else do they want? Parrot talk!' Dormitories were scarce and the refusal of most women to sleep unless with their own regional groupings made the task of the organizers a nightmare. Worse still was perhaps when at Shäno, a few miles away from Addis Ababa, three of the convoy trucks of the first group were put out of action, attacked by the Ethiopians. The settlers were unhurt but others were slightly injured.[189] These incidents though minor and unimportant *per se* seemed to be an ominous foretaste of things to come.

Otherwise the new land appeared to present an opportunity to realize a dream. With the benevolent and paternal help of the ONC, farms had been prepared for them, complete with a house; all the supplies and facilities they needed for their new life were being furnished: tools, seeds, livestock and machinery, finance and technical advice. Most important of all, there was the promise that with patience and diligence the prize of land and independence could be theirs. Yet for many the imperfections of the promised land soon manifested themselves. The ONC's houses were not spacious enough to accommodate even an average Italian family. At Holäta, for example, more than ten families of between nine and fourteen members had to be crammed into a house with only two small rooms and a kitchen which was also used as an entrance hall. In some cases, the situation was so bad that the MAI had to intervene and directly request their enlargement, only to be rebuffed by the ONC's president. Di

[188] Ibid. [189] ACS ONC 3/A/2, Taticchi to SC ONC, AA 8 Mar. 1939.

Crollalanza argued that the ONC had neither anticipated a presence of families larger than five people nor had the resources to build additional rooms.[190] Yet the question resurfaced when the problem of promiscuity and outbreak of epidemics of conjunctivitis and scabies became serious. At Bishoftu, some houses were built close to a bog, where malarial mosquitos bred, and many fell victim. Some farms lacked drinking-water, and only after a strong protest from the authorities was the ONC forced to install a purifier. Others lacked an indoor toilet and the families had to rely on an outdoor pit.[191]

The ONC's settlement programme, when brought to a halt by war in 1941, was not promising. In both farms, there were only about 93 families, a fraction of what ONC had anticipated. Of these, 81 were at Holäta and 12 at Bishoftu, in all numbering about 400 people.[192] Not all of these matched the ideal: 4 were young bachelors unwilling to marry locally, 11 married only immediately after their arrival and of these 6 by proxy, there being no suitable adult females nearby. The ONC had employed 53 artisans and 25 technicians.[193]

The cost of such settlement was high. In fact, the ONC had increased its budget from L.250,000,000 to L.400,000,000. It had claimed to have spent about L.55,000,000, with an average yearly investment of L.15,000,000–20,000,000.[194] Of this L.25,000,000 was spent on public works, such as housing, and interfarm communication. Each hectare was brought into production at an estimated cost of L.1,000–2,000—an average cost of L.98, 200–109,560 per farm, excluding the expense of transporting families, developing the social and economic infrastructure of the area, the cost of the land, and the military protection of the farms which lived under the constant threat of Ethiopian attack.[195] Holäta alone, where incursions from the Ethiopian patriots were not as significant as

[190] ACS ONC 15: Amedeo di Savoia to MAI, Comprensorio dell'ONC, AA 4 Sept. 1939; Di Crollalanza to MAI, 'Colonizzatori AOI', Rm 27 Oct. 1939. ACS ONC 3/B/2, Mazzuccato to SC AOI, AA 16 Oct. 1939; ACS ONC 27, Dati riguardanti l'AH, AA 15 Sept. 1939.

[191] ACS ONC 3/B/1: Cesare Curti, ONC AH: Relazione sulle condizioni salutarie delle famiglie coloniche, Holäta 3 Aug. 1939; Pasquale Pistone, ONC AB: Stato di salute e morale dei colonizzatori nel mese di luglio, Bishoftu 6 Aug. 1939; Sergio Fornari to SC, Relazione mensile, AA 13 July 1939; Relazione mensile mese di agosto, AA 26 Sept. 1939. Del Boca, *Caduta*, 204–5; Sbacchi, 'Italian Colonialism', 426.

[192] Sources on the number of families actually settled are not unanimous. Some give only 56 and others 81 families, all of them at Holäta Farm [Cf. Montefoschi, 'Centri agricoli'; *CI*, 20 June 1939]; Berretta, *Amedeo*, 259.

[193] *CI*, 16 May 1939, 20 June 1939.

[194] ACS ONC 17, Taticchi, Relazione.

[195] 'Notiziario agricolo commerciale AOI', *AC* 32/2 (Feb. 1938), 90; Sbacchi, 'Italian Colonialism', 443, 446.

those in Bishoftu, was garrisoned by about 500 armed soldiers, stationed at strategic positions on the farm.[196] Most importantly, the settlements involved the forceful removal of a large number of Ethiopians from lands rightly theirs. Neither the number nor the quality of settlers justified such an expenditure.

The ONC colonists enjoyed a privileged status. They were given their farms on credit and were considered owners. They paid no interest as it was paid by the Addis Ababa government. Such favourable conditions enabled a number of peasants to establish some credit towards the payment of their farms. Of these, one family at least entirely discharged its debts; another ten were on the verge of being assigned land ownership. Thirty-seven others had made considerable progress towards this same end.[197] Such individual instances could not alter the fact that by 1941 the project of military settlement had foundered on a welter of economic and social contradictions. The few settlers who were able to redeem their land, did so chiefly aided by their easy access to the burgeoning Addis Ababa market where a very high price was fetched for their crops, and by the availability of cheap Ethiopian labour. Their activities provided the Italian newspapers with many interesting photographs illustrating their sacrifices and hard work as pioneers in developing virgin Ethiopian soil. But their meagre contribution to the army's food supply did not greatly reduce imports from Italy.[198]

The ONC only realized a minute proportion of what it once hoped to achieve. With it the dream of creating a steady independent yeoman peasantry failed. At conceptual level, the scheme seemed perfect and few doubted its plausibility. But a number of factors militated to make the plan more of a complicated nightmare than a reality, thus frustrating the fruition of the ONC's reportedly well-planned settlement dream. Most of them were of the Italians' own making. Of course, the very short period that IEA lasted will not warrant one to make a conclusive judgement. And yet there were strong forces at work that would have thwarted the materialization of the plan.

The coming of Ethiopian independence delivered the final crushing blow to the ONC's plan when on 3 April 1941 the colonists were hurriedly evacuated to Addis Ababa with gleefully advancing Patriot fighters

[196] ACS ONC 15, Conrado di Domenico to ONC, Roccamonfina (Napoli) 26 Dec. 1942. Father di Domenico was ONC's chaplain at Holäta.

[197] Sbacchi, 'Italian Colonialism', 452.

[198] E. Weise, 'Ethiopia Now', *Harper's Magazine*, 115 (1937), 415–16; *New Times And Ethiopia News*, 24 Mar. 1945.

at their heels. The settlers' presence in the Empire's capital and how to deal with them divided the new British Administration of the Enemy Properties. Proposals to move both the settlers and the agency, first to Čärčär in Härärgé and then to Kenya, failed, opposed by the powerful Supply Office led by Colonel Stockpoole. Negotiations with the Ethiopian Ministry of Agriculture to recuperate the two farms foundered. Only in January 1942 did the settlers' agonizing wait came to an end with the unceremonious deportation of the first batch of women and children to Italy while the remaining 200 children, detained at Deré Dawa in Härärgé, were threatened by a measles epidemic.[199] Still more demoralizing was the plight of these returnees. 'Please help me to overcome the dramatic and pitiful situation in which my family presently finds itself',[200] a woman who had eight children and whose husband and elder son were left behind in a concentration camp in Rhodesia wrote to the ONC. Some held hopes of reviving their farm. 'My relative Luigi Contini has left . . . a box of clothes, four pair of shoes, a trunk of underclothes and a case of carpenter's tools. He asks if he will be able to get back whatever he left at the end of the glorious victory of our army.'[201]

But the situation was irreversible and the condition was tragic for those who remained behind and for others as well. Four of the ONC's personnel lost their lives in battle, one died of malaria, and another one was taken captive to India with some settlers, while the vast majority of the colonists languished in detention camps in Kenya, Rhodesia, and South Africa.[202] Most were deported at the end of the war to Italy where, as with their womenfolk before them, squalor and unemployment compounded their plight. Unable to re-establish themselves, they made passionate appeals to the ONC for succour: 'My situation is disastrous. I am at home engulfed in an extremely desperate situation with numerous family and without any possibility of re-establishing myself as a farmer.'[203] This was not an isolated situation but rather the painful reality that many ex-settlers experienced for a long time.[204]

[199] ACS ONC 17, Taticchi, Relazione.

[200] ACS ONC 15, De Luca Benvenuto to ONC, Rovvedo 25 Feb. 1943.

[201] ACS ONC 15, Giovanna Contini to Presidente ONC, 23 Jan. 1943.

[202] An exception to this was the head of ONC farms, Taticchi. Sentenced to fourteen months' imprisonment, he was bailed out by Mäsfen Seläshi, one of the resistance leaders who appropriated some of the most profitable Italian farms and became one of the wealthiest landowners in independent Ethiopia. He was employed on a two-year contract as head technician by the brain-starved Ministry of Agriculture until Feb. 1946 when he left for Italy on a USA petrol-carrier ship (ACS ONC 17, Taticchi, Relazione).

[203] ACS ONC 15, Costantino Antonio to Direttore ONC, 22 Mar. 1947.

[204] Later the Italian Government agreed to provide a package of assistance to various refugees returning from Africa but the implementation was slow and inefficient and many

With the evacuation of the settlers the farms collapsed, ransacked by the angry Ethiopians and with houses ruined by fire. However, there was no direct war damage for in that area no major military engagement had taken place. The lands and most of the auxiliary industries returned to their former owners while the Ethiopian Ministry appropriated the machinery, tractors, storehouses, and buildings and the Addis Ababa Municipality the trucks. But even before the evacuation the feasibility of the scheme was being questioned. The severe setbacks in production of wheat on which the agency's economic foundations rested and informative lessons from Kenya where scientific farming of wheat was conducted for thirty years with poor results had hinted that the ONC's entire programme needed a complete overhaul.[205] But most importantly the government's decision in 1938 to embark on other settlement projects that would cater for the needs of the destitute army veterans and peasant militia was largely symptomatic of its awareness of and dismay at the inadequacies of the ONC's performance.

cases were still being processed in 1970s (*CS*, 26 July 1970, p. 2; ACS ONC 8/A, Bertoni Giacomo to ONC Presidenza, Roncone 30 Oct. 1970).

[205] ACS ONC 3/A/1, Taticchi, Concorso degli indigeni alla colonizzazione demografica, [AA] 24 Oct. 1938; V. Harlow *et al.* (eds.), *History of East Africa*, ii (Oxford, 1965), 218.

# 5
# *Demographic Colonization: Regional Settlements*

## HISTORICAL SKETCH

Another scheme through which the Fascist regime attempted to develop a stable White population in its East African Empire was regional settlement. As we saw in Chapter 3, this was the brainchild of Lessona. With a few exceptions, it won wide support among the Italian élite, and from the outset Mussolini himself showed great enthusiasm for it. As its central objective, the programme sought to populate specific zones in Ethiopia with colonists from the more densely inhabited provinces of Italy, region by region, reproducing the atmosphere of the mother country from which they sprang. The new regions of the Empire would mirror those of the mother country which bore the same name.[1]

This scheme not only endorsed the Italian claims to the uniqueness of their colonization techniques, but also held certain features that seemed to guarantee its feasibility: the settlement of those with common habits and traditions in a region identical in topography and climate to their Italian home would foster social cohesion; and conflicts with regional roots would be avoided. In the long term, a stable link would be maintained between the metropolis and the Empire through a continuous flow of ideas and men between the regions.[2]

High on the list of benefits the regime hoped to reap from the programme of regional settlement was national and international publicity. The Fascist press played up the scheme as a new chapter in the methodology of empire-building. As the Rome correspondent of *The Sunday Times*

[1] FO 371/22020/J368/40/1, Lord Perth to Anthony Eden, Rm 22 Jan. 1938; A. Maugini, 'Primi orizzonti della valorizzazione agricola, dell'Impero', *Autarchia Alimentare* (June 1938), 19; E. Cerulli, 'La colonizzazione del Harar', *AAI* 6/1 (1943), 71; CD IAO 1921, Incartamenti e relazioni varie relativi all' ECRE 1944, p. 8; M. Pomilio, 'Problemi e realtà dell'Impero nel pensiero di Attilio Terruzzi', *Autarchia Alimentare* (Oct. 1938), 39; G. Pini, 'La terra ricchezza dell'Impero: Romagna d'Etiopia', *Azione Coloniale*, 22 June 1939; Lessona, *Memorie*, 283–4.

[2] Lessona, *Memorie*, 284; G. E. Pistolese, 'Problemi dell'Impero', *EI*, July 1938.

reported (23 January 1938) with its 'constant discussion' and '*press-agenting*', popular interest in the Empire appeared overwhelming. Indeed, emphasis on regional settlement was the most important departure from previous planned settlement plans and it has no parallel in any colonial history and in this one can see Lessona's innovation.

But approval of the scheme was by no means universal. In fact, wide divergence of opinion was quick to develop within the metropolis and the Empire as well. Initially weak and scattered, in the metropolis the dissension increasingly gathered momentum and by 1939 was strong enough to challenge the entire concept of regional settlement openly in Parliament. As appears in the notes of the meeting held on 24 October 1939 by the IEA Affairs Legislation Committee of the House of the Fascists' Second Corporations, the opposition's view was that regional settlements, in the long term, tended to be nationally divisive and racially degenerating. Accordingly, regional groupings would foster endogamy which, in the opposition's view, militated against Fascist racial doctrine. Effects from a language point of view were considered to be even more damaging. With regional settlements local dialects would flourish at the expense of a national language. As a result, national integration would be impaired and communication between different groups become difficult: different regional groups, when dealing with each other, would find Amharic the best means of common expression. In view of such a dreadful prospect, the opposition, with considerable sympathy from the rest of the House, called for a policy review, urging regional settlements to be replaced by nationwide groupings as in Libya.[3] Notwithstanding the considerable sympathy it had, the opposition was not able to reverse the situation.

International reaction to the government's publicity was mixed. Some journalists with Fascist sympathies were enthusiastic. In her admiration for the plan, Diel seemed to echo the sentiment shared by most: it was 'the first time in the history of Empire building' that such 'far-reaching methods' had been employed which 'in favourable circumstances . . . enable hundreds of thousands of Italians, notably small farmers, to make their homes in IEA'.[4] Other parties were more cautious, among them the British Foreign Office. One official remarked that the scheme was 'an interesting experiment' but emphasized the practical difficulties militating against its success, noting that 'the lot of the pioneers will remain far from easy'.[5]

Despite its enthusiasm for the scheme, in practice the government's behaviour was characterized by much publicity and little action. To the

[3] CD IAO 1921, Meregazzi to MAI: modificazioni al DL 6-12-1937, no. 2300, sulla costituzione dell'ECRE. [4] Diel, *Behold*, 70.

[5] FO 371/22020/J368/40/1, Perth to Eden, Rm 22 Jan. 1938.

annoyance of its supporters and despite strong pressure, the government failed to outline how regional settlement would be translated into practice, not encouraging mass settlements other than that of the ONC.[6] Obstacles such as proof of financial and technical fitness discouraged many potential applicants. State support was refused for destitute emigrants on the grounds of financial stringency.[7] A set of provisos meant to preserve economic order and stability in the colony forbade land transactions[8] and angered many willing settlers.[9]

Practical steps towards the formation of the much-awaited and much-publicized regional agencies proceeded at slow pace as internal squabbles handicapped the work of the Agricultural Council and only towards the end of June 1937 was discussion on the matter at an advanced stage.[10] The deadlock ended on 13 August with the announcement of the formation of three regional colonization agencies, with rough details of their areas of operation.

Each agency was promoted by a prominent Fascist personality: the ECRE (Romagna Colonization Agency of Ethiopia), was Mussolini's. Having welcomed from the outset the regional idea, Mussolini wanted people from Romagna, his own native province, to be the first settlers. So Fossa, a person in his confidence, was given orders to make preparations for the dispatch of colonists.[11] The ECPE was the brainchild of the Secretary of National Fascist Party, the Apulian Achille Starace who distinguished himself as a leader of Colonna Celere, a blackshirt militia whose so-called 'legendary march on Gondar' during the Italo-Ethiopian War was described as being one of the great epics of colonial history.[12]

[6] CD IAO AOI 1936: Lessona to GG, Rm 10 Mar. 1937; Graziani to GG AOI, AA 8 Apr. 1937.

[7] CD IAO AOI 1854: Lessona to GG AOI, Direttive sui problemi della colonizzazione agricola e dell'agricoltura indigena, Rm 29 May 1937, p. 4; MAI to GG AOI, Rm 9 May 1937.

[8] DGG 4-10-1936, no. 95: Norme che evitano nell'attuale periodo di sviluppo industriale e commerciale dei territori dell'AOI ingiustificati accaparramenti di terreni, in *GU* 2/20, 16 Oct. 1937.

[9] The decree was unbanned within a month (cf. DGG 13-11-1937, no. 805: Abrogazione dell'art. 1 del DGG 4-10-1936, no. 95, riflettente le norme intese ad evitare ingiustificati accaparramenti di terreno, in *GU* 2/23, 1 Dec. 1937) to be reimposed a few months later in modified form (DGG 3-3-1938, no. 155: Abrogazione dei decreti vicereali no. 95 del 4-10-1937, no. 805 del 13-11-1937, ed emanazione di nuove norme disciplinanti la materia delle alienazioni dei terreni, sia a scopo agricolo che a scopo edilizio, in *GU*, 3/10, 16 May 1938).

[10] CD IAO AOI 1936, IA to GG, Rm 22 June 1937; *The Times*, 20 Oct. 1937.

[11] CD IAO AOI 1936, Lessona to GG, Rm 13 Aug. 1937.

[12] Ibid.; *Gerarchia*, May 1936, p. 351. Starace's exploit is described in his book *La Marcia su Gondar della Colonna Celere* (Milan, 1936). Despite the Fascist's appetite for bombast and

The ECVE was presided over by Bergamese aristocrat Giacomo Suardo (1881–1947), a man with a colourful political background and then vice-president of the Upper House, Il Senato.[13] Other agencies never went beyond the planning stage.[14]

With the Royal Decree of December 1937, whose enactment was delayed largely due to sluggish bureaucracy and interdepartmental fighting over finance, the agencies gained legal status. In addition to defining operational areas, the Decree allocated resources and set out programmes for each agency. The ECRE's sphere of activity was the governorship of Amara and the region of Wägära was allotted as its specific field of operation. The ECPE was given the governorship of Härär and its site of settlement was the Čärčär highlands. The ECVE, whose activity never went further than the planning stage, was allocated the governorship of Galla and Sidamo, but without any defined site of settlement.[15]

The chief purpose of each agency was to settle in the shortest possible time a maximum possible number of colonial families as small farmowners.[16] The initial target was 1,000 families within a period of six years.[17] The official view was that within a few years the colonists, with State support

gullibility, the army leaders knew that it was nothing of the sort but rather an idle and uneventful march through a safe and empty route (cf. Del Boca, *Conquista*, 611–13; G. G. Farina, *Follie delle Follie* (Rm, 1945) 94; A. Mockler, *Haile Selassie's War* (Oxford, 1984), 109.

[13] CD IAO AOI 1936, Lessona to GG, Rm 13-8-1937.

[14] Ibid.; other planned agencies of regional or national character included those of Sicilia d'Etiopia, financed by Banco di Napoli (cf. [C. Giglio], 'Il finanziamento degli enti di colonizzazione demografica in AOI', *REAI* 27/2 (Feb. 1939), 155); Piemonte d'Etiopia and Liguria d'Etiopia (Pistolese, 'Problemi dell'Impero'); Aosta d'Etiopia and Marche d'Etiopia (*Azione Coloniale*, 28 Mar. 1940; *Impero e Autarchia*, 9 May 1940). But practical steps were taken only towards the Italian Overseas Colonization Agency or Ente di Colonizzazione degli Italiani all'Estero. Even though the agency gained legal status, practically no progress was made with exception of the transfer of about 100 workers who remained at May Habar, in Eritrea (cf. Legge 25 Aug. 1940, no. 1415, *GU* 249, 23 Oct. 1940; *GUGS* 5/48, 27 Nov. 1940; *Agenzia Le Colonie*, 9 July 1940; *Azione Coloniale*, 12 Aug. 1940; CD IAO 1916 A. Maugini and P. Nistri, Rapporto sulle possibilità di colonizzazione con italiani provenienti dall'estero nel territorio dei Gs Rm 18 May 1940; *CE*, 2 Dec. 1939; *PI*, 4 Jan. 1940; *GV*, 9 Jan. 1940; *Regime Fascista*, 24 Oct. 1940; *Rivista delle Colonie*, Aug. 1940; 'Notiziario Agricolo Commerciale: AOI', *AC* 34/10 (1940), 437).

[15] CD IAO 1921, Decreti costitutivi degli enti di colonizzazione Puglia, Romagna, Veneto d'Etiopia; see the same in: *GURI*, 25 Jan. 1938; 'Atti e Documenti: no. 3200, Costituzione di ECRE, 249–51; no. 2314, Costituzione di ECVE, 252–4; no. 2325, Costituzione di ECPE, 255–7; *REAI* 26/2 (Feb. 1938), 249–57.

[16] DIM 18-4-1939, Approvazione dello statuto dell'ECPE, in: *GURI*, 27 Apr. 1939; *REAI* 27/6 (June 1939), 769–74; *GU* 3/22, 16 Apr. 1939. DM 10–1–1941, Approvazione dello statuto dell'ECVE, in *GURI*, 7 May 1941.

[17] Maugini, 'Primi orizzonti', 19; *The Times*, 18 Jan. 1938.

and favourable economic conditions, would be strong enough to counterbalance the local population and make the Empire Italian in fact as well as in law.

On the domestic front, the political authorities used the settlement programme as a safety-valve to eliminate unemployment and quell political discontent. Since 1927 rural unemployment in Italy had been rapidly increasing. A downturn coincided with the mobilization for the Ethiopian campaign. During the winters of 1931–5 between 10 and 15 per cent of the agricultural workforce was unemployed, a figure above that of industrial workers.[18] It was in the regions where the companies recruited their settlers that there was a high rate of peasant unemployment and *braccianti*, day/seasonal-labourers. These traditionally had been centres of social discontent and radical politics.

A notable example was the region of the Veneto. According to a report by the local party official to the Secretary of the National Party in Rome, in 1937 the conditions in the region were explosive, with incidents involving 'singing of subversive songs'. By 1939 crowds of unemployed peasants and workers were reported to be hanging around government buildings and Party offices 'with an impatience and restlessness that cannot be further ignored'.[19] Not surprisingly, the Veneto had been one of the largest suppliers of colonists for the settlement programme in Libya. In 1938 57 per cent of the settler families, and in 1939 62 per cent, were immigrants from that region.[20] 'Tens of thousands of peasants and workers' were thought to be eager to migrate to the Empire from this same region.[21]

## SETTLEMENT AREAS
## PHYSICAL LANDSCAPE

Fascist propaganda claimed that demographic colonization was carefully directed from above and regulated in the smallest detail.[22] The settlement zones were chosen for their healthy climate, high yield, and proximity to

[18] Segrè, *Fourth Shore*, 82.

[19] ACS PNF: B11, Situazione politica delle provincie (Padua 1928–42); B26 (Treviso). For similar situations in Aquila and Regio Emilia, regions targeted for the recruitment of the settlers, see ACS ONC 15: Carlo Perrone [Presidente Federale] to Di Crollalanza, Aquila 17 Jan. 1938; Eugenio Bolondi [Segretario Federale PNF] to Di Crollalanza, Regio Emilia 27 Apr. 1939.

[20] 'Notiziario agricolo commerciale', *AC* 34/11 (1940), 483; Segrè, *Fourth Shore*, 140–1 where he gives the statistical data of emigration to Libya from each region.

[21] ACS PNF B11, Situazione Politica.

[22] *The Sunday Times*, 23 Jan. 1938.

markets and major communication centres.[23] With its rolling countryside and a salubrious climate likened to that of May in Italy, Wägära, the ECRE's settlement area, appeared to perfectly meet these requirements. As defined in the Royal Decree of 6 Dec. 1937, the boundaries of Wägära were marked to the North by the eastern and western high plateau of the Semén province, to the East by the regions of Bäläsé, and to the West by the westerly slopes of the great Bära valley and the plains of Kosoyyé, the subdistrict of Amba Giyorgis. It consisted of an area 80 km long and 25 km wide.[24] Its rich and black soil was compared by Almagia to the wheatlands of Russia;[25] it had plenty of water, rainfall averaging 1,200–1,300 mm a year, and excellent communication via the Gondär–Asmära road, then being asphalted. Though timber was scarce, there were numerous wild pine trees, while lime was available in the valley of Bära, West of Dabat, an important market on that road. It was ideal for the cultivation of both grain and vegetables, and crop failures were almost unknown. For years it served as the breadbasket for the Italian colony of Eritrea. Moreover, there was good pasturage and cattle, goats, donkeys, horses, mules, and sheep were in abundance. The highlands were well suited for the rearing of long-haired sheep, and the lowlands for cultivation of cotton, rice, peanuts, and coffee. The existence of wild grapevines suggested that vineyards could be established in warmer, more sheltered places. It was also pointed out that such spots were favourable for mulberry trees, and hence for raising silkworms, while the abundant rainfall afforded possibilities of hydroelectric development in the Semén area.[26] The head of the Amara government's agricultural service, Mario Pavirani, summed up: 'The advantages deriving from such initiative (demographic settlement) should be regarded as of supreme importance, and, thus, of immense benefit for the function and influence in the Empire of the colony of the Amara. Wägära on account of its happy location would become a reservoir of resources and foodstuff products of absolute importance to all future problems.'[27]

[23] CD IAO AOI 1936, Ministro to GG, Rm 4 June 1937; [C. Giglio], 'Importanza dell'ambiente fisico-agrologico-economico nella colonizzazione demografica', *REAI* 27/1 (Jan. 1939), 29–38.

[24] RDL 6-12-1937, no. 2300, Costituzione dell'ECRE, in *GURI* 19, 25 Jan. 1938; CD IAO: 1806, Dr. Mario Pavirani, Uogherà, Apr. 1937; 1111, G Am DAF Commissione Indemaniamenti, Accertamenti di terreni demaniali: rilievi di campagna eseguiti nell'Uogherà, Gn 14 Dec. 1937.

[25] Fossa, *Lavoro*, 503.

[26] Ibid. 504–6; MAI, 'Valorizzazione', 294; UA Am, 'Aspetti generali e Zooteenici del Lago Tana', *AC* 32/6 (1938), 263–4.

[27] CD IAO AOI 1806, Pavirani, Uogherà, p. 7.

Wägära was estimated to have approximately 200,000 ha of colonizable lands. The ECRE was to make use of between 50,000 and 60,000 ha for the settlement of 1,000 Italian peasant families with an almost equal number of Ethiopians as labour reserve.[28] Wägära was not chosen in a casual fashion. As Giuseppe Pini, the head of the Supreme Council for the Public Work Division (Sezione del Consiglio Superiore dei Lavori Publici) pointed out, Wägära had a number of characteristics in common with Italy's province of Romagna, with strikingly similar scenery, gently undulating land, and vast horizons.[29] Most of the construction workers on the Gondär–Asmära motorway were from Romagna, and the remarkably good state of their health suggested that the climate suited peasants from the region.[30]

The land for settlement by the ECPE, the Čärčär highlands, consisted of two types: to the East there was broken country with an average altitude of about 2,200 m and a cool but temperate climate; to the West was warmer undulating land with an average altitude of 1,750 m. In all it covered a total area of 285,000 ha, a quarter of it cultivable. Initial settlement was to start in the western part as it was judged to offer a better climate, and more abundant supplies of construction materials, particularly lime and limestone. It was sparsely populated thus necessitating less expropriation of Ethiopians. Yet all the land was considered fertile. The luxurious vegetation and the variety of crops under cultivation, including wheat, barley, oats, maize, *ṫef*, *durrah*, chick-peas, lentils, beans, peas, onions, garlic, as well as coffee and bananas, citrus fruit, castor and linseed oil, tobacco and cotton were evidence of immense productive potential. As the area was located close to both the Addis Ababa–Härär road and the Addis Ababa–Djibuti railway, communications were favourable. It was also within easy access of the nearby markets of Bädässa and

[28] CD IAO 1921, Fuzzi, Programma d'azione per il primo anno d'attività, 7 Jan. 1938; Pini, 'Terra ricchezza'.

[29] Pini, 'Terra ricchezza'.

[30] Ibid. The original plan was to begin settlement in the nearby Achäfär region where it was thought that about 180,000 ha of State land with excellent agricultural potential was available. Achäfär was a guerrilla stronghold, and as punishment the entire region was declared public domain. In addition to being malaria-infested, the area was unsuitable for crop cultivation, with poor communications and difficult access to market. Owing to these factors the plan was dropped. (DGov 5-6-1937, no. 46144: In demaniamento del terrilorio dell'Achefer, *BUGA* 3/4, 16 Apr. 1938; CD IAO Ex-AOI 600, Relazione sull'attività svolta dalla sezione della colonizzazione dalla data dell'occupazione al 28–2–1939, Gn 4 Mar. 1939, p. 8; UA Gn, 'Acefer', 56, 59; *Colonie*, 19 Oct. 1938; *CE*, 27 Sept. 1938; MAI, 'Valorizzazione', 294.)

Gälämso, two centres that had been considerably developed after the occupation.[31]

Galla and Sidamo, the vast south-westerly governorship, allotted for settlement to Italians from the Veneto, had excellent agricultural conditions. It included areas of rich cultivation yielding all kinds of cereals as well as coffee and cotton. Mario dei Gaslini described it as 'perhaps the most suitable in all IEA for numerous and gradual national settlement',[32] a view shared by the journalist Polson Newman, who saw it as 'the best region for settlement' where the soil was of 'exceptional quality and the climate most suitable for Europeans'.[33] And yet colonization in the Galla and Sidamo area proceeded at a slow pace. This was largely because the territory was the last to be occupied militarily and was little known to the Italians despite the presence for several decades of the Consolata Mission who played a significant role in the occupation.[34] There were few roads and it was far removed from the sea.

The situation was further compounded by the ECVE's internal organizational problems and political complications between Rome and Addis Ababa. As a result the agency never set foot in Galla and Sidamo.[35] Instead, two different forms of demographic enterprise emerged locally: the EDR, set up by the government of Galla and Sidamo, operated at Borä, about 10 km from Jimma, on 2,000 ha. On the outskirts of Jimma

[31] CD IAO Hr 1847, Proprietà terriera nel Cercer, n.d.; G. Giannoccaro, 'Prime tappe dell'Ente "Puglia d'Etiopia" in AOI', *Africa Italiana*, 1/1 (Nov. 1938), 25–6; [Giglio], 'Importanza', 7–8.

[32] CD IAO Hr 2214, Il Cercer Orientale, n.d. 'Colonizzazione del GS', *RSAI* 2 (1939), 672.

[33] *Abyssinia*, 104.

[34] Del Boca, *Caduta*, 26–34.

[35] Not only did the agency fail to win the co-operation of the colonial authorities inside the Empire but its President Renzo Morigi, ex-secretary of the Fascist Party, who gained national prominence for the murder of the well-known communist Massaroli at Ravenna in 1927, had poor entrepreneurial skills and hardly any tangible programme. He is alleged to have been appointed to the post largely because of his friendship with the minister of agriculture. Perhaps these facts were sufficient enough to allay the confidence of the agency's financing institution, INFAIL (cf. ibid. 207; Stella, Qualche note). The Fascist media on the occasion of the President's visit to Ethiopia continued trumpeting for a month with the news of him obtaining between 20,000 and 25,000 ha of land on the Jimma–Bonga road, which was sufficient to set up the first 500 farms (cf. *Tribuna*, 31 Jan. 1940; *Corriere Padano*, 28 Mar. 1940; *Azione Coloniale*, 4 Apr. 1940). But Morigi's report to the Presidential Council points out that his attempt to secure a meagre 4,000 ha around Jimma was frustrated by the local authorities. By the end of 1941 one million lira had been spent, mostly on travel, on a project that hardly moved beyond the drawing board. Cf. CD IAO 1918, Documenti e relazioni relative all'ECVE: Verbale della seduta del Consiglio di Presidenza in data maggio 1941, p. 9.

along the road to Bonga, the CAG was set up by the local Fascist Party.[36] These two agencies operated in a quite different framework from the other three regional bodies.[37] The areas they operated were acclaimed as 'healthy with abundant waters, extremely fertile with good topography and extensive irrigation potential'.[38]

Contrary to official claims, the choice of demographic colonization centres was not preceded by an accurate and serious study because of inadequate financial, technical, and personnel resources. The planners were lured by effortlessly growing vegetation rather than realistic edaphic and climatic assessment. Like the ONC's settlement centres, the soil of the areas was predominantly vertisol, which required management skills of the type that contemporary technology could not offer, and extensive use of labour.[39]

Wägära was relatively densely populated by the Amharas, who were dedicated agriculturalists and cattle-breeders.[40] So attached were these people to their lands, that even the tangible gains offered by road construction were unable to lure them, thus forcing the authorities to recruit Sudanese labourers. Public land in the area accounted for a few thousand

[36] The war stopped a number of agencies promoted by the Fascist Party in other parts. Two such were Arnaldo Mussolini at Asmära and Azienda Agricola Adigrat in Tegray province (cf. *Colonie*, 18 Mar. 1939; *CM*, 21 Jan. 1939).

[37] Even though a particular area predominated, the composition of EDR's and CAG's settlers had more a national rather than regional character. As its name suggests, the CAG was a co-operative to help those with limited technical and financial resources to develop their farms (cf. Berretta, *Amedeo*, 181–3).

[38] In the sources the two agencies are dealt with simultaneously and periodicals often confuse them. EDR was originally known as 'Cavallieri di Neghelli', a cavalry unit led by Thesaurus De Rege who died as war hero in May 1936 while fighting at Mälka Gulba, in GS. CAG was set up in Apr. 1938 to man Pattuglie del Grano (Wheat [Production] Task Force), a farm established by 30 demobilized Blackshirts in May 1937 with the view of promoting wheat cultivation (CD IAO GS 1822 GS UA, Attività agricole esistenti nel territorio del G GS, Jan. 1940, pp. 2–3). For a detailed account of the development of EDR and its initial tormented life see: CD IAO GS 1821, Esperienze della gestione di una azienda di colonizzazione demografica, Jan. 1940; CD IAO Gm 2919, Comprensorio di EDR in Gm n.d. [1940]. See also: Gm GS 845, Brevi Notizie sul comprensorio 'De Rege', Gm 1948; *Azione Coloniale*, 30 Nov. 1939; *CE*, 3 Dec. 1939; *GI*, 27 Nov. 1938; *GP*, 9 May 1940; *IOM*, 20 Jan. 1940; *Mattino*, 24 Nov. 1939; *Messaggero*, 4 July 1939; *PI*, 23 June 1938, 24 Nov. 1939]; *Tribuna*, 23 Dec. 1939; Villa Santa *et al.*, *Amedeo*, 235–6.

[39] Interview with Mr Yaqob Edjamo, crop-physiologist and agronomist 21 May 1988; Giuseppe Rocchetti to the author, Fl 17 May 1984.

[40] The population, predominantly Christian with some Muslims and Fälashas, was estimated to be more than 35,000 at about between 20–5 persons per sq km (cf. CD IAO 1111, p. 12; this conflicts with the figure given by another source according to which the population was no more than between 15,000 and 20,000 with density of 7–10 per sq km, see [Giglio], 'Importanza', p. 36).

hectares, scattered in plots of various sizes located at a considerable distance from each other.[41] Italian peasant settlement in the area could not be effected without massive evacuation of the existing owners from their land. Without such measures, settlements confined exclusively to the existing public domain would not be economically viable.[42] In addition, Wägära was one of the strongholds of national resistance. At the centre of the ECPE's settlement, the presence of unhealthy marshlands demanded extensive reclamation.[43] As with the case of the ECRE, the settlement of the EDR required massive displacement of the existing landowners.[44]

The settlement centres were scattered over a wide area. For the colonial administration in Ethiopia, who had the task of building the social and economic infrastructures and providing protection, this was a heavy burden on its already extremely tight human and financial resources.[45]

These factors were the focal points over which a major clash developed between the Duke of Aosta and the MAI. For the Duke of Aosta, the peasant settlement scheme as an answer to Italy's superfluous labour force was a commendable enterprise, but it was far too ambitious for the available finances. He saw capitalist initiative as the best option for initial development; with such a scheme the government would shoulder less responsibility and be in a position to direct its resources to more urgently needed areas; it would also allow a breathing space until complex problems of land tenure were studied and legally regulated, and give enough time to establish domanial lands and to draw up realistic plans. Yet the Duke of Aosta's appeal for a moratorium on demographic colonization was rejected by the MAI. Then he urged that, in line with the early plan, Shäwa be declared the first zone of colonization with the aim of forming around the capital of the Empire a belt of Italian population, and that the three agencies of Romagna, Puglia, and the Veneto should

> concentrate their work in Shäwa where I'll be able to provide them with further facilities, but for the future, unless instructed otherwise, I will restrict myself to the above criteria which, understandably, do not preclude small private initiatives in the agricultural fields at the periphery of the main urban centres of each governorship or commissariat; but I repeat, no dispersal of efforts and scattering of our farmers over different and remote regions.[46]

[41] CD IAO 1111, Lodi, accertamenti, p. 11.

[42] Ibid., p. 13; CD IAO 1806, Pavirani, Uogherà, p. 4.

[43] MAI, 'I servizi sanitari', *AAI* 3 (1940), 783; *CS*, 27 Jan. 1940.

[44] CD IAO Gm 2919, Comprensonio di EDR, pp. 1–2; CD IAO 1916, Maugini and Nistri, Rapporto, p. 13.

[45] Villa Santa *et al.*, *Amedeo*, 225–6.

[46] Ibid.; CD IAO AOI 1936, Amedeo di Savoia to MAI, AA 5 Jan. 1938.

The MAI acknowledged the merits of directing all colonial activities to Shäwa; yet it was by no means prepared to countenance a major shift of policy on the grounds that:

> the programmes of the demographic colonization agencies of Romagna, Puglia, and the Veneto—wanted by the Duce and organized by this Ministry in collaboration with the National Party—are already in an advanced stage of elaboration and implementation as experts have already studied on the spot the possibility of cultivation and the settlement of the colonists and the respective zones have been so far allotted; thus it seems impossible to turn back and concentrate the activities of the agencies in Shäwa. As to the political-military situation demanding the formation of a strong national nucleus in Shäwa, it should be borne in mind that this need equally exists in the other regions, which, in fact, have far better agricultural and climatic conditions. And therefore it is important not to divert from the guidelines of March marked by the Duce, and in the mean time colonization in Shäwa, as the Governor-General rightly proposes, should be stepped up at its fullest strength, within the framework of the general needs of the Empire. It seems that the most suitable agency for such an objective is the ONC which is locally better equipped and whose activity at present is the subject of so warm commendation by H.E. the Head of the Government.[47]

On the whole, the Ministry's position prevailed, but only after it had agreed to cut the size of the settlers of both the ECRE and ECPE, and defer indefinitely the activities of the ECVE.[48]

## OLD IDEAS IN NEW CLOTHING
## STRUCTURAL CHARACTERISTICS

In carrying out their mission, the colonization agencies were assisted by a substantial legacy of ideas and institutions. The classic point of reference in formulating their policies, in overall practice although not in detail, was the work of the colonization agencies in Libya. In this respect, they were less of an innovation than an imitation of a number of mass settlement agencies operating in Libya. What made them different was their regional character and the nature of their finance. Other variations owed much to the different political and environmental settings in which they operated.

The most obvious sign of the close relationship between the Libyan colonization companies and those of the Empire was in their bureaucratic

[47] CD IAO AOI 1936, Terruzzi to GG, Rm 10 Jan. 1938. The hard experience of ONC in obtaining land in Shäwa rather confirms the Ministry's position.

[48] Ibid., Amedeo di Savoia to MAI, AA 15 Jan. 1938; Villa Santa *et al.*, *Amedeo*, 253.

structures.[49] The organizational pattern of each agency rested on the legal framework laid down by Royal Decree on 6 December 1937. The decree established, in addition to the office of a president, a board of directors, an advisory committee, and a board of auditors, leaving to the by-laws of each agency the definition of their respective fields of action. The President was responsible for all the agency's affairs and decided matters of ordinary administration.[50] But supreme power was vested in the Board of Directors, known as Consiglio del Presidente, or the Presidential Council, in which were represented, in addition to the Ministries of Italian Africa, Finance, and the Interior, all the organizations that were concerned directly or indirectly with colonization and agricultural development.[51] The Council set the general guidelines of colonization, appointed higher officers, decided on budgetary matters and, whenever consulted by the President, expressed its views on specific issues. It operated from Rome headquarters.[52] As a legislative body, the glaring shortcoming of the Council was the absence of representatives from the colonies themselves. This was rectified only in early 1940. Such failure to take into account the needs of the colonies was one of many stumbling-blocks in the smooth development of the programmes.

The Advisory Committee resided in the area of the concession and as such it was meant to fulfil the shortcomings of the Council, based at a remote distance from the centre of colonization.[53] The Advisory Committee was a consultative body of government officials and experts living in the colony. Its task was to aid the President on a wide range of issues related to colonization. Thus it was not a decision-making body nor had it any power on budgetary matters.[54] The Board of Auditors, composed of three members, was responsible for the agency's finances. At the end of

[49] For the Libyan experience see Segrè, *Fourth Shore*, 89–101.

[50] DIM 18-4-1939: Approvazione; CD IAO 1921, Incartamenti: Statuto giuridico e trattamento economico; DM 10-1-1941, Approvazione.

[51] In addition to the said ministries, these included: PNF, General Command of MVSN; Commissariat for Internal Immigration and Colonization, CFA, CFLI, and an expert on matters of colonization.

[52] While the composition of the Advisory Committee remained constant in all the agencies, that of the Presidential Council varied. So we have 9 members of the ECRE, 11 for ECVE, and 12 for ECPE in the Presidential Council, and 7 for all three in the Advisory Committee (cf. 'Atti e Documenti').

[53] Ibid.

[54] CD IAO 1921, Incartamenti, 1944, pp. 5–9. In this incomplete document the opposition's view is briefly described; the MAI is challenged to explain clearly the financial implications that the agencies would have on the State treasury; excessive bureaucratic presence in the Presidential Council and the reasons for exclusion in other boards of institutions that are more closely affected; how the difficult issues of land shortage and expropriation were going to be solved.

each financial year, it presented the annual budget for the examination of the Council who, together with the report of the auditors, passed it on for MAI approval.[55]

The agency official most directly in touch with the day-to-day technical and administrative activities of the agency in the colony was the Director, or *direttore*, or *vice direttore tecnico*. With the help of his immediate staff, he kept the President informed on the agency's conduct and, whenever necessary, proposed to him a programme of action for the agency in its concession.[56] The exceptions to these general rules discussed so far were the EDR and CAG, who operated under the management of their local agricultural office with rules laid down by the Viceroy.[57]

In the Presidential Council and its advisers were merged various political, civil, military, and labour organs. This demonstrates both the regime's predilection for bureaucracy and its desperate attempt to involve the whole nation and divergent interests in the enterprise. Such a scheme was seen as a showcase of the harmonious working of the corporative state. In practice, however, the relationship between different office-holders and institutions towards the colonization agencies remained ill-defined and was often a source of serious friction. The President of the ECPE, Giambattista Giannoccaro, ran the agency as if it were his own personal fief and delegated no authority.[58] Until his forcible removal from office, the director of the ECRE was engaged in an acrimonious struggle with its president.[59] The government of Galla and Sidamo had often to intervene to uphold the tottering authority of the management undermined by the continuous interference of the local Fascist Party officials in the administrative affairs of the EDR.[60]

If the division of responsibility between officials was vague, there were a good many obligations and not much security for the settlers. The rules governing the conduct between the agencies and the colonists were fairly similar and, in substance, did not differ greatly from the guidelines followed by the ONC. Such similarities are identifiable in the rules relating

[55] CD IAO 1921, Incartamenti: Statuto dell'ECRE; DM 18-4-1939.

[56] CD IAO 1921, Incartamenti: Ordinamento degli uffici; CD IAO Hr 1992, ECPE, Ordinamento tecnico amministrativo contabile, allegato b, n.d..

[57] DGG 16-10-1939, no. 987: Istituzione presso il G GS della gestione di colonizzazione De Rege per la valorizzazione agraria, in *GU* 4/26 (suppl.), 19 Oct. 1939; FO371/24635/J376/18/1, CG Gibbs to PSSFA, AA 14 Dec. 1939; Villa Santa *et al.*, *Amedeo*, 235–6.

[58] CD IAO Hr 1992, Maugini, Appunti per l'Ing. Giannoccaro, Fl 27 May 1940.

[59] CD IAO 1921, Incartamenti: ECRE 1938, Appunti per il Duce, 23 July 1939.

[60] CD IAO GS 1821, Esperienze.

to the selection of families, the treatment of the colonists, the development and distribution of farms.[61]

It was understood that the agencies were temporary institutions who undertook to develop the project on behalf of the State and who were expected to dissolve once their goals were attained. Yet the agencies took a qualified view of the ideal of an autonomous peasantry. The future picture they had was that of a co-operative community of farmers capable of managing their affairs through a form of landowning settlers' consortium. Then the agency's role would be to assist them by providing managerial and technical know-how, credit facilities, marketing, and processing services.[62]

Before assuming ownership, the Italian peasant remained tied to the agency by two contracts: salary and redemptive. During the salary phase the peasant was employed as a wage labourer and a member of the workers' militia. He alternated communal work as defined by the agency with semi-military duties. The salary was attractive.[63] It was supplemented by costs for accommodation, discomfort, and family allowance, including medical assistance, insurance fees, and *Dopo Lavoro* subscription. But a number of deductions at source and the high cost of living left very little in the pocket of the peasant.[64] Part of the deducted money was directly sent to his family in Italy. It was argued, however, that the prospect of becoming a landowner compensated for low wages, making the peasant's salary more rewarding in the long run.

The salary phase was a trial period and thought to be beneficial to both the agency and the settler. For the agency, it offered a breathing space to

[61] Cf. CD IAO AOI 1937, CFA, Schema di norme generali per il regolamento dei rapporti fra enti di colonizzazione e coloni in AOI, 1938. Owing to the farmers-union disagreement with the original draft, the official approvement of the scheme took place only in Feb. 1940 by the newly formed Agricultural Council after laborious work and a number of modifications. Cf. *Azione Coloniale*, 19 Feb. 1940.

[62] CD IAO AOI 1937, Schema; Cerulli, 'Colonizzazione', 72.

[63] For more detailed duties imposed upon the settlers see [C. Giglio], 'Da colono all proprietario in AOI con la colonizzazione demografica', *REAI* 27/4 (Apr. 1939), 401–5. The salary was L.18 per diem for unskilled, L.20 for a semi-skilled, and L.23 for skilled (cf. [id], 'Alcuni problemi']. ECPE paid slightly higher wages which was L.19.8 per diem for an unskilled, L.23.1 for semi-skilled, and L.25.3 for skilled (cf. CD IAO Hr 1992, ECPE, Relazione ECPE, 20 Nov. 1940).

[64] Nor were such wages comparable to those paid for others within IEA. The scale was not constant but that of 1939 was as follows: unskilled L.30, semi-skilled L.35, skilled L.40, unskilled foremen L.40, semi-skilled foreman L.45, skilled foreman L.50, chauffeurs and motor mechanics L.1,200 (cf. FO371/23380/J575/296/1, CG Bird to PSSFA, AA 20 Jan. 1939; *CI*, 15 Jan. 1939).

implement reclamation work and a housing programme. More importantly, it was a convenient way to assess the peasant's credentials as a worker and potential landowner and to eliminate unsuitable candidates. For the peasant, it gave him the opportunity to acclimatize himself; should he decide to leave he lost nothing. If he chose to stay, he was joined by his family to start the period of redemption. But critics castigated the salary phase for the unnecessarily huge financial cost it entailed to the agency and the psychological trauma that the peasant had to endure without his family, making him work badly. They maintained that it was better to settle the whole family at once.[65] For money-starved agencies these economic arguments held great sway. In fact the two home-grown agencies, the EDR and CAG, abolished the salary period altogether, adopting the policies advocated by the critics.[66]

At the beginning of the redemption phase the colonist received a family farm as his own personal property and shouldered a number of responsibilities: he had to upkeep the premises and see that all farming implements and household goods were put at the exclusive use of the farm. In addition to his duty to execute the agency's orders, he had to keep the family labour unit intact. This meant that he had to prevent his sons from drifting into temporary high-paying construction jobs or stop his daughters from seeking work as domestics or worse in the city. It was particularly important that he groomed the eldest son as the natural leader and successor to the farm.[67]

At the redemption phase, the existing agreements were replaced by *contratto colonico*, a labour code or pact between the colonist and the agency, regulating the conditions of rescheduling of debts before assuming a full title to the ownership of the farm. In this respect, the redemption phase differed from the Libyan experience where it was preceded by between five and ten years of sharecropping. The new contract consisted, as detailed in an account book, *libretto colonico*, of three sections; in the introduction there was a summary description of the farm and of its boundaries with the aid of a small map; a record of livestock, household goods, and farming equipment provided by the agency, with the services and other effects received in cash or in kind; finally, the debts and the credits accumulated over the years. The debts included the following expenses: housing and implements believed to be necessary by the

[65] [Giglio] 'Da colono', 402. [66] *CC*, 25 Dec. 1939.

[67] CD IAO Hr 1992, Patto di ECPE; CD IAO 1921, Incartamenti: Statuto dell'ECRE; DIM 18-4-1939, Approvazione; [Giglio], 'Da colono', 405; E. L. P. Newman, *Abyssinia*, 102; Quaranta, *Ethiopia*, 49–50.

administration to the settlement; deforestation and reclamation including machinery, tools, animals and seeds; improvements and extension to the farm; transfer of the peasant and his family from Italy; any additional financial assistance, in cash or in kind. Added to this was a 5 per cent interest charge.[68]

No definite time-limit was set for the settler to discharge his debt. The only stipulation was that before freehold ownership of the land was granted the amortized cost had to be cleared. It was thought that this would take less than half the time (i.e. ten years) required in the Libyan situation where the settler had to contend with a more hostile environment.[69] In their efforts to force the settler to discharge his debts, agencies like the ECPE imposed a number of 'production incentives'. But in actual fact the agencies ran the farms autocratically, controlling every phase of activity. The peasants had little say either in the disposal of their labour or the expenditure of funds derived from it.[70] Complaints by Maugini of wastage due to carelessness and lack of proper storage reinforce this point.[71]

Theoretically, the subsidies were provided only for the trial period. Upon his admission to the farm, the colonist was obliged to pay the money back. But this was never the case because there was no year in which the gross income was greater than the annual subsidy. This was due partly to the agencies' failure to develop an adequate number of hectares as they had originally planned. As a result a considerable sum was spent in family allowances until the farm became fully productive and this ultimately served only to increase running costs of the farm. A typical case was the ECPE, which provided its fourteen families with six hectares of reclaimed land. In 1939 the agency paid an allowance of L.169,611. Despite expansion of the cultivated area and improvement in the farming, in 1940 this increased to about L.180,000.[72] This was a

[68] CD IAO AOI 1937, Schema di norme della convenzione per il finanziamento, annex 2; DIM 18-4-1939, Approvazione; CD IAO Hr 1992, Patto di Colonizzazione, p. 7; DGG 16-10-1939, no. 987, Istituzione; CD IAO Gm 2919; F0371/24635/J376/18/1 Gibbs to PSSFA, AA 14 Dec. 1939; [Giglio], 'Da colono', 402–5; Quaranta, *Ethiopia*, 49–50; E. L. P. Newman, *Abyssinia* 102.

[69] [C. Giglio], 'Finanziamento', 156–7.

[70] As regards labour, ECPE's labour code shows that the agency could resort to the peasant's labour at any time. As regards disposal of the produce, the proceeds from the sale of this was divided into three parts: a fixed quota was put aside towards amortization credited with 3% interest; 20% of the remaining was kept as reserve against the effects of freak years, generating 5% interest; the remainder was left at the settler's disposal (cf. DIM 18-4-1939, approvazione; CD IAO Hr 1992, Patto di colonizzazione, pp. 8–9).

[71] CD IAO Hr 1992, Situazione dell'ECPE al 31-12-1939, Rm 13 May 1940; F. Pierotti, *Vita in Etiopia, 1940–1941* (Bologna, 1959), 36–8.

[72] CD IAO Hr 1992, ECPE: Bilancio consuntivo anno 1939 e bilancio preventivo per l'anno 1940, Rm 13 May 1940, p. 19.

financial embarrassment for the agencies and an additional burden on their slim resources.

## THE QUESTION OF FINANCIAL ARRANGEMENTS

The success of colonization largely depended on good financing and the question of the best way of doing it had been a source of controversy among experts since early days.[73] Unlike in Libya where the State was almost the sole investor in demographic colonization, in Ethiopia the regime shared responsibility with other public or para-statal organizations. Each agency was given an advance of L.50,000,000 payable in six annual instalments. The payments progressively decreased.[74]

The ECRE, owing to its connection with Mussolini, received privileged treatment: it was directly financed by the State. The ECPE's quota was divided equally between the Banco di Napoli and the INFPS, which was involved in a similar programme in Libya. Each paid L.25,000,000. The ECVE was supported by the INFAIL. In addition, substantial sums were to be provided by the public administrative corporations of the provinces of origin, payable in six consecutive financial years. The ECRE was allocated L.2,000,000 from public institutions of Ravenna and Forli, the ECPE L.5,000,000, and the ECVE L.3,000,000.[75] Of the two home-grown agencies, the EDR had a budget of L.6,000,000 and CAG L.1,000,000.[76]

As agents of the State, the companies had legal standing and enjoyed tax exemption on any financial contributions made by welfare organizations, banks, or acquired through bequests and donations. In areas of their operation, they had preferential rights to any new land incorporated into the public domain, to the use of pasturage, and exploitation of natural resources. To minimize the cost of colonization, the agencies could combine farming with industrial and commercial activities. This enabled them to gain an exclusive monopoly over the trade and most resources in their area, such as cattle-breeding, marketing, and building contracts.[77]

[73] CD IAO 1921, Incartamenti, 1944, pp. 5–9; [Giglio], 'Finanziamento', 154–6; RDL 5-9-1938, no. 1607, in *GURI* 241, 20 Oct. 1938.

[74] The first year's payment was L.20 million; for the remaining four years: L.15 million, L.8 million, L.4 million, and L.1 million (cf. 'Atti e Documenti').

[75] Ibid.; DIM 18-4-1939, Approvazione.

[76] Of this L.750.000 was paid by GG and L.250.000 by the Bank as a loan (cf. Villa Santa *et al.*, *Amedeo*, 236).

[77] CD IAO Hr 1992, ECPE: Relazione, Bari di Etiopia 20 Nov. 1940, p. 9; CD IAO 1921, ECRE, Relazione del Presidente, Rm 7 Feb. 1939, p. 21, and 11 Jan 1940, pp. 13, 32;

The money was given in the form of a loan, repayable within a fifty-year period starting ten years after the formation of the agency. But no specific arrangements were provided as regards securities and interest rates, issues that were of vital importance.[78] Such shortcomings caused friction between the ECPE and its creditors until accommodation was reached with the Royal Decree of 5 September 1938 and the special convention of 6 October of the same year which brought substantial amendments to earlier arrangements. In the new agreements the interest rate was fixed at 5 per cent and the period of repayment was reduced to twenty years, with amortization starting in the sixth year.[79]

As collateral, the creditors were given wide-ranging special privileges. This involved rights to all kinds of State lands at the agency's disposal, to works of improvements and agricultural developments accomplished on it, to livestock and all types of assets including the eventual State contributions to land reclamation. The prerogative was to remain valid so long as the loan was not entirely settled. Any failure by the agency to keep the agreement, even the payment of only one year's mortgage, would dissolve the contract and lead to an immediate payment in full of both the outstanding debt and interest.[80] Support for the agency's ability to meet the terms of the agreement was drawn from the findings of the Libyan experience where, it was alleged, the colonists, working in environmental conditions far less favourable than in Ethiopia, were able to pay their debts within twenty years. Yet some critics pointed out that this comparison overlooked a number of important differences. In their view, in Ethiopia unlike in Libya, practically no institutions were available to verify the state of landownership and the rules of its transaction. It was true, they argued, that the Ethiopian soil was much more fertile than in Libya; yet it was equally true that the colonist had much less to fear in Libya than in Ethiopia in terms of competition from Ethiopian production and manpower. The critics focused mainly on the interest rate which in the context of the Ethiopian agricultural environment—virgin soil with everything to be built from scratch—was excessive. With high interest rates, the cost of each farm increased and made the production cost too high against Ethiopian competition.[81]

CD IAO 1792, Walter Caravelli and Luigi Zampighi, Lineamenti dell'ECRE a norma del decreto costitutivo e dello statuto, Rm 10 Oct. 1938.

[78] [Giglio], 'Finanziamento', 155.

[79] RDL 5-9-1938, no. 1607; RDL 13-5-1940, no. 823: Modificazione dell'art. 7 del RDL dicembre 1937, no. 2314, costitutivo dell'ECVE, in *GUGS* 5/34, 21 Aug. 1940, and *GU* 166, 17 July 1940.

[80] CD IAO AOI 1937, Schema di norme.

[81] [Giglio], 'Finanziamento', 155, 157.

What was to be done? How could the situation be altered? Once again the old question of State subvention was debated. Originally, the agricultural council recommended that the task of financing all forms of agricultural activities be vested in one agency, mentioning the Institute of Mortgage Loan as an ideal choice. Basing itself on the Libyan experiment, it warned of the dangers and expenses of extensive State-subsidized settlement.[82] Even for the State interventionists, the less resort made to State coffers, the better; yet for them a total lack of State intervention was fraught with dangers. They asserted that demographic colonization was a symbol of great power overseas and the means to create a large Italian population abroad. Because of this it had, in addition to economic, the highest political and social goals. Thus, if State subsidy from a strictly commercial standpoint was unremunerative, from the political point of view it was not unduly extravagant expenditure.[83]

But as the cost of colonization became apparent, the agencies invariably complained of financial stringency and its limitation on their activities. They discovered that the early optimistic estimated unit cost of settling a family, L.55,000–70,000 per farm, was unrealistic. Where the most economical means were employed, such as extensive use of Ethiopian labour and inexpensive material, the cost of a farm amounted to between L.120,000 and L.140,000, excluding investment in infrastructure.[84] A number of factors were blamed, including shortage of labour, rising wages, turbulent political conditions, and the scarcity of building materials. With little fresh capital at their disposal, the agencies found their original capitalization of L.50,000,000 ludicrously inadequate for the task before them. The ECRE estimated the minimum financial requirement for settling 1,000 families to be double of its actual grant.[85] The feeling was stronger with the agencies that paid interest on the loan. Giannoccaro, the President of the ECPE, realized that, after interest deduction, the ECPE's entire loan in real terms amounted to only L.34,000,000. As it appears in a sketchy memorandum for the ECPE's Board of Directors, Giannoccaro regarded this as 'too inadequate and absurd' to finance a programme on such a scale.[86] Costs may decline as colonization

[82] CD IAO AOI 1775, Seduta: Rm 15 Apr. 1937, pp. 11–19; Rm 22 Apr. 1937; Rm 26 Apr. 1937. Consulta, 'Avvaloramento', 1563–4.

[83] The upholders of such a view called on the government to pay 3 per cent of the interest due to the creditors and leave the remaining 2 per cent to the agencies themselves (cf. [Giglio], 'Finanziamento', 158).

[84] CD IAO Gm 2919, Comprensorio di EDR, p. 6.

[85] CD IAO 1921, Fuzzi, Programma, p. 3.

[86] One of the solutions suggested by ECPE was for the government's intervention by paying about 4 per cent of the interest. CD IAO Hr 1992: Relazione Ing. Giannoccaro al

proceeded, but serious doubts had already been cast on the practicality of the scheme.

## PLANS VERSUS REALITY

Settlement agencies did not operate uniformly. Need to accommodate local demands gave rise to administrative and organizational differences. The common factor was that, owing to financial constraints, the settlements were run on a tight budget and, as much as possible, using local resources; and owing to the unsettled military situation, strategic considerations were given overriding importance in the pattern of house construction and farm organization.

The settlement plan of the ECRE was drafted by the agency's future director—Ing. Savini, under explicit orders from Mussolini.[87] Savini began his work in the early part of 1937 and in substance if not in detail, his settlement programme was identical to that of the ONC. He planned to settle 1,000 families within a maximum period of six years; this meant that each year 10,000 ha would be put under cultivation and 200 houses built. According to the plan, the first 200 houses would be completed by the end of 1938.[88] Each family was to be given 50 ha of land, divided into three sections: 2 ha would be used for artificial and natural pasturage for cattle which would play an important part in the development of the farms; the remaining land would be ploughed, with the exception of 1 ha for vegetable-growing. The size of the farm was justified on the grounds that it responded to the needs of an increasing family demanding its further partition, and the tendency to rapid deterioration of tropical land under production requiring crop rotation.

Initially the plan was to build isolated homesteads, but for reasons of security this was abandoned. Instead Savini designed a single-storey house with three rooms and a kitchen and an annexed courtyard enclosed by a solid wall. Four houses were situated in the corners of their plot and joined together by a vast communal wall, forming a square with only one entrance. The courtyard was to serve as a shelter for the colonist's livestock, forage, and tools. As their original architect Tito Piccialuti claimed, the geographical set-up of the houses with their solid walls, in addition to facilitating the sharing of communal services and relatively reducing the

Consiglio della Presidenza, 24 Mar. 1940; ECPE to Maugini, Situazione amministrativa contabile anno 1940, Rm 11 Apr. 1940.

[87] Detailed information on the plan is given by Fuzzi in his report to the meeting of the Presidential Council (cf. CD IAO 1921, Fuzzi, Relazione, 7 Jan. 1938; CD IAO 1806, Pavirani, Uogherá, p. 6).

[88] Fossa, *Lavoro*, 497; Diel, '*Behold*', 70.

road network, gave its inhabitants, 'specially at night, once the gate was closed, a sense of total safety as if they were in a small but well protected fortress'.[89]

The ECRE began its operation in April 1938, seven months after schedule. Unlike other agencies, the ECRE started from an advantageous position. It purchased four construction sites from the departing Asmära–Gondär highway construction agency Ragazzi.[90] All the sites were equipped with basic services and stone-built premises, and at one site there were well-furnished houses for the management. Each site was to serve as a settlement centre, each bearing district names similar to those of Romagna province. Eventually, work was started in only three of these. The main administrative and settlement headquarters was Romagna Centre located at Dabat, 70 km north of Gondär.[91] The settlement was cut in two by Dib-Dibit River which served for irrigation.

Activities were centred to the West of the river. In addition to the administration and social amenities, there were stables and mechanical workshops; in a nearby irrigated garden all sorts of vegetables were grown. The remaining two districts were located at Dära; they bore the names of Lugo of Ethiopia (Lugo di Etiopia) and Cesena of Ethiopia (Cesena di Etiopia) and were located 50 km and 40 km from Gondär respectively.[92] The first group of colonists who disembarked at Mešwa numbered 124 men, 300 less than the original plan envisaged. They included, in addition to 10 agency officials and their assistants, 50 heads of settler families; the rest were workers with different skills, predominantly builders. In December 1938 and 1939 65 and 340 individuals respectively were added, consisting exclusively of technicians, artisans, and craftsmen.[93]

The ECRE's achievement was hailed as the pride of the Italian provincial schemes. But the reality challenged such claims. Even though the

[89] 'Progetto per la formazione di una legione di lavoratori agricoli per l'AOI', *REAI* 26/2 (Feb. 1938), 1999.

[90] G. B. Lusignani, 'L'ECRE', *Autarchia Alimentare*, 1/7 (15 Dec. 1938), 36–7.

[91] As the main motorway-construction site Dabat had premises that could accommodate 600 people (cf. CD IAO 1921, ECRE, Relazione del Presidente: 7 Feb. 1939, pp. 3–4).

[92] CD IAO 1921, ECRE, Relazione del Presidente 7: Feb. 1939, p. 19, and 11 Jan. 1940, pp. 13, 16–7; *Azione Coloniale*, 22 June 1939. As the names of the districts reflect the settlers' origin, it seems that the first settlers were natives of the respective district of Lugo and Cesena. Of the two, Lugo was at an advanced stage and work had started in early 1939. Contrary to the report of some sources (cf. Pankhurst, 'Italian Settlement', 151, who quotes P. Monelli's *Mussolini: An Intimate Life* (London, 1953), 19), Predappio d'Etiopia, though planned very early to be set up in the memory of Mussolini's birthplace, it appears that it never came into light.

[93] They were mainly tractor-drivers, farm-mechanics, builders, and carpenters (CD IAO 1921: Relazione, 7 Feb. 1939, p. 12; Fuzzi, Programma 1940, 7 Feb. 1939, p. 4).

exact number of the workforce can only be approximated, by the end of 1940 there were about 450 people consisting of workers and day agricultural labourers, including about 150 masons and 30 agency officials. The peasants amounted to less than 100 and none of them were united with their families.[94] Of these some were settled at Lugo, but the majority were concentrated at the main headquarters, while Cesena was occupied almost exclusively by workers.[95]

When assessing the agency's work, the role of Ethiopian labour which the agency employed 'on a generous scale' with a normal intake of between 500 and 1,000 day-labourers must be taken into account. On the housing side, at the beginning of 1940 a total of about sixty houses were built including storage area and stables. Of these about forty were completed and twenty were in progress.[96] The pictures are equally meagre if the developed land area was examined. In the first year 240 ha were reclaimed. Of these 110 ha were worked exclusively by Ethiopians. Land sown amounted only to 105.45 ha. For 1939 the agency set out a 'realistic' programme and intended to put a minimum of 2,500 ha under cultivation. Eventually, the total reclaimed land area, including the previous year, amounted to 1,013 ha and of these 873 ha were cultivated: 10 ha by the Dabat Agricultural Office and 125 ha by Ethiopians on a sharecropping basis.[97] In 1940 the agency concentrated all its efforts on agricultural activities but it was able to sow a mere 1,250 ha.[98]

Colonization agencies had an integral approach to their settlement programme. In this respect the ECRE's efforts to make its settlement self-sufficient could boast a number of achievements. By the end of 1940 the settlement had its own clinic, a recreation centre, and a post office. In addition to this there were a wide range of agricultural machinery with spare parts and various implements, a well-stocked mechanical workshop, a carpentry shop, an electric generator, a brick furnace with annual production potential of a 500,000 pieces, and two wells with a daily capacity of about 20,000 l.[99] Investment in cattle, albeit modest, was promising if

[94] CD IAO: 1921, ECRE: Situazione febbraio 1940; 1793, Attilio Tomassini, ECRE, Promemoria, Rm 15 May 1944, p. 2; 1792, Caravalli and Zampighi, Lineamenti, pp. 3–4.

[95] CD IAO 1921, ECRE, Relazione del Presidente, 11 Jan. 1940, pp. 17–18.

[96] CD IAO: 1921, Situazione 1940; 1792, Caravalli and Zampighi, Rineamenti; Quaranta, *Ethiopia*, 48.

[97] CD IAO 1921, Relazione, 11 Jan. 1940, pp. 6–7.

[98] CD IAO 1921, Fuzzi to Maugini, AA 16 Jan. 1941.

[99] The machinery consisted of tractors (14), mechanical sowers (18), threshers (3), reapers (2), binders (2), fodder press (1), trucks (5), harrows (23), ploughs (27), incubators (1), and a wide range of minor implements. CD IAO 1921, Situazione 1940; Berretta, *Amedeo*, 80–1. For slightly different fig. see Pankhurst, 'Italian Settlement', 151.

the difficulties of purchasing them locally were taken into account.[100] By the end of 1940 the animal stock was 429 consisting of 317 head of oxen, 37 horses, 62 pigs, and 13 goats in addition to about 1,000 poultry.[101]

The key stumbling-block for the ECRE's activities was the perilous political climate and the regime of land tenure, further compounded by internal administrative shortcomings and lack of co-operation from both local and central governments. Owing to policy differences between Rome and Addis Ababa, the ECRE was forced to start its activities with a manpower cut to less than half the size of its original plan, from 400 settlers to 130. Colonists were waiting for five months as the agency's efforts to start in September 1938 were deferred twice; the first, because of the insistence by the Ministry of Finance that demographic colonization would not constitute any burden to the State treasury; then by the long debate between the MAI and the colonial administration in the Empire, who demanded the delay of the agency's activities until enough land was secured and the turbulent political climate of the area settled down. Further postponement was avoided only by the forceful intervention of Mussolini who ordered an immediate start, notwithstanding the realization that the economic consequences of such a move were damaging.[102]

After a short burst of enthusiasm the saga of disappointments began, confirming the fears of the colonial administration in Addis Ababa. According to plan, the agency was to carry out its work in 50,000 ha of land that would be formed by conversion of small domanial holdings into large plots through the principles of expansion and amalgamation as laid down by Mazzocchi Alemanni. But the government of Amara, apprehensive of the political situation and under pressure from the central government, withdrew from its original commitment to provide the agency with 20,000 ha immediately, and 50,000 ha within a few years. In the first year the agency hoped to cultivate 2,500 ha of land with an expected yield of 20,000 ql which, at the current market price of L.200 per ql, would have given the agency a revenue of L.4,000,000. Eventually the agency cultivated

[100] Despite the reported highly lucrative offer from the agency, the Ethiopians sold their best animal stock only when forced (cf. CD IAO 1921: ECRE, Relazione del Presidente, Rm 7 Feb. 1939, pp. 17–18; and Rm 11 Jan. 1940, pp. 13–14; Fuzzi, Programma, pp. 12–13). Such difficulty was equally experienced by other agencies (cf. CD IAO Hr 1992, A. Maugini, Promemoria Per l'Ing. Giannoccaro, Fl 27 Sept. 1938).

[101] Owing to the difficulty of Italian poultry to adapt to high altitude (2,400–2,500 m), Italian fine-quality chickens gave mediocre results, both in the field of pure- and cross-breeding. Cf. CD IAO 1921: ECRE, Relazione del Presidente, Rm 11 Jan. 1940, p. 13; Fuzzi, Programma 1940, pp. 13–14; Situazione 1940. Pini, 'Terra Richezza'.

[102] CD IAO 1921, Fuzzi, Relazione, 7 Feb. 1939, pp. 6–10.

only 105 ha from which it made a very modest revenue of L.53,128.85. The agency's explanation for its failure to keep pace with its planned target was lack of adequate machinery to plough a land that was almost virgin. It successfuly pleaded with the Ministry of Finance for the purchase of foreign-made tractors but the proposal was practically blocked by Mussolini. Even though he agreed with the agency's plan, Mussolini allowed only a fraction of the original fund to be released for the purpose.[103]

Difficulties in obtaining land were symptomatic of the turbulent political climate. Wägära, the ECRE's operational area, was a stronghold of national resistance.[104] Dabat, the ECRE's centre, was as a garrison under siege. At one stage the centre was ransacked and supplies had to be provided by air. The worsening situation led the Governor, General Mezzetti, to establish a task force at Dabat. Each group of four houses was protected by a barbed wire fence, and heavy artillery and machine-guns, manned by Italian peasant militia. Every Sunday the farmers received military training and exercise in pitched combat, the use of hand grenades, and automatic weapons.[105] The fluid military situation frustrated the work of agricultural offices which were one of the oldest and had a conspicuous presence in each major urban centres of the Amara governorship. As a result, the knowledge of the land tenure system of the area remained poor and the exchange of land slow.

In 1939 there was a remarkable improvement in political conditions largely due to a number of military offensives conducted by the new Governor, General Frusci, which led to the submission of a number of chiefs and allowed a degree of control over surrounding areas. However, the possibility of attaining total disarmament of the population and asserting effective control remained remote. And yet the Amara government took advantage of the lull to evacuate lands destined for demographic colonization and promote share-cropping between Ethiopians and the ECRE. In the process a combination of force and the local chiefs were used to convince the Ethiopians. Deceptively, the initiatives seemed to proceed smoothly.[106] But soon the government was in retreat when it became clear that the land exchange measures worsened the delicate political situation. In the end, the agency, out of its planned 50,000 ha of

[103] Ibid., pp. 1–8, 14–15; 'Rassegna agraria coloniale', *AC* 33/7 (1939), 446–7.

[104] On the operation of the patriotic movement in this area cf. Gärima Täfärra, *Gondäré Bägashaw* (AA, 1949 EC).

[105] *CS*, 1 Aug. 1939; Sbacchi, 'Italian Colonialism', 349.

[106] CD IAO 1921, Fuzzi, Relazione, 11 Jan. 1940, pp. 18–19.

land, was able to obtain about 5,600 ha and of this only 4,000 ha were useful for demographic colonization.[107]

The strategy of sharecropping with its twin purpose of peaceful penetration and solving the problem of land scarcity had modest success. Earlier Ethiopian farmers agreeing to this form of contract were discouraged by the government's imposition of *asrat*. This was done when no such obligation was imposed on other Ethiopian farmers. The ECRE interpreted this as a deliberate policy meant to undermine its work. With the change of the Amara government at the end of 1939, these strictures were removed. But, despite the agency's use of intensive political pressure, only 350 contracts were signed. Although the harvest was good, the land covered only 95 ha.[108]

Scarcity of land was not the agency's only problem. Since the early days it had experienced a leadership crisis that remained throughout its life-span. The conflict between the General Director of the agency, Savini, and the President, Fuzzi, was so serious that Mussolini himself became worried by the publicity it generated. An urgent *ad hoc* committee led by the General Finance Inspector, Mola, attributed the conflict to a combination of personality and policy differences. Following the subsequent dismissal of Savini, the administration was restructured. But his departure left the agency with a leadership crisis from which it never recovered.[109]

The difficulties exposed the ECRE to the mercy of competing political interests. For the Duke of Aosta, the situation offered an opportunity to fit the agency's programmes within his own colonization paradigm. With its programmes badly trimmed and its strength weakened by the internal crisis, ECRE was left with little option but to succumb. The developments forced the agency to divert a significant part of its actions to other areas. In February 1938 it was given a once-flourishing farm confiscated from a Russian adventurer, Babitchev, in Shäwa. Later in May 1938 and November 1939 two other farms were given, located in the Upper Awash Valley at Täfqi and at Tullo Bullu.[110]

[107] CD IAO 1921, Fuzzi, Programma, 19 Jan. 1940, p. 2. [108] Ibid.

[109] Savini was removed in Aug. 1939 and by the beginning of 1940 four directors were appointed and then left or were removed. CD IAO 1921: De Rubeis A., Appunti per il Duce, 23 July 1939; Relazione del Presidente, Rm 7 Feb. 1939, pp. 13, 29, and 11 Jan. 1940, pp. 2–3; Timò Mansueto, Rm 9 Feb. 1943. *CS*, 1 Aug. 1939.

[110] Villa Santa *et al.*, *Amadeo*, 230; MAI, 'Valorizzazione', pp. 269–70, 273; *New Times and Ethiopia News*, 1 July 1939.

Having confined Ensign Babitchev to Rhode Island the Italians confiscated his farm, as they did most other foreign businesses.[111] The farm was located at Ada, within 50 km from Addis Ababa, at an altitude of 1,800–1,900 m, protected by the surrounding Yärär mountains. As in the pre-Italian period, in post-occupation Ethiopia Ada was one of the most important modern agricultural centres. It is rich in water resources from the Wädächa River and a spring situated in the lower slopes of Yärär mountain ranges. Apart from the threat of malaria from the palustral zones and stagnant waters along the Wädächa river, it has a healthy climate. With the completion of reclamation work and reorganization of the Wädächa river by the ONC, it was hoped that the area would be suitable for demographic colonization. And yet the scarcity of land forced the agency to confine the settlement only to the families of agricultural managers and technicians, not peasants.[112]

The farm extended over about 180 ha made up of four plots varying in size and intersected by other properties. Both dry and irrigated farming was used. Although limited to a mere 20 ha, irrigated areas could be doubled through rational use of water resources. Cultivation took place using a mixture of sharecropping and direct management with contracted labour. The total labour force consisted of 120 peasants and their families who lived within the farm. Each family gave four days' work to the farm in exchange for an irrigated plot of *c.*1,000 sq m, known as *carta*. Both forms of management involved part irrigation and part dry farming. Under sharecropping, dry farming was made up of *c.*15 ha and irrigated *c.*10 ha, while under direct management they consisted of 10 ha each. Crops cultivated in dry farms were cereals consisting mainly of *ṭěf*, *durrah*, barley, wheat, and chick-peas. In the irrigated farms under sharecropping were sweet potato and sugar cane, while on the plots directly manned the key crops were coffee, papaw, bananas, and citrus fruits.[113]

Irrigation was shared with two other bordering groups: three days were allocated to the Oromo peasants of nearby Debasso area, who tilled about 120 ha, and Qäñazmach Täsämma, an owner of a 40-ha farm. The

[111] DG 24-3-1938, no. 672, Espropriazione dell'azienda agricola appartenente al sig. Giovanni Babiceff sita nel territorio della Residenza di Adda, in *GU*, 3/10 (suppl.), 21 May 1938; DGG 26–9–1938, no. 819, Indennità da corrispondere al sig. Giovanni Babiceff per l'espropriazione dell'azienda agricola sita nel territorio dell R. Residenza di Adda, in *GU* 4/1, 1 Jan. 1939. The Commission assigned a compensation of L.400.000 (cf. FO371/22020/J641/40/1, CG Bird to PSSFA, AA 4 Jan. 1938).

[112] CD IAO 1921, ECRE, Relazione del Presidente, Rm 7 Feb. 1939, pp. 23–4, and 11 Jan. 1940, p. 34.

[113] Ibid., 7 Feb. 1939, pp. 22–4, and 11 Jan. 1940, pp. 21–3.

Babitchev's farm had a four-day quota from both the river and the spring: twelve daytime hours were allocated to the farm and twelve night-time hours were used by the *carta*-holders.[114]

Despite its reliance on traditional techniques, the Babitchev farm had all the features of contemporary modern farms established in the second half of the twentieth century.[115] It was equipped with a considerable number of residential buildings made of a mixture of stone and *čeqa*, animal shelters, and silos. It also had a water-mill on Wädächa River.[116] Its great disadvantage was the presence of brigands and resistant fighters in the surrounding Yärär mountains. In fact the farm had been the theatre of a number of military operations. This, combined with two years' neglect, had left it in a pitiful condition. When the ECRE took possession, the farm animals were reduced to only ten head of oxen, two donkeys, and a she-mule; the activities of the *carta*-holders were restricted to their own individual plots.[117]

Soon after the take-over, the farm, renamed after one of Mussolini's daughters Villa Anna Maria, was reorganized under the direction of C. M. Battaglia, a surveyor, and his assistant agriculturalist, Savorelli. Battaglia's work was facilitated by the fact that he was also Vice-Resident of Yärär district, a political office that gave him considerable weight in his dealings with Ethiopians. Battaglia had at his disposal fourteen Italian workers, composed of a market-gardener, tractor-driver, two agricultural workers, a storekeeper, and eight road-builders and labourers working under a superintendent on the 10-km-long Villa Anna Maria–Bishoftu Road construction project. Later with the addition of an accountant, an agricultural technician, a botanist, a horticulturalist, and a herdsman, the farm could boast a well-qualified, though modest, agricultural team.[118]

According to the plan, Anna Maria Farm was to be a model estate and an excursion venue for the inhabitants of Addis Ababa because of its picturesque landscape, medium altitude, abundant water resources, and location a few miles from the capital. The scheme envisaged a total restructuring of the farm to make cultivation more rational and surveillance easy. This demanded the removal of the *carta*-holders occupying the most fertile areas of the farm and the owners of the criss-crossing or adjoining tracts of lands, including Oromo peasants. Such measures would

[114] CD IAO 1921, ECRE, Relazione del Presidente, Rm 11 Jan. 1940, p. 23.

[115] Pankhurst, *Economic History*, 208–9.

[116] CD IAO 1921, ECRE, Relazione, Rm 7 Feb. 1939, p. 23, and 11 Jan. 1940, pp. 21–2.

[117] Ibid., 7 Feb. 1939, pp. 24–5, and 11 Jan. 1940, pp. 22, 33, 35.

[118] Ibid., 7 Feb. 1939, pp. 24, 26, and 11 Jan. 1940, pp. 34–5, and 19 Jan. 1940, pp. 34–5.

ensure that the farm was a compact and unbroken unit and had little competition for water resources.[119] Other operations entailed the organization of the plants in a row and terracing of the soil to make the plots easily accessible; the introduction of new domestic and imported plants and citrus fruits; the renewal of old plants; the extension of vegetable-growing areas; the elimination of not easily marketable crops; finally, improvement of the irrigation system by a more rational use of available water resources.[120]

To do this Battaglia relied on a handful of tractors manned by Italian personnel, and extensive use of Ethiopian labour. To counter eventual shortage of labour, he maintained the existing contractual arrangement between the farm and the *carta*-holders. Those *carta*-holders removed to make room for the farm were settled at its margins. In addition, an average of seventy-five Ethiopians were employed as day-labourers, cattle-keepers, guardsmen, in water regulation, in the services of mill and *gebbi*. The size of the farm was gradually expanded to about 350 ha through the usual practice of appropriation and confiscation of bordering lands. The measures sparked some revolts. In consequence, at the end of 1939 and the early part of 1940 the farm was the scene of a major military operation.[121] Even though the fighting caused grave damage to the farm, the revolt was ruthlessly crushed. The agency was paid an indemnity with which it was able to replenish its depleted livestock.[122]

Improvement in the use of water was achieved by either reducing the quota allocated to the Ethiopians or, according to the initial plan, removing them to other areas. The most important operation in the irrigated farms was the successful transplantation of a number of coffee and fruit trees; sugar-cane plants and sweet potatoes were eliminated on the grounds of low market demand by the Italians and high water requirement. On the nursery side, ornamental plants both of local and foreign variety were extensively cultivated partly to meet the demand from Addis Ababa inhabitants and the Town Hall.[123]

Actual progress, however, was slow. The 1938 annual report did not fail to point out the modest nature of the achievements. It blamed the

[119] Ibid., 7 Feb. 1939, p. 9, and 19 Jan. 1940, pp. 33–4.

[120] Ibid., 7 Feb. 1939, pp. 24–6.

[121] Ibid., 11 Jan. 1940, pp. 24–5, 35.

[122] Despite the preventive treatment by the Zooprophylactic Institute of AA, a 1939 epidemic killed most of the animals, esp. oxen. The agency's total livestock amounted to 86 animals—30 oxen, 16 cows, 6 calves, 16 goats, 10 donkeys, 7 mules, 3 sheep, and 2 horses (ibid., 11 Jan. 1940, pp. 11–12).

[123] Ibid., 11 Jan. 1940, pp. 26, 29–30; CD IAO 1921, Fuzzi to Maugini, AA 16 Jan. 1941, p. 5.

unsettled political situation and the scarce resources of men and material, but at the same time emphasized the valuable lessons learnt. The statistics of 1939 were not comforting. With the help of two tractors and extensive use of local workforce, 20 ha had been reclaimed and 75 ha sown—65 ha dry farming, the rest irrigated.[124] In 1940 the total cultivated area amounted to 136 ha, with 26 ha irrigated and 110 ha dry farming.[125]

The Upper Awash farms of Täfqi and Tullo Bullu were located along the Addis Ababa–Jimma highway and within close range to Awash, one of Ethiopia's principal rivers. This, combined with their high productive potential, gave the two farms one of the most promising sites for successful demographic colonization. In addition they were to serve as an experimental centre for the study of the agricultural potential of the region as well as water resources. Yet once again it proved difficult to obtain land. The *residenti* contacted were slow to act and only after long and laborious negotiations was a total of *c*.700 ha obtained, approximately 350 ha at each place.[126] As most of it was virgin land, work proved difficult. Using 10 agricultural labourers, 6 tractor-drivers with 5 Caterpillar tractors under the supervision of a hard-working agricultural expert, Giacomo Fiumana, 400 ha were reclaimed at Tullo Bullu. At Täfqi, 150 ha were sown with *ṭéf* (100 ha), chick-peas (40 ha), and *nug* (10 ha). As the farm was experimental, emphasis was laid on the cultivation of different varieties of local and imported crops. Wheat, fodder, oilseeds, vegetables, and citrus fruit were cultivated on the remaining land.[127] In 1940 the cultivated areas were below the planned target. Against 1,200 ha planned, 1,600 ha were reclaimed and 1,049 ha sown and, contrary to early expectations, the average yield, as in most of Shäwa lands, was poor.[128] As in other farms, the agency combined farming with cattle raising. By 1940 it had a modest but promising animal population of 16 cows for breeding, 12 draft oxen, 4 bullocks, 8 mules, and 3 horses.[129]

[124] CD IAO 1921, Relazione, 11 Jan. 1940, pp. 27–30.

[125] CD IAO 1921 Fuzzi to Maugini, AA 16 Jan. 1941, pp. 5–6.

[126] The initial concession was at Busa, 24 km from the AA–Jimma highway but because of lack of accommodation and difficulty in getting supplies, Täfqi was chosen where the agency's personnel occupied part of the military garrison. The change entailed the evacuation of the Ethiopian owners from their *rest* lands at Täfqi to domanial lands at Busa. The process proved very slow. At Tullo Bullu the work started immediately and accommodation was not a problem as the agency purchased the road-construction sites from Parisi company (cf. CD IAO 1921, Relazione, 7 Feb. 1939, pp. 11–12, and 11 Jan. 1940, pp. 37–8, 44).

[127] Ibid., 1 Jan. 1940, pp. 39–43, 45.

[128] CD IAO 1921 Fuzzi to Maugini, AA 16 Jan. 1941, pp. 3–4.

[129] Ibid., pp. 5–6; CD IAO 1921, Relazione, 11 Jan. 1940, pp. 24–36; 19 Jan. 1940, pp. 30–6.

The ECPE fared no better. The ECRE was credited with being the first regional demographic agency, but it was the ECPE that first began actual settlement. Using funds advanced by the government of Härär, it started work even before its legal constitution had been approved. In many respects, the ECPE's scheme was identical to that of the ECRE and ONC, but differed in the pattern of house construction and organization. To its advantage, it had a relatively peaceful area with plentiful water and a sympathetic colonial administration. Its main problem was obtaining a sufficient indigenous workforce.[130]

As no bank of information existed on the area's agricultural potential, the agency set out a number of rules of thumb to maximize economic resources and keep risk to a minimum. It planned to combine cereal farming on land directly manned by the agency with cattle husbandry.[131] To reduce labour costs and remedy its shortage, mechanical farming was to be dominant. Animal husbandry was to involve both draft and—to meet the increasing demand of the growing settler community—industrial animals. Maugini emphasized the need to resort to political pressure to overcome Ethiopian reluctance to sell animals.[132] But as was so often the case, plan differed from practice. The agency showed little stock-raising interest and its animal population amounted to only about seventy oxen and a few poultry and horses.[133]

The agency planned to allot each settler family a large farm of 50 ha or more. But as a short-term strategy, it allocated between 15 and 35 ha. On the remaining land the agency, using exclusively Ethiopian labour, carried out directly mechanical farming and cultivated crops of home and foreign variety. Emphasis was laid on lucrative industrial crops that could offset the financial cost of the project. After a while these plots were to be allotted to the settlers, while the agency explored and put new land under cultivation.[134]

The decision to allocate plots ranging between 25 and 35 ha hinged on a number of assumptions. On the productivity side, since Wächo Valley's agricultural and livestock resources were immense, if intensively cultivated the plots would be highly remunerative. On the settler's side, the farms were within the means and abilities of the colonist families. It was

[130] As regards administrative structure see CD IAO Hr 1992 Schema di norme: Statuto giuridico and 'Patto di Colonizzazione'.

[131] CD IAO Hr 1992, Maugini, Promemoria, Fl 27 Sept. 1938.

[132] Ibid., pp. 1–2.

[133] CD IAO Hr 1992: ECPE, Bilancio consuntivo, p. 24; Agostino Volpi, ECPE: Relazione, Bari d'Etiopia, 20 Nov. 1940, pp. 9–10.

[134] [Giglio], 'Alcuni problemi', 1877–8.

**FARM SIZE** 25 ha

| | |
|---|---|
| Buildings, roads ditches | 1 ha |
| Citrus orchard, arboreta | 2 ha |
| Rotation exempt grazing field | 2 ha |
| Rotating herbal crop | 16 ha |
| Fallow during rotation period | 4 ha |
| | 25 ha |

(Fallow land may stay bare or covered)

| Year | Semester | I(Ha. 4) | II(Ha. 4) | III(Ha. 4) | IV(Ha. 4) | V(Ha. 4) |
|---|---|---|---|---|---|---|
| I | 1<br>2 | Renewal<br>Wheat | Herbs<br>Oilseeds | Durrah | Fallow | Durrah |
| II | 3<br>4 | Herbs<br>Oilseeds | Durrah | Fallow | Durrah | Renewal<br>Wheat |
| III | 5<br>6 | Durrah | Fallow | Durrah | Renewal<br>Wheat | Herbs<br>Oilseeds |
| IV | 7<br>8 | Fallow | Durrah | Renewal<br>Wheat | Herbs<br>Oilseeds | Durrah |
| V | 9<br>10 | Durrah | Renewal<br>Wheat | Herbs<br>Oilseeds | Durrah | Fallow |

A. *RENEWAL*: Potato, corn, broad beans, beetroot.

B. *INDUSTRIAL*: Oil plants, cotton, flax, castor, groundnut, sunflower.

C. *HERBAL*: Fodder-crop.

D. *LEGUMINOUS*: Lentils (subject to prior manuring), chick-peas, beans.

(A,B,C,D depending on the outcome of the experimentation in progress)

TABLE 10. Crop rotation at Ente Puglia Farm
*Source:* [Giglio], 'Alcuni problemi' (1879).

pointed out that in an average family of eight the maximum working members would be five. Such a family could not cultivate effectively a farm of 50 ha without the aid of Ethiopians. But owing to the scarcity of local labour and, above all, the desire to obtain as much land as possible, the settlers were not allowed to engage Ethiopians. The agency made allowance for the gradual enlargement of the settler's plot either after a five-year period when the rest of his farm had become fully productive, or a part of it could even be given to the colonist's eldest son when he was ready to farm on his own.[135]

Another strategy was extensive use of sharecropping contracts with Ethiopians. This system, in addition to ensuring revenue, had political advantages as it was thought to favour peaceful penetration into areas where the Ethiopian population was dense.[136] Indeed, as indicated in land directly cultivated, the agency made generous use of Ethiopian labour.[137] But in no place do its records attest that it practised any form of sharecropping.

Construction of the colonists' dwellings proceeded hand in hand with the creation of the farms. But unlike the ONC and ECRE, the houses were scattered along the road, in pairs with 50 m between them. Facing these were another pair at a distance of 80–100 m, separated by a road. The initial plan was to build five-roomed stone houses with a central passage, a kitchen and small oven, and a lounge at the centre flanked at its right and left by two separate rooms plus storage. Eventually only two-bedroom houses were built with a lounge and kitchen. Unlike both the ONC and ECRE who built communal bakeries and wells, the ECPE equipped each farm with a baking oven with an overhanging henhouse and a well of its own, ostensibly to avoid bickering among the housewives. The farm's set-up, as a whole, was intended to reduce interfarm tracks, enhance close co-operation, and guarantee security.[138] At the end of 1940 about 120 houses were under construction, and only 27 were complete.[139] Their inadequacy in design and size soon became apparent. The oven with the overhanging henhouse made the rooms smoky, hot, and filthy. There were not enough rooms to accommodate large families, and storage for food and tools was lacking.[140]

But the ECPE should be credited for at least having settled a few families. The first 105 settlers, who embarked on 17 January 1938 from

[135] Ibid.; CD IAO Hr 1992, Situazione amministrativa, p. 2; 'Rassegna Agraria coloniale', *AC* 33/12 (1939), 687.

[136] CD IAO Hr 1992, Maugini, Promemoria.

[137] No data exists on the number of Ethiopians employed by the agency but records for 1939 and 1940 indicate that wages paid in 1939, including board, amounted to L.97,434.25 and those for up to 30 Nov. 1940 consisted of a total of L.54,411 (wages) and L.854 (board). Allowing that pay consisted of official L.2 per diem, it can be concluded that in 1939 *c.*130 and in 1940 *c.*78 Ethiopians were employed daily (cf. CD IAO Hr 1992; CD IAO ECPE, Bilancio, p. 14; Ufficio Presidenza, Conto costi e redditi, Rm 11 Apr. 1941).

[138] G. Giannoccaro, 'Prime tappe', 25–7; 'L'ECPE', in *Africa Italiana*, 7 (suppl.), 9 May 1939; *Colonie*, 7 May 1939; *GP*, 14 Jan. 1939; MAI, 'Valorizzazione', 281–4.

[139] CD IAO Hr 1992, Volpi, EPE: Relazione, Bari d'Etiopia, 20 Nov. 1940, p. 3.

[140] CD IAO Hr 1992, Situazione, p. 2; 'Rassegna Agraria Commerciale', *AC* 33/12 (1939), 687.

Brindisi, arrived at Wächo on 2 February. They came from Foggia, Lecce, Taranto, and Brindisi and the plan was that each settlement was to bear the name of these districts and to be inhabited by people of that area. More than a year later in June 1939 they were joined by an additional 99 settlers. The settlement was concentrated at Wächo and took the name of Bari di Ethiopia. Towards the end of 1940 it had a population of 276. About 80 were construction-agency workers, others were employed by the agency as mechanics and tractor-drivers. The rest were peasant settlers. Among these, 14 were with their families composed of 74 children aged between 1 and 18 years. In all, they were 104 individuals. There had been three deaths and two births among the settler families.[141]

These modest accomplishments were not matched on the farming side. In an attempt to overcome the prevailing uncertainties, the ECPE relied on diversified cultivation. With the exclusion of one ha that was to be occupied by ditches, wells, and building, the remaining 24 ha were partitioned into six sections of four ha each (see Table 10). Of these, five sections were set aside for rotating cereal crop cultivation. The type of crop subject to rotation each year varied according to climatic conditions as well as the crop cycle. In the remaining section, two ha were allocated for arboriculture and the other 2 ha for grassland. Out of 6,000 ha planned for cultivation in the first year, 1,100 ha were reclaimed and 650 ha sown. In the following year an additional 1,000 ha were reclaimed, but sowing remained confined to 200 ha. There was a substantial increase in land reclaimed in 1940, but land actually cultivated remained quite modest in size. Against 3,000 ha reclaimed, a mere *c.*260 ha were sown, 103 ha by the fourteen settler families, and 157 by the agency itself.[142]

The EDR's scheme was to farm large plots removing its occupants to a nearby area, far enough away to avoid social contact but close enough to guarantee a flow of labour supply. In this way the agency secured 2,500 ha. In the process 682 Ethiopian families were relocated and given compensation for their property, part in cash and part in kind, and 700 houses and 2,730 animal shelters were destroyed. Each family was given an average of 1.02 ha of land and compensation of L.366.5, with a total payment amounting to L.250,000. They were settled within 10 km from the farm.[143]

[141] Giannoccaro, 'Prime Tappe'; id., 'ECPE'; *Colonie*, 21 Nov. 1939; *CI*, 24 Jan., 21 June, and 25 June 1939; *Messagero*, 24 Jan. 1939; *PI*, 3 Apr. 1939. Accordingly they came from: Lecce (29), Bari (18), Taranto (18), Brindisi (17), Foggia (10). CD IAO Hr 1992, Volpi, Relazione, pp. 1–3, 6; 'Le prime due nascite a Bari d'Etiopia', *CM*, 21 Nov. 1939.

[142] [Giglio], 'Alcuni problemi', 1878.

[143] CD IAO: Gm 2919, Comprensorio di EDR, pp. 1–2; 1916, Maugini and Nistri, Rapporto, p. 13. Villa Santa *et al.*, *Amedeo*, 236.

In the construction of the houses and organization of the farms, the EDR followed the ECPE, but the labour code was quite different from other regional agencies. Owing to financial considerations, it excluded the salaried period and the peasant started with the redemptive phase. Its housing programme envisaged two types: thirty-five two-roomed houses for small families and five three-roomed houses for larger ones. Both had a kitchen and looked on to a vast porch. Some were isolated and others grouped in twos or threes, each within their own plot. Each group shared a communal oven and well.[144]

By the end of 1940 the EDR was able to settle forty-one families. Thirty came from Italy while the heads of the remaining eleven were chosen from Jimma among the veterans of the Italo-Ethiopian war. Twenty-five were from Bergamo, five from Abruzzi, and the rest mostly from central Italy.[145] The EDR's major shortcoming was in developing adequate land for the settlers. The agency targeted to provide each family with 35–40 ha. At the moment of settlement the family would receive 10 ha already reclaimed and ready for cultivation and the rest as a pasture and woodland. Actually, EDR developed a total of 120 ha—far short of its planned target.[146]

The CAG followed a quite different approach. Unlike the EDR, it left the Ethiopians on part of their original land, as their labour, particularly in heavy-duty jobs, was of vital importance.[147] But like the EDR, it introduced the settler and his family only after accommodation was ready and land sufficient enough to support them. Each settler received 5–6 ha of reclaimed land. Upon payment of a lump sum of between L.10–15 as part of its amortization, the settler was transferred on to the land. The settler assumed the status of an ordinary concessionaire and, with regard to the anticipations received by the way of housing expenses and start-up capital, his relation with the CAG was like that of a debtor *vis-à-vis* his creditor. As a result, the settler shouldered all the responsibility of the farm, including management. There was no time-limit for the discharging of debt, but he was expected to do so as soon as he could. In case of

[144] CD IAO Gm 2919, Comprensorio di EDR, pp. 2–4.

[145] The regions involved included Alessandria, Catanzaro, Cuneo, Littoria, Parma, Reggio Emilia, and Napoli. They came in two batches. The largest contingent arrived on 9 Dec. 1939 and was 162 strong. It included 120 children aged between 1½ months and 10 years. The second group came in Apr. 1940. (Cf. *Azione Coloniale*, 23 Nov. 1939; *CC*, 25 Dec. 1939; 'L'arrivo a Massaua dei coloni diretti nel Gimma', *Corriere Eritreo Sportivo*, 8 Apr. 1940; *GI*, 24 Nov. 1939; *IOM*, 20 Jan. 1940; *Mattino*, 23 Nov. 1939, 30 Mar. 1940; *Popolo di Sicilia*, 12 Dec. 1939; *PR*, 9 Apr. 1940; *Tribuna*, 12 Dec. 1939.)

[146] CD IAO: Gm 2919, Comprensorio di EDR, p. 5; GS 1822, Attività agricole, pp. 2–3. *CI*, 23 Nov. 1938; *GP*, 9 May 1940.

[147] CD IAO GS 1822, Attività agricole, pp. 4–5; Villa Santa *et al.*, *Amedeo*, 236.

extraordinary works, the CAG rented the necessary equipment. In this way, at the end of 1940 the CAG was operating on 700 ha and had settled fourteen families. The head families were selected locally, and each was provided with about 50 ha.[148]

The picture that emerges is that regional demographic colonization agencies' achievements in transplanting Italian rural life to rural Ethiopia were modest. The fruition of the scheme was thwarted by a number of factors, some of them common to the Eritrean and Libyan experience. The agencies suffered from administrative confusion and managerial incompetence and financial stringency. The conduct of some agency officials bewildered the colonists. Giannoccaro was known for his dictatorial behaviour, avarice, and irresponsible conduct towards the peasants; this was also the case with De Rege's authorities. As a whole, the leadership lacked clear vision and a sense of commitment.[149]

Equally important was the meddling of local organizations, ranging from the Fascist Party to the trade union and colonial administration, in the internal administration of the agencies. Each group expressed conflicting views on the plight of the settlers and on their administrators. For some critics, the agencies' sole purpose was to provide cushy jobs for a few Party and government officials, medical officers, and agricultural experts.[150] The agency officials naturally saw the settlers' growing grievances in a rather different light. They believed that many peasants were simply fortune-hunters who came to Africa lured by absurd promises made by irresponsible people, i.e. Party officials.[151]

Close examination of the settlers' background suggests that the selection procedure was far from ideal. The official view of the peasant settler was of a hard-working man and a father of a large family, aged between 25 and 40, an ex-fighter with 'robust physique, the Fascist spirit, and good character',[152] or as the Minister of Italian Africa described, 'simple and honest men attached to bountiful turf by deep-rooted racial instinct, . . . people of resolute heart and loyal who have rushed from the motherland to remote Africa and who in Roman style bring fertility to those lands barren for centuries'.[153]

[148] CD IAO GS 1822, Attività agricole, pp. 4–5; *Azione Coloniale*, 16 Nov. 1939; *CI*, 14 Apr. 1940; *GP*, 9 May 1940; *Mattino*, 24 Nov. 1939; *Messaggero*, 17 Nov. 1939; *Pattuglia*, 30 Dec. 1939; *PR*, 17 Apr. 1940.

[149] CD IAO: Hr 1992, Appunti per l'Ing. Giannoccaro, Fl 27 May 1940, pp. 1, 7; GS 1821, Esperienze.

[150] Pierotti, *vita*, 36–8; FO371/23380/J1776/296/1, CG Bird to Halifax, AA 1 Apr. 1939.

[151] CD IAO GS 1821, Esperienze.

[152] *The Sunday Times*, 23 Jan. 1938.

[153] Pomilio, 'Problemi e realtà', 39; *Colonie*, 28 Feb. 1939.

Few met such an idealistic description. A random survey of the settlers of the EDR and ECPE reveals the presence of social misfits and trouble-makers.[154] There were unskilled workers who jumped between jobs as miners, farm-labourers, and road workers, hoping to find in the Empire an easy life; pedlars and craftsmen who lacked elementary agricultural skills; former soldiers turned university students; adventurers who came to the Empire to escape domestic unemployment and accumulate fortune before returning home. A few had a farming background but before they came to the Empire, they had changed to non-farming employment. There were quinquagenarian men and women whose ability to be successful farmers was questionable.[155]

An overview of the health of many belie repeated claims of strict vetting. Out of 42 repatriations to Italy by the ECPE, 28 were for ill-health.[156] A number of EDR settlers had tuberculosis and heart trouble.[157] The higher authorities, embarassed by these developments, appealed for strict vetting.[158] Most of the settlers were far removed from being the ideal Italian Mussolini depicted—ready to 'Believe, Obey, and Fight'.[159] Many were repatriated for indiscipline.[160] On occasion troublesome families successfully exploited the competing authorities and ignored the agency's rule with impunity.[161]

Excellent farmers were in the minority. Of course, all of them had a biblical vision of Africa where land and food were in abundance.[162]

[154] The technical officers were unequivocal in their complaints about these groups. The case in point is Giuseppe Brescia's family, described as 'lacking any dignity, apathetic, indolent, and of poor morals' (CD IAO Hr 1992, Volpi, Relazione, 20 Nov. 1940, p. 6).

[155] *Azione Coloniale*, 30 Nov. 1939; *PI*, 24 Nov. 1939; *CE*, 3 Dec. 1939; CD IAO Hr 1992, Volpi, Relazione, 20 Nov. 1940, pp. 6–7.

[156] These included chronic bronchitis, dyspepsia, haemorrhoids, rheumatism, chronic fever, organic depletion (cf. CD IAO 1992: Relazione Ing. Giannoccaro, 24 Mar. 1940; Volpi, Relazione, 20 Nov. 1940, pp. 4, 6).

[157] CD IAO GS 1821, Esperienze.

[158] ACS ONC 15, Carlo Severini to Giuseppe Lombrassa (Commissario Generale per le Immigrazioni e la Colonizzazione Interna), Rm 4 July 1940.

[159] ACS GRA 45/41/5, Notiziario Radio [suppl.], 29 Sept. 1937.

[160] Out of 105 settlers of ECPE, 42 were repatriated, 12 of them for indiscipline. CD IAO Hr 1992: Relazione Ing. Giannoccaro, 24 Mar. 1940; and Volpi, Relazione; CD IAO GS 1821, Esperienze.

[161] What happened to the authorities of EDR is one of many cases that stands out to illustrate the constantly growing colonist's psychology. In Jan. 1940 the supply of meat for the settlers was delayed by one day and the arrival, clearly following complaints by the colonists, of the local Fascist Party leader prompted one of the colonists to remark to the management: 'Ah well, here is the Federale. Run to do as you are expected!' (CD IAO GS 1821, Esperienzes).

[162] They migrated because they saw better prospects in the Empire. Their enthusiasm shines through in their interviews, obviously written for propaganda purposes. (Cf. *PI*, 24 Nov. 1939. Luxuriant growth of the first harvest prompts an elderly peasant to make a

Disillusionment came with hard work, harsh climate, privations of all kinds, and squalid living conditions. The optimism and confidence which had characterized their journey to Africa evaporated once they came in contact with reality—a sharp contrast to the fanfare that marked their trips and was glowingly reported by the media. It was routine before leaving as pioneers of the 'second Roman Empire' for the colonists to be lavishly entertained. Equipped with the khaki cotton uniform of the Fascist colonial militia and supplied with some farm necessities including seed grain, they paraded chanting 'Giovinezza' ('youth') and at each stop on their journey were greeted by enthusiastic crowds meticulously organized by Party cadres and government officials.[163] The glowing picture which the leadership painted, combined with the ease and speed of the Italian victory over Ethiopia, lent weight that these promises could be realized.

To some extent, the settler was cushioned from the full rigours of frontier life by the facilities and services provided for him. He paid no tax, and received free medical care. Yet the colonial life was also full of problems. There was the sluggish bureaucracy that distributed necessary tools and supplies only after long delays. Others found the materials and livestock delivered inadequate or inappropriate. Where the grain was distributed, much of it had been ruined by poor storage. As in the case with the ECPE, food was insufficient and poor nutritionally, and accommodation depressing. The settler, originally in good spirits, was demoralized by bad weather, drought, and insects that plagued his crops. He lived in constant fear of being attacked by Ethiopians.[164]

Already by the end of 1939 the problem of recruitment had become acute.[165] Indeed, the settlement had gradually lost the popular enthusiasm they had at the initial period of the conquest. During this same period the Italian authorities had begun to express doubts as to the viability of their colonization techniques. The achievements were mediocre but the costs were enormous. How much the agencies spent on their colonization projects during the years 1937–41 is not known. But the Empire contributed only a small fraction of the expense. Out of a total budget of

remark for a visiting Duke of Aosta: 'It seems to me like being in the promised land. The potato is harvested three times. Even the most wretched grass has a height of no less than three metres.' (Berretta, *Amedeo*, 182.)

[163] *The Sunday Times*, 23 Jan. 1938.

[164] Reports are full of locust invasion and drought that brought in havoc in the settler's as well as the agency's confidence. See *Daily Telegraph*, 21 Feb. 1939.

[165] According to the agencies, the shortage was due to large settlements that took place at the Pontine Marshes and Libya. See CD IAO 1921, Programma 1940, p. 4.

TABLE 11. Land allocated and developed by demographic agencies (ha)

| Agency | Land | | | | | | | |
|---|---|---|---|---|---|---|---|---|
| | Planned | Granted | | Reclaimed | | Cultivated[a] | | |
| | | 1938 | 1939 | 1940 | 1941 | 1938 | 1939 | 1940 |
| ECRE | 60,000 | 5,600 | 380 | 1,013 | — | 148[b] | 1,227 | 2,431 |
| ECPE | 60,000 | 6,000 | 1,100 | 1,100 | 3,000 | 602 | — | 261 |
| EDR | 2,350 | 1,600 | — | — | — | — | — | 160 |
| CAG | 700 | 700 | — | — | — | — | — | 65 |
| TOTAL | 123,050 | 13,900 | 1,480 | 2,113 | 3,000 | 750 | 1,227 | 2,917 |

[a] No figures available for 1941.

[b] Inclusive of lands worked by the Ethiopians under sharecropping.

*Source:* CD IAO 1921: Fuzzi, Relazione del Presidente, Rm 7 Feb. 1939, pp. 5, 25–9, 48; Fuzzi to G Am DCL, Dabat 2 Sept. 1940; Fuzzi to Maugini, AA 16 Jan. 1941. Giannoccaro, *ECPE*, 19; CD IAO RGGS UA, Attività svolta nei vari settori dalla costituzione dell'Ufficio fino ad oggi, Gm, Jan. 1940, p. 2; Villa Santa, *et al.*, *Amedeo*, 2.

L.50,000,000 allocated to each of them, the ECRE had spent more than L.35,000,000 and the ECPE about L.29,000,000.[166]

Hopes and speculations vastly exceeded actual results. The enterprise which had begun earlier in a mood of optimism had lately become a more sober affair, beset by incertitude and anxieties. Demographic colonization was no longer regarded as an effective way to solve Italy's population problem. The scheme was questioned, and in its place the capitalist methods, so strongly advocated by the Duke of Aosta, began to gain wider currency.

[166] CD IAO 1792, Caravalli and Zampighi, Lineamenti, p. 1; CD IAO 1793, Tommasini, ECRE: Promemoria, pp. 1–2; CD IAO 1992, Bilancio patrimoniale, 30 Dec. 1940; Sbacchi, 'Italian Colonialism', 442–6.

# 6
# *Commercial Farming*

## THE POLICY IN ITS HISTORICAL PERSPECTIVE

Commercial farming was introduced into the overall plan for Italian colonization of Ethiopia in response to the pressing needs of a sudden conquest of an empire that was remote from the motherland. The chief challenge was presented by the influx of Italian military and civilians, dramatically increasing consumption demands. As locally available resources were quantitatively and qualitatively inadequate, practically everything had to be imported, and much of it purchased from non-Italian sources. This imposed a steadily increasing burden on the already strained Italian finances and became economically unacceptable.

Strategic considerations also called for a contingency plan that would guarantee the Empire's survival in the event of an internal uprising or a major war implicating Italy. In case of external aggression significant contributions towards such an objective were expected from demographic colonization and native agriculture. But an exclusive reliance on both these sectors was judged economically unwise. Native farming was thought to require rapid transformation in order to meet production targets, achievable only if age-old habits were discarded. It would be years, perhaps centuries, before the Ethiopians could grow cash crops and participate effectively in an exchange economy.[1] However, in the event of widespread internal insurgency, dependence on the native production was seen as a liability.[2] Equally, a great deal of time, finance, and organization was needed before demographic colonization could be sufficiently established.[3]

In the face of such shortcomings and determined to make self-sufficiency in foodstuffs the first objective in the economic development of IEA, the

[1] CD IAO AOI 1854, Lessona to GG AOI, Direttive, p. 2; ATdR 24/99, De Benedictis, Relazione sull'autonomia alimentare dell'Impero, Nov. 1937.

[2] P. Bono, 'L'impresa agricola privata nella colonizzazione dell'AOI', *GAD*, 13 Nov. 1938; *CI*, 11 Sept. 1938.

[3] CD IAO AOI 1990, Tassinari to Duce, pp. 7–9; ATdR 24/100, Relazione Maugini, pp. 14–15.

Italian government pursued an agrarian policy based on commercialization of agriculture, and encouragement of agrarian capitalism. The key objective was import substitution both in the colony and the metropolis. As Mussolini summed up, 'in the battle for the attainment of autarky both the territories of the metropolis and the Empire formed an inextricable unit'. This was within the framework of Italy's view of the Empire as a natural extension of the metropolis.[4] In the light of this, commercial farms were expected to fulfil twin objectives[5] which Lessona summed up as follows:

What must be clear from the start is the fact that the fundamental objective in the economic field must be *to secure the self-sufficiency of IEA in the shortest possible time and in the most comprehensive forms that can be effected.* Alongside with this but in as far as it does not conflict with the attainment of this primary objective, *East Africa must develop those trends of overseas trade predating our occupation and be able to produce raw materials that our country needs.*[6]

The need to rely on this sector became even more compelling when Mussolini set the end of 1938 as the deadline after which any kind of metropolitan-foodstuff subsidy to the Empire would cease.[7] Since early 1936 there had been an understanding within the Italian colonial administration that such an objective was not beyond reach. An annual production of between 800,000 and 1,000,000 ql of foodstuff was calculated to be sufficient to free the Italian settlers and their local collaborators from dependence on imported food and on Ethiopian peasant production.[8] This assessment optimistically predicted that the commercial farms, aided by demographic colonization, would bring into cultivation *c.*20,000 ha per

[4] *CE*, 11 Nov. 1938; *CC*, 25 Nov. 1940; *IC*, 19 Jan. 1940.

[5] CD IAO AOI 1936, Promemoria per S.E. Il Capo del Governo, Agricoltura e colonizzazione nell' AOI, Rm Nov. 1937; ATdR 24/99, De Benedictis, Relazione. Autarky was a theme recurrent throughout the occupation. Cf. 'La politica autarchica e le sue ripercussioni sui traffici commerciali', *EI* (July 1938); 'L'Impero nell'economia di guerra', *EI* (May 1940); 'L'AOI nel quadro dell'autarchia nazionale', *II* (Apr. 1940).

[6] CD IAO AOI 1854, pp. 1–2; FO371/24635/J412/18/1, Sir P. Loraine to Visc. Halifax, Rm 29 Jan. 1940, 'Notiziario agricolo commerciale: AOI', *AC* 32/3 (1938), 136.

[7] ACS GRA 44/36–2, Mussolini to Graziani, Rm 13 May 1937, and Rm 30 July 1937. Mussolini's deadline, originally set to 1 July 1938, was later brought towards the end of that year. The immediate effect of the order was the rise of feverish activities that were largely disorganized and abortive. The most significant among these was the formation of a Civil Mobilization Committee and Co-ordinated Supreme Defence Commission whose main task was to put the whole country on a war footing and requisition foodstuffs; yet both these institutions existed only on paper. Cf. ibid., AA 20 May 1937.

[8] FO371/23380/J1776/296/1, Bird to Halifax, AA 1 Apr. 1939, p. 7.

year. With such a pace of development it was thought that autarky would be a reality within a five-year period.[9]

The development of commercial farming was complicated by the struggle between the MC in Rome and the colonial administration in Ethiopia. The conflict was reflected in a wide range of issues concerning the type of farms that should be encouraged, the areas in which land should be granted, and the mechanisms by which this should be accomplished. However, no well-defined position existed as policies and attitudes shifted with changes of personalities. In the early stages, specifically under the government of Graziani, the Ministry's enthusiasm for large-scale commercial farms was viewed inside the Empire with some apprehension. The main focus of the regime within Ethiopia was on the industrious small cultivator; its belief was that the Empire should be an open field of activity for all Italians and not for a privileged group, whether private capitalists or corporate bodies. Underlying such a position was the fear that the encroachment of large companies would lead in the long term to a *de facto* monopoly. The outcome would be the concentration of land and agricultural products in the hands of a few companies at the expense of the small-scale investor and of the organic and healthy development of the Empire's agriculture.[10]

The Ministry's argument rested on a realistic appraisal of the socio-economic condition of the Empire. The most significant consideration was that, as a backward economy on the extreme periphery of the world capitalist system, the Empire needed a massive infusion of capital and infrastructure. But the economic resources at the command of the State were meagre and skilled personnel scarce. Therefore, its ability to carry out such enormous pioneering work was lamentably limited.[11] In the light of this, the Ministry found the expediency of partnership with big farms the only viable alternative. Underlying this belief was the view that, with sufficient backing in Italy, the companies should be able to invest enough capital in their holding, and furnish the necessary technical knowledge and managerial skills to ensure that the basis of agricultural development was properly laid.[12]

[9] CD IAO AOI 1774, G. Pini, Attività economica dei nazionali in AOI prima ed all'inizio della guerra—febbraio, 1941 AA 15 Dec. 1941, p. 2. In fact, the MAI Terruzzi claimed that 700,000 ql would be produced in 1939. Cf. Pomilio, 'Problemi e realtà', 39.

[10] Villa Santa *et at.*, *Amedeo*, 221–2.

[11] Del Boca, *Caduta*, 144–5.

[12] CD IAO AOI 1936, DGACL (Lessona) to GG Produzione di frumento per fabbisogno AOI, Rm 12 Oct. 1937, p. 5. According to Lessona private initiative would bring three advantages to the economy: (1) Ensure the transfer of sizeable agricultural machinery. (2)

With this approach the Ministry aimed to use large companies at least as a time-buying strategy until the basic organizational and developmental infrastructures were laid down. But its position appeared rather baffling, as in the mean time it reiterated its commitment to upholding the paramountcy of the social goals of colonization. Thus, according to the Ministry's plan a number of small- and medium-sized private farms whose primary scope was to populate the land were to be promoted alongside large-scale commercial farms. Families with adequate skills and capital were to be encouraged to take up farming.[13] Tight control by the government, it was thought, would ensure promotion of such mass settlement initiatives and curb the eventual concentration of land in the hands of a few commercial farms. The strategies used for such control were mainly careful processing of applications for land concessions and imposition of stringent conditions for the development of the lands. As we will explore later in this chapter, such a system strengthened the hands of bureaucracy. It became impossible to get anything done without going to innumerable offices in the capital of the respective governorate who all referred the matter to their superior in Rome. From the point of anti-monopolistic control and promotion of genuine investors, the measures also proved inadequate. Companies and individuals obtained generous land grants for their past services and, in some governorates, a system of extravagant land concessions was taking place which, in due course, would have encouraged speculative interest.

As initially envisaged, commercial farms were to be confined to a few carefully selected areas unsuitable for settlement but economically viable. Viability was measured in terms of the presence of abundant labour and the potential for large returns for capital invested.[14] But the plan was never translated into action and most commercial farms operated close to demographic colonization. In fact, the promoters of the scheme failed to see that from the start the plan was wrought with internal inconsistencies. The areas that were singled out included such scattered and remote regions as those surrounding Addis Ababa to the Awash River basin in the South and South-West, from Érär Valley and the JeJega planes in Härär in the South-East, up to the planes of Qobbo and Qorbätta in the far North.[15]

Enable the transfer of experienced and skilled farmers with their own equipment to start agricultural activity immediately. (3) Furnish useful experience in the relationship between commercial farms and local native farmers.

[13] Villa Santa *et al.*, *Amedeo*, 221–2.

[14] CD IAO AOI 1936, Lessona to GG IA, Rm 9 June 1937.

[15] Ibid.; ATdR 24/100, Relazione Maugini, p. 17.

But more importantly, some of these areas were beyond the Italian authorities' immediate political grip. Secondly, an abundance of vacant land did not necessarily mean that the land was conducive to commercial farming. In fact, as the Italians themselves discovered, labour tended to be extremely scarce, the climate unhealthy and communication difficult. In addition, despite their public pronouncements the political authorities in Rome and Ethiopia made only spasmodic attempts to implement the policy and were more conspicuous in undermining than encouraging it. As we will see, the most glaring example was the allowance for a governor to issue certain concessions in his respective territory.[16] In some governorates this policy was pursued vigorously to the chagrin of the Viceroy's government in Addis Ababa which was often forced to intervene to correct excessive abuses.[17]

## ORGANIZATIONAL PATTERN

At the eve of Ethiopian Liberation in 1941, there were farms of all dimensions and combinations differing in structure, organization and orientation. Towards early 1939 these were divided into three categories.[18]

1. There were small homestead farms developed by farmers with modest starting-up capital. As in demographic colonization these were largely composed of individuals entitled to the nation's gratitude who, in a sense, were social casualties: war veterans, particularly the demobilized soldiers of the Ethiopian campaign, fathers of large families, and unemployed workers. Each farm ranged between 1 and 60 ha.[19] Located at the outskirts of an urban centre, the farms had as their immediate objective the provision of fresh vegetables and fruits to urban populations and military garrisons. However, like demographic colonization their ultimate aim was mass settlement and because of this they were treated favourably. But unlike demographic colonization, the concessions were given directly to a head of a household and not to a purposely formed State-funded agency.[20]

[16] CD IAO AOI 1936, Graziani to MC, AA 14 Apr. 1937; Graziani to GG AOI, AA 9 May 1937; Lessona to Graziani, Rm 14 May 1937.

[17] ACS GRA 43/34 22: Graziani to Lessona, AA 2 Nov. 1937; Lessona to Graziani, Rm 4 Nov. 1937. CD IAO AOI 1936: GG AOI (Petretti) to MAI, AA 25 Sept. 1937; GG AOI (Petretti) to G Hr, AA 30 Sept. 1937. CD IAO AOI 1926, Terruzzi to Duce, 27 Apr. 1938.

[18] The division roughly reflected Lessona's ideas as developed in his master plan of 5 Aug. 1936, where he envisaged four types of colonization (cf. Ch. 3).

[19] According to some sources they covered an area between 1 and 15 ha, while according to others they consisted of between 1 and 50 ha.

[20] MAI, 'Valorizzazione', p. 210; *La Stampa*, 9 Aug. 1940.

2. Medium-sized farms included three distinct forms of concessions:

(*a*) 'Concessions directed towards agricultural development with or without obligation requiring peasant land settlement;'

(*b*) Agro-pastoral;

(*c*) Forestry.

The first group of medium-sized farms (2a) were divided into first and second categories. The first was oriented to mass land settlement. In this case the land was granted as either a concession transferable into a definite ownership upon the payment of, *in toto* or *pro rata*, the price of the land, or permanent concessions pending discharge of specific obligations. The size of the land ranged from 61 to 500 ha and its price was assessed at market value; and so, unlike small-scale commercial farming, the concessions were not subject to special treatment. But like small-scale concessions, they were given to the head of a household. In the second category, the lowest holding was 500 ha and the maximum 2,000 ha. Yet cases of 5,000 ha were not uncommon. An element common to this and other categories of concessions was that the land was not transferable into permanent ownership; they were leased for a thirty-year period and renewable for a maximum of ninety-nine years.

The holdings of the agro-pastoral concessions were of variable size with a maximum limit of 10,000 ha. The forestry concessions included companies charged with either reforestation or exploitation of existing forestry resources and their by-products. As for size, reforestation entailed an upper ceiling of 5,000 ha and for the first twenty years was rent-free. For the concessions connected with forestry exploitation, there were no clear conditions set on the size, the duration, or lease. Both size and lease were computed on 'the value of the wood and secondary products, putatively extractable from the forest'. Upon such evaluation also depended the time-scale which, however, 'will not in any event exceed a twenty-year period'. Medium-sized farms originally were made up of those members of the CFA who had as their mission imperial self-sufficiency in food-crops, particularly wheat. Non-members, for whom wheat was optional, engaged themselves in other crops assumed to be more lucrative and less risky. This development, combined with the continuous failure of wheat, forced the CFA at a later stage to diversify production and involve themselves in farming of industrial crops.[21]

3. Large-scale concessions were to produce industrial crops such as cotton, sugar cane, indiarubber, coffee, tea, oilseeds, and textile fibre,

[21] MAI, 'Valorizzazione', p. 210; *Messaggero*, 2 Nov. 1937.

with a twofold target: to provide the raw material required by the metropolitan industries and to conquer the foreign market. Most of the farms were *ad hoc* creations, and some were subsidiaries of large metropolitan companies. The companies held concessions of variable sizes, with each holding extending between 500 and 30,000 ha. Practically, they were beyond the direct control of local authorities as they were directly accountable to the MAI. This arrangement was a frequent cause of political friction between local authorities and companies.

Among large-scale farms the most important were the Cotton Districts, controlled by the ECAI. This agency was established on 7 January 1937 by Royal Decree with a capital of L.25,000,000 partly paid from the State Treasury and partly by the ICI. Its task was 'to oversee the development, regulation, and exploitation of cotton in IEA'. Within a year the agency had set forth its programme, establishing a number of Cotton Districts, each measuring on average one million hectares. The underlying concept of the organization of these districts was to boost the production of cotton in preference to other cultivations.[22] Among other large companies worthy of consideration were SIMBA which operated in Härär governorate—mainly Asälla, Fadis, Villa Baka—and Società Nuova Africa. These two companies employed considerable capital in both men and material.[23]

This threefold division, though inadequate, was considered to meet the existing administrative conditions and serve as point of reference for future concessions. But the Italians were not able to give theoretical and practical justification of such categorization.

Three censuses of 1938, 1939, and 1940 summarily deal with the status of commercial farms in each governorate.[24] Like any survey, these censuses are limited in their range, chronology, and consistency of data,[25] but they are valuable for the fragmentary data they give. A remarkable common

[22] For a more comprehensive discussion on Cotton Districts and bibliographical ref. see Ch. 7.

[23] MAI, 'Valorizzazione', 210; G. Pesce, 'Disciplina giuridica delle concessioni agrarie: Gli obblighi e le facoltà del concessionario', *Azione Coloniale*, 7 July 1939; G. Mondaini, 'Tradizione e innovazione nell'incipiente ordinamento fondiario dell'AOI', *Georgofili*, 7/7 (1941), 378–83.

[24] CD IAO: AOI 1808, Am UA, Relazione sulla colonizzazione e i servizi agrari dell'Am, Gn June 1940; GS 1822, Attività agricole; Hr 1813, L'attuale organizzazione dei servizi dell'agricoltura nel G Hr all'inizio del 1940, n.d; Sc 1802 (this document has no title and contains an incomplete list of concessions up to 1939); Sc 1801, Attuale.

[25] Owing to these limitations, it is not possible to give an exact picture of the annual rate of expansion in each province, the activities engaged, or the amount of capital at the concessionaries' command.

feature is the staggering growth of commercial farms and their steadily increasing concentration in southern regions.

When considering the difficulties under which they operated, the pace of development was rapid, even if far below official expectations. Towards the end of the 1940s there were approximately 224 farms holding about 166,639 ha. According to some sources, 50 per cent of these were engaged in cattle-breeding, 38 per cent in cereal production, 8 per cent in industrial cash crop, and 4 per cent in banana- and coffee-production.[26] In terms of their size, about 61 per cent were made up of small farms which owned over 2 per cent of the total land area. Medium-sized farms consisted of 35 per cent, holding a total of 21 per cent in terms of land area. The remaining 77 per cent of the area belonged to large concessions which formed only 4 per cent of the total farms. The combined holding of mass-settlement-oriented farms—small and medium first category—totalled 16,478 ha or about 10 per cent of the total land under commercial farms. Thus most of the land was concentrated in the hands of large companies (see Tables 12 and 13). This reveals Rome's claim that monopolistic tendencies would be tightly controlled to be a hollow pronouncement.[27]

The census throws additional light on geographical aspects of commercial farming. Regional distribution showed that most farms were centred at Härärgé and Shäwa Provinces. Despite their great agricultural potential, commercial interest in Sidamo and Käffa developed late, whereas in Illubabor, Wälläga, and Gämu Gofa it was conspicuously absent. A number of reasons contributed to this. Primarily, the effective occupation of the regions took place comparatively late. Illubabor, Wälläga, and Gämu Gofa were isolated by lack of communications, and Sidamo and Käffa were the last to have a motorway to Addis Ababa. Another deterrent was their distance from the port of Meŝwa, 1,600–2,000 km away. Even though the opening of the Asäb–Däsé road reduced such distance substantially, it still remained considerable (1,200–1,600 km). This weighed heavily on the transport costs of machinery, seed, and other products, and encouraged many farmers to settle in agriculturally less favourable but otherwise more convenient areas.[28] The situation was aiding to make Sidamo and

[26] The data do not clarify whether these figures are inclusive of the activities in the old colonies. Cf. *CC*, 25 Mar 1940. For somewhat inflated official figures, see 'Rassegna Agraria Coloniale', AC 36/5 (1942), 113.

[27] M. Pomilio, 'I problemi attuali dell'Impero nel pensiero di Attilio Terruzzi', *Alimentazione Italiana* (13–31 Aug. 1938).

[28] CD IAO: GS 1985, Attivitá agricole esistenti nella regione di GG al 1939, n.d. pp. 1–2; GS 1822, Attivitá agricole, pp. 1–2.

TABLE 12. Commercial concession areas, 1941

| Territory | Area | | | | | |
|---|---|---|---|---|---|---|
| | 0–60 (ha) | 61–500 (ha) | 501–2,000 (ha) | 2,000 (ha) | Total | %age |
| Shäwa | 550.5 | 3,096 | 10,110 | 4,500 | 18,256.5 | 10.9 |
| Härärgé | 1,185.7 | 2,620 | 3,600 | 40,000 | 47,405.7 | 28.4 |
| Arsi | 606 | 300 | 700 | 13,000 | 14,606 | 8.8 |
| Balé | — | — | 4,000 | 7,500 | 11,500 | 6.9 |
| Bägémder | 515 | 255 | — | | 770 | 0.5 |
| Wällo | 291 | 2,134 | 2,200 | — | 4,625 | 2.8 |
| Käffa | 270 | 3,100 | — | — | 31,370 | 18.9 |
| Sidamo | 72 | 760 | — | 28,000 | 35,832 | 21.5 |
| Tegray[a] | 152.2 | 571 | 1,550.7 | 35,000 | 2,273.9 | 1.3 |
| Wälläga | — | — | — | — | | — |
| | | | | — | | |
| TOTAL | 3,642.4 | 12,836 | 22,160.7 | 128,000 | 166,639.1 | |

*Note:* Figures are elaborated by the author.

[a] Figures available only for 1939.

*Source:* CD IAO: Sc 1801; AOI Sc 1802; 1808, Am UA, Relazione sulla colonizzazione e servizi agrari dell'Amara, Gn June 1940; Hr 1813, L'attuale organizzazione dei servizi dell'agricoltura nel governo del Hr all'inizio 1940, n.d.; GS 1822, RGGS UA, Attività agricole esistenti nel territorio dei GGS, Jan. 1940; 1984, Elenco delle aziende agricole e dei piccoli coloni di Asella, n.d.

TABLE 13. Concessionaires according to their farm size

| Territory | Concessionaires | | | | | |
|---|---|---|---|---|---|---|
| | Farm size (ha) | | | | Total | %age |
| | 0–60 | 61–500 | 501–2,000 | 2,000+ | | |
| Shäwa | 19 | 15 | 9 | 2 | 45 | 20.1 |
| Härärgé | 45 | 13 | 3 | 2 | 63 | 28.1 |
| Arsi | 15 | 1 | 1 | 2 | 19 | 8.5 |
| Balé | — | — | 3 | 2 | 5 | 2.2 |
| Bägémder | 23 | 2 | — | — | 25 | 11.2 |
| Wällo | 13 | 12 | 2 | — | 27 | 12.1 |
| Käffa | 7 | 10 | — | 1 | 18 | 8.0 |
| Sidamo | 5 | 3 | — | 1 | 9 | 4.0 |
| Tegray[a] | 9 | 2 | 2 | — | 13 | 5.8 |
| Wälläga | — | — | — | — | — | — |
| TOTAL | 136 | 58 | 20 | 10 | 224 | |

*Note:* Figures are exclusive of Cotton Districts.

[a] Only 1939 figures are available.

*Source:* CD IAO: Sc 180, Attuale; Sc 1802, AOI Concessions; AOI 1808 Am UA Relazione; Hr 1813, L'attuale organizzazione; GS 1822, Attività agricole.

Kaffa the field of monopolistic concessions which, unlike small farms, had resources to absorb these difficulties. Despite the predominance of small- and medium-sized farms, in Härärgé and Shäwa there was a relatively balanced presence of all forms of enterprises (see Tables 12 and 13).

The geographical distribution of the farms reflected concern for economic viability and political conditions of the regions. Almost all the concessions were situated along highway routes and in areas with relatively abundant water resources and at 2,000 m above sea level where there was little threat from malaria or endemic diseases. Economically, proximity to a highway meant easy access to markets was guaranteed and transport costs reduced; politically, an Italian presence along highways was thought to ensure their continuous functioning and, in an emergency, military assistance could be easily provided.

TABLE 14. Tractors owned by various groups in IEA, 1940

| Governorate | No. of Tractors owned | | | | | | | |
|---|---|---|---|---|---|---|---|---|
| | Agricultural offices | Demographic colonization agencies | | | | | Commercial farms | Total |
| | | ONC | ECRE | ECPE | EDR | CAG | | |
| Amara | 3 | — | 14 | — | — | — | 22 (14 owners) | 39 |
| GS | [a] | — | — | — | 7 | 2 | 27 (10 owners) | 35 |
| Härär | 38 | — | — | 15 | — | — | 71 (14 owners) | 124 |
| Shäwa | 4 | 44 | 7 | — | — | — | 85 (26 owners) | 140 |
| Eritrea | 1 | — | — | — | — | — | 21[b] (9 owners) | 22 |
| Somalia | | — | — | — | — | — | 344 (3 owners) | 344 |
| TOTAL | 46 | 44 | 21 | 15 | 7[c] | 2 | 570 (76 owners) | 704 |

[a] Tractors at its disposal are those same indicated under EDR.
[b] Includes Tegray where two companies had 7 tractors.
[c] Inclusive of companies of Cotton District.

*Source:* CD IAO: 1800, Promemoria per Maugini; Gm 2919, Comprensorio di EDR; Hr 1813, L'attuale organizzazione, p. 8. (Totals as in the original.)

TABLE 15. Ambo: Annual production at a medium-sized farm (*Scagliarini Roberto & Sons*)

| Year | Cultivated areas (ha) | Production (ql) |
|---|---|---|
| 1938–9 | 150 | 1,000 |
| 1939–40 | 350 | 3,000 |
| 1940–1 | 300 | 300 |

*Source:* CD IAO Documenti AOI, Concessione agricola 'Scagliarini & Figli'—Ambo, Rm 8 Mar. 1941. Scagliarini had 2 Lanz tractors of 45HP, 2 of 35HP, Balilla tractor of 16HP, and a substantial array of implements. Cf. CD IAO: AOI 1936, Macchine ed attrezzi agricoli per la lavorazione del terreno esistenti presso le aziende agricole costituite nella zona di Ambo, AA 11 Apr. 1938; AOI 1800.

In Härärgé, Sidamo, Käffa, and Arsi, where more peaceful circumstances prevailed, farms were scattered over a relatively wide area. Where the political situation was precarious, they were concentrated in a few areas: on the high plateau of Mount Säqälti, between 15 and 30 km from Gondär in Bägémder; Boruméda and Čaffa in Wällo; Mäqälé and Adwa in Tegray.[29] In Shäwa, they were restricted to the West and South-West along the Addis Ababa–Jimma and Addis Ababa–Härär motorways and along the belt of Awash River basin. The vast majority operated at Ambo where they represented 24 per cent of the total farmers holding more than 50 per cent of the total land area under commercial farming in the region. The dispersion showed inconsistencies in the government's policy which, as we mentioned earlier, aimed to channel commercial farming to a few carefully selected areas that were unsuitable for mass settlement.[30]

The concessions were expected to develop primarily as enclaves, independent of traditional Ethiopian agricultural farming techniques. Indeed farms like those of SIMBA and, to a lesser degree a number of those belonging to CFA members, were considerably mechanized. In addition to an extensive array of agricultural implements, SIMBA had 30 tractors

[29] CD IAO: Ex AOI Gn Am 600, Relazione sull'attività svolta dalla sezione della colonizzazione dalla data dell'occupazione al 28-2-1939 pp. 1–3; GS 1820, IL problema della colonizzazione capitalistica, nei GS, Jan. 1940, p. 6; GS 1822, Attività agricole, pp. 5–14. *CI*, 27 Dec. 1938; *Colonie*, 14 Dec. 1939; 'Rassegna agraria coloniale,' *AC* 33/7 (1939), 446–7.

[30] More importantly, the developments particularly in these regions had a historical significance that went beyond the period of occupation. In fact, it was largely in these same areas that the phenomena of commercial capitalism appeared in 1960s and 1970s.

TABLE 16. Ambo: annual production at a medium-sized farm (*Bisacchi Ugo*)

| Year[a] | Cultivated area | Production (L.s) | | |
|---|---|---|---|---|
| | | Revenue | Cost[b] | Deficit |
| 1938–9 | 180 | 218,623 | 673,560 | 455,937 |
| 1939–40 | 425 | 738,195 | 1,363,571 | 625,376 |
| 1940–1 | n.a. | [c] | [c] | [d] |

[a] The agricultural year runs from 1 Apr. to 31 Mar.
[b] The cost does not include capital investment. For more detailed information see Table 17.
[c] The report states that the production was much lower than that of 1939–40 because of an unexpected smut attack just before harvest.
[d] The concessionaire claims that the deficit is higher than that of 1939–40 but does not give details.

*Source:* CD IAO AOI 1837, Azienda agricola Bisacchi Ugo, Comprensorio di Ambo, n.d.

(see Table 19) and the 21 farmers of CFA had 85 tractors (see Table 20) between them. But according to a census carried out at the end of 1939, there was a total of *c*.210 tractors, and of 224 farmers only 65 had a tractor (see Table 14). Over 70 per cent of the concessionaires, who had no modern agricultural machinery, relied totally or partially either on traditional methods of farming, i.e. use of animals[31] or a few hired tractors,[32] or were dependent upon government assistance.[33]

[31] Some of the concessions operating under such a method produced exceptionally good yields, as was the case with the concession of De Zorzi in Käffa where an average production of between 15 and 20 ql per ha was obtained. Cf. CD IAO GS 1822, Attività agricole, p. 11.

[32] The government tractor-hiring scheme was well organized only in Härär where the UA, in addition to a large variety of agricultural implements, possessed 38 tractors, 13 threshers, and a fully equipped repair and maintenance workshop. In other regions, such facilities were uncommon at either private or governmental level. Although the press reported the formation of a special body to hire out agricultural machinery, it does not appear that such a body became operational (cf. CD IAO Hr 1813, L'attuale organizzazione, pp. 8–11; annex 1, G Hr DCL IA, Norme per il noleggio di macchine agricole agli agricoltori; D Gov Hr 13-7-1938, no. 439, Noleggio di macchine ed attrezzi agricoli dell'amministrazione a favore degli agricoltori, in *GU* 3/17, 1 Sept. 1938; *Colonie*, 24 May 1938; FO371/22020/J371/40/1, Lord Perth to FO, Rm 28 Jan. 1938).

[33] This was particularly the case with small farmers around Gondär who had the initial ploughing work done gratis by the tractors of the UA (cf. CD IAO Ex AOI Gn Am 600, Relazione sull'attività, p. 6).

TABLE 17. Cost of production and revenue

| Type of Cost | 1938–1939 | | | | 1939–1940 | | | |
|---|---|---|---|---|---|---|---|---|
| | Area (ha) | Revenue | Expenses | %age | Area (ha) | Revenue | Expenses | %age |
| Capital[a] | 180 | 218,623 | 915,000 | 60.5 | 425 | 738,195 | 333,000 | 20.98 |
| Labour | | | 232,020 | 15.3 | | | 735,221 | 46.32 |
| Operational | | | 336,900 | 22.3 | | | 477,430 | 30.08 |
| Management | | | 29,240 | 1.9 | | | 41,566 | 2.62 |
| TOTAL | 180 | 218,623 | 1,513,160 | 100.0 | 425 | 738,195 | 1,587,217 | 100.00 |

[a] This indicates capital investment.

*Source:* CD IAO AOI 1837, Azienda agricola Bisacchi Ugo, Comprensorio di Ambo, n.d.

What proportion of this land was actually developed at the end of 1940? With the exceptions of Käffa and Sidamo, information on regions is incomplete. Neverthless, the answer depends on a number of factors, including age, size, and location of the concession. As regards age, the majority of the concessions belonged to the land-rush period of 1937 and 1938. From the end of 1938 onwards there was a gradual but considerable decrease. Exceptions were Käffa and Sidamo where there was a sustained interest in contrast to the restrained attitude of the earlier period. One of the most remarkable post-1938 phenomena was the rapid growth of profit-oriented, large-sized farms at the expense of mass-settlement-oriented farms.

As regards areas developed, the fragmentary nature of the available data makes nationwide assessment difficult. But sources attest that there was not a single farmer who succeeded in cultivating all his holding. Even the achievements of the old and relatively well-equipped concessionaires was far from satisfactory. Table 15 shows the example of the farm of Scagliarini Roberto & Sons. The concessionaires belonged to the CFA group and began work in early 1938, making use of modern machinery. But out of 1,000 ha of highly fertile land at Ambo, they cultivated a mere 150 ha in the first year, 350 ha in the second year, which in the third year was reduced to 300 ha.

Of course, there were regional variations. In the governorates of Shäwa and Härär, where the farms had been established for a longer period and, in the case of Härär, under more peaceful circumstances, they were considerably developed. But the limited development by farmers of the CFA in Wällo and those of Käffa and Sidamo regions suggests there was no spectacular success both at individual or regional level. The six CFA members of Wällo began their operation in early 1938 on 3,170 ha.[34] They had a sizeable stock of modern machinery and a variety of agricultural tools. Yet the total area cultivated at the end of 1940 amounted to about 1,500 ha.[35] Regionally, in Sidamo and Käffa provinces, of the total concession area of 67,202 ha belonging to the commercial farms, a paltry sum of 805 ha was developed.[36]

Similarly, the level of production was lamentable. With a few exceptions, in a good year the average yield of wheat totalled between 4 and 6 ql per ha. For example, the 30,000 ql of wheat raised from CFA farms in autumn 1937 were hailed as the best. But to gauge the strength of this

[34] In 1940 it had increased to 3,794 ha (cf. ibid., p. 7; CD IAO AOI 1808, Am UA, Relazione, p. 18). [35] CD IAO AOI 1808, Am UA, Relazione, p. 18.
[36] CD IAO GS 1822, Attività agricole, pp. 5–14.

statement one has to bear in mind that this was obtained in 8,000 ha of land with 12,000 ql of seeds and using solely Ethiopian labour.[37] Each year the farmers had to contend with climatic hazards attendant upon any agricultural enterprise, and marauding insects. In 1939 crops at Härärgé and Arsi uplands were considerably damaged by the combined assault of excessive rainfall, invasion of locusts, and depredation of pests, in particular the larvae of the Noctaid moth.[38] In 1940 the harvest was so ridiculously poor that many of the farmers did not even bother to gather it.[39] Reduction in production affected even coffee. This staple export crop had plummeted to its lowest level in 1938. In September 1937 its export to Italy was severely curtailed. Exporting firms were allowed to send to the metropolis consignments equivalent to 30 per cent of their exports to other countries. In 1931–4 inclusive, total export averaged 17,000 tons per year. But at the end of 1936 export had practically stopped to the extent that a rich coffee-producing country had to import from Brazil.[40] With the recovery of 1938, coffee export totalled 4,533 tons. Of this 1,360 tons (value L.9,620,000) went to Italy. But in 1939 production was reduced to 2,000 tons.[41] The government undertook a number of special measures to boost production but the results remained limited.[42] Härär governorate, where the government conducted a concerted campaign and where a number of European concessions existed before the occupation, provides the best illustrative example. In 1931 a total of 10,000 tons of coffee was produced. Production for the whole of 1935–7 amounted to 8,000 tons, reduced to about 3,000 tons in 1938–9. One European plantation which

[37] Villa Santa, *et al.*, *Amedeo*, 223; FO371/22020/J371/40/1, Lord Perth to FO, Rm 28 Jan. 1938.

[38] FO371/24635/J5/5/1, Gibbs to Halifax, AA. 27 Nov. 1939, p. 3.

[39] CD IAO: Documenti AOI, Concessione Scaglairini; AOI 1837, Azienda agricola, 6; AOI 1838, Vincenzo Valducci to Prof. Maugini, Fl 10 Jan. 1942, p. 5; AOI 1929, Pavirani to Testa, Aselle 4 Nov. 1940; GS 1328, Ciccarone, Relazione sulla missione alle concessioni cerealicole del comprensorio di Asella, AA 16 Oct. 1940; Ex AOI Sc 1329, Ciccarone, Relazione; E. Tischer to De Bendictis, Togonà 3 Nov. 1940; Castellani, 'Problemi', 10–11; Maugini, 'Appunti', 406–7.

[40] FO371/23380/J1992/296/1, De Clermont and Donner Ltd. to FO, 12 May 1939; 'Notiziario agricolo commerciale: AOI', *AC* 32/7 (1938), 380–1; E. Bartolozzi, 'Il commercio del caffè nell'AOI', *AC* 32/7 (1938), 316–19.

[41] ACS MCP De Palma B7/74, 9 May 1943; FO371/23380/J1776/296/1, Bird to Halifax, AA 1 Apr. 1939, p. 9; FO371/24635/J5/5/1, Gibbs to Halifax, AA 27 Nov. 1939, p. 3; R. Ciferri, 'Problemi del caffè nell'AOI', *AC* 34/4 (1940), 144.

[42] CD IAO Hr 1813: annex. II, G Hr DCL IA, Provvedimenti a favore degli indigeni coltivatori di caffè, Hr 27 Apr. 1938; annex. no. 9, UA RG Hr, no. 13357, Vivai di caffè, Hr 20 Feb. 1940. MCP De Palma B7/74, 9 May 43, p. 6.

before the occupation had an annual production of between 7,000 and 8,000 ql by 1937–8 had an output of only 2,000 ql, and over 900 ql, by 1938–9.[43]

Low output was matched by soaring production costs which seriously shook the enthusiasm of many farmers (see Tables 16 and 17). The situation made their need to resort to State subsidy more acute. This was an embarrassment to the colonial administration, for the shortcomings of the farmers exposed the inadequacy of its control aimed at preventing the establishment of unprofitable farms.

## STATE ASSISTANCE AND CONTROL

Recognizing that the growth of commercial farming was not possible without State intervention, the colonial government undertook a number of measures to ensure rapid but planned development, aided by a host of learned institutions within Italy. Assistance came through the formation of a network of agricultural offices and experimental stations and in the form of a variety of incentives. Control was exerted through a careful processing of applications for land and imposition of an agreed development programme. But the government's objectives were seriously limited by the combination of inadequate official support, random and ill-defined directives, and meagre financial commitment.

The agricultural institutions in Ethiopia provided technical assistance, propaganda, and surveillance. But their achievements were far more limited than officialdom claimed. Their involvement began with the enactment of the decree of 16 June 1936 which saw the creation in the Reale Academia d'Italia of a research centre, known as Study Centre, whose task was to

1. Promote and carry out through its own missions scientific research in IEA with a view to supplying precise information necessary for development by the State and private companies;
2. Co-ordinate and set guidelines for all private initiatives that have similar objectives and capital to invest.
3. Examine requests and proposals that would be submitted to it by the MC.

[43] DGG 4-9-1939, no. 803, Norme per la salvaguardia e disciplina delle piantaggioni di caffè nel territorio del Gov del Hr, in *GU* 4/23, 27 Sept. 1939; CD IAO, Relazione sulla situazione caffearia Hararina alla fine della campagna 1937–1938, AA July 1938, pp. 1, 6.

The Centre was jointly financed by the Reale Academia d'Italia and the MC and had its headquarters in Rome with a branch in Addis Ababa.[44] The Director-General of the MAI participated in almost every activity that the Centre carried out in its Rome headquarters. The Centre despatched a number of exploratory missions which collected valuable scientific and agricultural information; but the findings offered no immediate solution to the problems facing the farmers.

The Centre's role was overtaken by the IACI of Florence which, alongside other para-statal as well as private agencies, carried out useful investigations through specialized commissions. Founded in 1903, the Institute was one of the leading bodies for the study of tropical and subtropical agriculture and served as a centre for the diffusion of colonial propaganda. Most of the technical directors, experts, and supervisors of the colonies were trained there and its ex-students and staff carried out a number of valuable missions. The Director, Professor Armando Maugini, was also the Head of the Department of Agricultural Services of the MAI.

As it had in the older Italian colonies, the Institute played a crucial role in the development and organization of the agricultural services inside the Empire.[45] In each of the four governorates an agricultural department, Ispettorato Agrario, was set up, complete with a veterinary department and an experimental station, to carry out experimental work on livestock and crops. The Ispettorato was divided into seven units specializing in areas such as 'native' agriculture, stock-raising, phytopathology, and plant-breeding as well as experimentation.[46] For the commercial farms the most important unit was the Land Reclamation and Agricultural Colonization Unit (Sezione Bonificamento e Colonizzazione Agricola). This unit had the task of allocating land and overseeing development, assisting and controlling activities, and providing fuel at a subsidized rate; in some instances, but particularly where other units were not operational,

[44] MAE, Avvaloramento, 342–4. Among such missions that of Lake Tana, led by one of the founders of the Institute, Giotto Danieli, is of considerable academic importance in terms of seriousness and the wide range of topics covered.

[45] Ibid.; ATdR 24/116, RD per l'ordinamento del RIA AI: Relazione al Consiglio dei Ministri, Rm 1937; ATdR 24/119, Schema di RD ordinamento dei servizi dell'agricoltura nell'AI, Rm 1937.

[46] On the work of these institutions see agricultural departments' reports: CD IAO: Sc 1801, Attuale; Hr 1813, 'L'attuale organizazione'; AOI 1808 Am UA, Relazione; GS 1819, Organizzazione dei Servizi dell'agricoltura e di esigenze per il prossimo avvenire, Jan. 1940. At the later stage the Experimental Department became independent from the UA and the areas of its action widened (cf. RD 29-7-1938, no. 2221, Ordinamento dei servizi dell'agricoltura nell'AI, in *GU* 63, 15 Mar. 1939).

it had an extremely wide remit. An important role was also played by the experimental stations, many of which carried out valuable research on local and foreign crops and animals. Most significant of these were those of Mälco in Galla and Sidamo, Amaräsa in Härär, Gudär and Mäkanisa in Shäwa, and Pirzio Biroli in Amara. The experimental work undertaken at these stations aimed to assist the farmers with new and lucrative crops.[47] The law envisaged the extension of each unit to every district and subdistrict; in the event, only in part of Härär was this the case.[48]

The agricultural offices had wide-ranging and demanding activities which often overshadowed their main tasks of technical assistance, propaganda, and surveillance of commercial farms.[49] These extra duties extended from land to topographical and agricultural surveys of the soil which, in case of small- and medium-sized farms, entailed parcelling out into plots of 60 and 500 ha and partitioning further into smaller units for the purpose of assessing the annual rent and operational capital; evaluation of a development plan with reference to irrigation, availability of drinking water, transport facilities, access to market, and labour supply; and land-exchange plans, involving eventual compensation for and resettlement of displaced inhabitants and study and composition of charters of concessions, or *disciplinare*.[50]

Despite their impressively elaborate appearance, the agricultural departments were largely underfunded, undermanned, and undermined. Considering that agriculture was the key area that the government wanted to promote and the lifeblood of the country, the budget devoted to agriculture was minute.[51] For example, Härär's annual budget amounted

[47] Archival material on the experimental stations and their works is vast: cf. CD IAO: AOI 1323, Relazione sull'attività svolta dalla sperimentazione agraria negli anni 1937 al 1940, 14 Jan. 1940; Hr 1325, Giovanni Piani and Carlo Carloni, Primi risultati di un biennio di ricerche agrarie sperimentali nel Hr, Hr 1 Feb. 1940; Hr 1335, Alinari, Relazione sull'attività svolta nel campo sperimentale di Neghelli nel 1936, n.d; Gm 1336, Relazione sull'attività dall UA nel vivaio e campo dimostrativo di Malcò, 1937–1938, n.d.; Sc 1337, Vasco Gatti, Attività del campo di orientamento di Olettà nel 1937, n.d.; Sc 1338, Campo sperimentale; AOI 1808, Am UA, Relazione; GS 1817, Relazione sull'attività svolta nel vivaio e campo dimostrativo di Malcò al 31-12-1938, Jan. 1939; Hr 3034, Giovanni Piani and Carlo Carloni, Primi risultati di un biennio di ricerche agrarie sperimentali nel Hr, Hr 1 Feb. 1939.

[48] RD 29-7-1938, no. 2221, Ordinamento dei servizi dell'agricoltura nell'AI, *BUGA* 4/7/1, 15 Apr. 1939.

[49] CD IAO GS 1820, Il problema, 6; Sc 1801, Attuale, 2.

[50] CD IAO: Sc 1801, Attuale, p. 2; GS 1820, Il problema, pp. 2–9.

[51] According to a sexennial plan, the total fund earmarked for agricultural development amounted only to L.200,000,000 or 1.8 per cent of the total budget. This was almost 30% less than that allocated to defence (4.5%) and higher only than that allocated to communication (0.5%) and mining (0.91%) (cf. ATdR 24/79, MAI, Piano sessennale delle opere e delle spese per l'AOI, Rm 1937; 'Notiziario agricolo commerciale AOI', *AC* 31/7 (1937), 301).

TABLE 18. Agricultural personnel, 1940

| Governorate | Personnel | | | | |
|---|---|---|---|---|---|
| | Professional[a] | Skilled[b] | Semi-skilled[c] | Unskilled[d] | Total |
| Amara | 9 | 13 | 1 | 3 | 26 |
| GS | 17 | 34 | 7 | 33 | 91 |
| Härär | 14 | 11 | 1 | — | 26 |
| Shäwa | 10 | 1 | — | 1 | 12 |
| TOTAL | 50 | 59 | 9 | 37 | 155 |

[a] Professional group includes agronomists (31), graduates (16), planners (3).

[b] Skilled consist of nursery workers (19), clerical workers (13), assistant agronomists (8), tractor-drivers (8), surveyors (3), gardeners (2), mechanics (2), apiculturalists (1), horticulturalists (1), carpenters (1), smith (1).

[c] Semi-skilled involves nursery assistants (9).

[d] Unskilled include workmen (27), storekeepers (4), footmen (2), stablemen (2), shepherd (1), messenger (1).

*Source:* CD IAO: Sc 1801, Attuale; AOI 1808, Am UA, Relazione; Hr 1813, L'attuale organizzazione; GS 1819, Organizzazione dei servizi.

to L.1,100,000 while the minimum it needed to maintain its basic performance was L.2,000,000. But owing to the large presence of commercial farms in the territory, Härär received relatively generous treatment.[52] The same can be said as regards personnel, which at the end of 1940 consisted of about 150 individuals (see Table 18). Of these only 50 had higher qualifications of any kind. Most of these were stationed in Galla and Sidamo, which by 1940 appears as the only governorate comparatively adequately equipped in terms of both number of departments and manpower, although still poor in financial resources.[53]

Lack of manpower and finance was compounded by inadequacy or total lack of premises. With a few exceptions, in the early stages departments

[52] CD IAO Hr 1813, L'attuale organizzazione, 5.

[53] CD IAO: AOI 1808, Am UA, Relazione; GS 1819, Organizzazione dei servizi; Hr 1813, L'attuale organizzazione; Sc 1801, Attuale.

operated from temporary shelters. Even as late as 1940 there were complaints of inadequate office space. Almost all the peripheral units functioned either from a temporary *harish* or from a room shared with other government organizations.[54] To further aggravate smooth functioning, there was a network of overlapping political and military jurisdictions and a relative lack of autonomy that 'made interferences inevitable and harmonious and profitable co-operation often difficult'.[55] The overall perception that the concessionaires had about these agricultural departments was negative. Most recurrent complaints directed against them were that they were dilatory and unsympathetic to their enterprise. They were blamed particularly for supplying seeds that were inadequate or inappropriate or after long delays.[56]

The technical officers were not only numerically inadequate but they lacked specialized training. As most of their experience was gained in the field, they lacked time to absorb the growing amount of information about the agricultural environment. Thus there was little wonder that even those devoted men among them did little to meet the expectations of most of the farmers. Reports talk of their enormous frustration with the rigours of frontier life. An early report by Härär agricultural department contrasts the conducive agricultural environment of the governorship with the depressed state of agricultural officers. On the one hand, there was one of the most peaceful regions ruled by a governorate interested in colonization and agricultural development. On the other, stood a relatively well-qualified but apathetic personnel, demoralized by the uncomfortable lifestyle. To further complicate matters, the Director was reportedly unpopular and ill-disposed towards the IACI and its graduates from where a substantial number of his staff came. The list of complaints from technicians were many. They included lack of accommodation and equipment, job dissatisfaction due to overwork, confusion on job boundaries, favouritism, and political interference. Reports attest how the work of these officers was hampered by lack of funds, the vast areas they had to cover, transport difficulties, a hostile indigenous population, and an intractable bureaucracy. Owing to this, propaganda and technical assistance remained largely limited to quick visits to experimental centres by concessionaires, or extension work by one of the overworked technical officers, trying in a hurried visit to persuade the farmer to use improved

[54] CD IAO: Hr 1832, IA Hr, Promemoria to Prof. Maugini, 12 Sept. 1938; Hr 1813, L'attuale organizzazione; Sc 1801, Attuale.

[55] CD IAO Sc 1801, Attuale, 2.

[56] CD IAO: Documenti AOI: Concessione Scagliarini; AOI 1837, Azienda agricola; AOI 1838, Valducci to Maugini, 5.

farming techniques such as hoeing or growing a particular crop. However, in 1940 in some areas even such practices were reduced to the exchange of a few words during one of those sporadic contacts between the concessionaires and the technical officers.[57]

As regards incentives, these were provided in the form of subsidies which largely consisted of reduced agricultural fuel, supply of seeds, free travel up to the port of destination for the personnel to be employed on the farms, and free transport of equipment and machinery. For some crops, such as wheat, a guaranteed price was offered which the colonial authorities thought would provide an adequate profit margin. To encourage extensive cultivation of virgin lands, a land reclamation fund was set up consisting of L.250 per ha of land put under cultivation.[58] Farmers were also exempt from payment of *asrat* to which all Ethiopian farmers were subject.[59] In addition, there were special aid packages of short-term measures adopted on a nationwide or regional basis either by the central administration or by the respective governorate in special circumstances. These included assistance in building a house, free access to a tractor, a special subsidy in case of crop failure or loss due to climate or locust invasion.[60]

Most of the subsidies were obtained through a sustained pressure that involved acrimonious dispute and long negotiations. This was particularly the case with the land reclamation fund and agricultural fuel.[61] The

[57] CD IAO: Er 1832, Appunti relativi sull'IA di AA As Hr Sm, 1938; RG GS UA, Attività svolta nei vari settori dalla costituzione dell'Ufficio ad oggi, Jan. 1940, pp. 3–4; GS 1820, Il problema, p. 6.

[58] CD IAO AOI 1936: DGACL (Lessona) to GG, Produzione frumento; Lessona to CFA, Trasferimento in AOI di imprese di coltivazione, Rm 20 Nov. 1937; GG AOI [Cerulli] to MAI (DGACL), Concessioni provvisorie per le semine di cereali, AA 25 Mar. 1938; Ugo Rossi *et al.* to Marcello Bonfondi [Segretario Federale di AA], Promemoria, AA 30 Mar. 1937; Carburante agricolo e disciplina della sua distribuzione 1937; CD IAO AOI 1800, Promemoria per Maugini; DGG 23-7-1938, no. 828, Prezzo dei carburanti agricoli, *GU* 3/15, 1 July 1938; DGov 31-12-1938, no. 206152, É accordata l'esenzione della decima agli imprenditori di Dessié e all'ECRE per l'annata 1938–1939, *BUGA* 4/6, 15–31, Mar. 1938; DGov Sc 13-2-1940, no. 201, Concessione di anticipazione di frumento da semina e di carburanti agricoli ad agricoltori nazionali che svolgono la loro attività nel territorio del G dello Sc., in *GUGS* 5/25, 19 June 1940; Villa Santa *et al.*, *Amedeo*, 240; 'Notiziario agricolo commerciale: AOI', *AC* 32/1 (1938), 44.

[59] FO371/23380/J1776/296/1, Bird to Halifax, AA 1 Apr. 1939.

[60] DGG 10-2-1938, no. 94, Esenzione dai diritti di confine del petrolio e dei residui della distillazione degli olii minerali destinati esclusivamente alle attività agricole dell'AOI, in *GU* 3/10, 16 May 1938; CD IAO Ex AOI Gn Am 600, Relazione sull'attività, 6; *Colonie*, 24 May 1938; *Tribuna*, 7 Feb. 1940; *GV*, 7 Apr. 1940; G. Piani, 'Rassegna di alcune attività autarchiche del Hararino nel campo agrario', *Autarchia Alimentare* (Oct. 1938): 38; 'Notiziario Agricolo Commerciale: AOI', *AC* 32/1 (1938), 44.

[61] For the Government in AA, price reduction in agricultural fuel entailed price increases in other sectors of the economy and tax exemption affected its revenue. Cf. CD IAO AOI:

Viceroy's government in Addis Ababa was impervious to the plight of the farmers who found themselves in the unusual position of having to persuade the sympathetic MAI to intervene on their behalf.[62] Addis Ababa suspected that the commercial farms wanted to capitalize upon the government's anxiety to increase agricultural output and put the blame for shortcomings in their performance largely on poor technical skill and managerial incompetence of the farmers. For the farmers such a view reinforced their own perceptions of the colonial authority's unsympathetic attitude towards their painful situation arising largely from the tremendous lack of knowledge about the physical, social, and economic conditions of the areas in which they operated.

Ambo's CFA group turned out to be the most vociferous critics. As pioneers of wheat production, they ridiculed government's suggestion of speculative intentions; they pointed out that most of their problems were the outcome of broken promises and misleading information on the part of the colonial authority. In addition to severe lack of extension services, they pointed to unacceptably slow land allocation, supply of poor seeds, and unsafe political conditions. Quite rightly, they claimed that although wheat prices were guaranteed in advance, in fact what they were paid was less than that they would have earned if they had sold it on the open market.[63] But they were in no way willing to face the rigours of constantly changing international market and Ethiopian competition. 'The concessionaires do not ask for either annual contribution nor any sort of bounties from the Government-General, but only a guarantee to purchase [at agreed price] at harvest—a guarantee that is in no way denied for a much higher amount to firms engaged in road construction.'[64]

1839, Di Crollalanza, Promemoria, July 1938, p. 19; 1936, Petretti to GHr; 1926, Terruzzi to Mussolini.

[62] CD IAO AOI: 1936, G. Pini to DS AE, Situazione primi concessionari, AA 19 Dec. 1937; 1929 Simba, Appunto dettato dalla S. E. Testa, 13 Mar. 1941.

[63] Initially the agreement was to purchase a quintal of wheat for L.200 provided that the production unit was 8 ql per ha. But this was modified after protracted wrangling that followed the failure by the farmers to produce the agreed target per ha. The second agreement took into account a production level lower than 8 ql per ha. and set graduated purchasing price in such a way that it increased in parallel to the decrease of output. In this way e.g. a quintal of wheat produced at the scale of 4 ql per ha was purchased for L.305. (Cf. CD IAO AOI 1936: MAI IA (Lessona) to GG Hr Gn Gm, Rm 25 Oct. 1937; Lessona to CFA, Trasferimento; Rossi to Bonfondi, Promemoria; Cerulli to MAI, Concessioni provvisorie.) At about this same period the cost of a quintal of wheat in AA was L.326, in Gm L.398; while the cost per quintal of wheat flour amounted to L.383 in AA, L.302 in Hr, L.353 in Gn and L.355 in Gm (cf. CD IAO AOI 1936, GG AOI DS AE (Graziani) to MAI, AA 3 Nov. 1937).

[64] CD IAO AOI 1936, Rossi to Bonfondi, Promemoria.

The colonial authority gave in to most of the pressures of the farmers' lobby, and the differences were largely settled towards the end of 1939. Nevertheless the farms continued to experience a number of difficulties. One such problem was the absence of credit institutions that specialized in lending money at reasonable rates of interest for productive rather than commercial activities. Though its need was strongly emphasized by the Agricultural Council as early as 1937 and despite persistent requests by the farmers, the Agricultural Loan Bank never came into existence. As a result the farms suffered from shortage of capital. Those seriously affected were small- and medium-sized farms most of whom were chronically undercapitalized and operated at a loss.[65] Only after a number of unfulfilled promises to meet the repeated pleas by the farmers for credit institutions, did the government seem determined towards the end of 1940 to create co-operatives and set up an agency that would facilitate their financing. In addition, it was thought that this same agency would take over the allocation of land in order to free the would-be applicant from undue bureaucratic delays.[66]

## MECHANISMS OF CONTROL
## MEASURES, EFFECTS, AND TRENDS

Delays in land allocation were largely caused by the government's desire to control each phase of agrarian development and transfer of land so that the Ethiopian cultivator was not sacrificed for the speculator, nor the genuine investor with capital discouraged from using surplus land. The antidote was sought in a thorough scrutiny of applications, enforcement of occupation and development conditions, and a strict control of transfers. However, these mechanisms proved neither efficient nor practical. The effect was rather to antagonize many bona-fide applicants. Archival

[65] CD IAO: AOI 1848, Gustato Fonti to G Hr, Concessione agricola Foschini G. Ravioli, Bisidimò (Hr), Hr 2 Dec. 1940; AOI 1837, Pier Bono, Promemoria sulla situazione attuale delle Imprese Agricole private in AOI, e sulla necessità divergenti provvedimenti, n.d. 'Cose e case dei produttori', *CE*, 24 June 1939.

[66] Equally CFA's plan to set up a Consortium or Assistential Agency to small- and medium-sized farms was not put into operation. Cf. Consulta, 'Avvaloramento', 1571; CD IAO AOI 1936: Rossi to Bonfondi, Promemoria; Pini, Considerazioni sull'attività agricola in AOI: Credito agrario, n.d; DGG 13-2-1940, no. 193, Estensione delle attività dell'Ente Assistenziale UMA in AOI, in *GUGS* 5/13, 27 May 1940. *Azione Coloniale*, 23 Mar. 1939; 7 Dec. 1939; *Colonie*, 2 Mar. 1939; *CC*, 25 Mar. 1939; *Piccolo*, 6 Feb. 1939; 'Per lo sviluppo dell'agricoltura nel GS', *Rivista delle Colonie Italiane*, 19 May 1938; *La Stampa*, 13 June 1939; Villa Santa *et al.*, *Amedeo*, 254–5.

material reveals denunciations, and sometimes even violent disputes, between the colonial administration and commercial farmers, most of whom had been enticed to the Empire by the government's own propaganda machine. They were indignant about the way the colonial administration tempered their aggressive enterprise while the latter viewed their behaviour as undisciplined and embarrassing.[67]

The procedure for obtaining land was complex and subject to much administrative and political confusion. As there was no land survey for allotment, the concessionaire was told to select his own. The only requirement was that he should be able to show proof of possessing the necessary means—farming skill and capital. Although the rule was the application be made in person, cases by attorney were not uncommon. As regards land the criteria for selection was that it had to be unoccupied or uncultivated or domanial. The administration quite erroneously assumed that there was a vast, if undefined, quantity of such land and that the prospective settler would be content to apply exclusively to it. The system was bound to cause confusion and infringe upon the rights of the Ethiopians. One of the natural results of settler prospecting was that applications were made in scattered localities. The vast majority of the applicants selected lands within the nearby urban centre while only a few bothered to prospect further afield. It was not also unusual that different settlers applied to the same land. Most importantly, selection did not proceed according to the administration's expectations. The settlers were quick to realize that lands avoided by the Ethiopians tended to be waterless, bush-covered, or unhealthy. As they saw the Ethiopian occupation indicative of productive land, a substantial number of them deliberately selected lands on a high state of cultivation on the pretext that it was domanial. When they were denied, tempers flared and the administration was accused of betrayal and treachery. Each application was registered by the land office and, after a preliminary survey and examination of the Ethiopian rights, the applicant was initially issued with a certificate of temporary occupancy and with proper titles afterwards.[68]

Regardless of the type of the concession or authorizing body, for all but a very few Fascist Party officials and large influential commercial farms,

[67] CD IAO AOI 1936: Pini, Situazione; Rossi to Bonfondi, Promemoria; A. de Rubeis, Promemoria per S.E. il SottoSegretario: Semine di frumento da parte degli agricoltori organizzati dalla CFA, Rm 12 May 1938.

[68] CD IAO 1926, Terruzzi to Mussolini; Villa Santa *et al.*, *Amedeo*, 226; ASMAI AS III, 2ª Guerra mondiale, P4, Mezzetti to Terruzzi, Däsé 13 Aug. 1938; 'Notizie agricolo commerciale: AOI', *AC* 32/1 (1938), 44.

acquisition of land took a long time. The processing of applications was tedious and totally unsuitable for the rapid settlement of claims as it involved lengthy bureaucratic procedures:

The applicant has to wait a minimum of six months to a year from the moment he presents his application until he gets possession of the land, naturally, under a simple temporary permit, and this without considering the time that he had spent in the selection of the land itself. Faced with such a situation many of them are discouraged and, as a rule, they give up; only those whose financial resources would allow them to live in the capital destined for the agricultural enterprise endure. Would so much be necessary for carrying out the measures necessary for the handing over of the concessions? In so far as the system currently followed stands, 'yes'.[69]

Three bodies were empowered to process an application depending on the amount of capital to be invested: the Agricultural Council, the CFA, and local governments. Farms with a capital in excess of L.100,000 applied directly to the Agricultural Council who made a preliminary investigation as to whether the concessionaire had sufficient capital and tested technology at its command, and examined the feasibility of the scheme. The criteria was to allocate land on the basis of L.1,000 per ha, the maximum size of the holding to be 500 ha.[70] But this rule was not always followed and, as we saw earlier, many companies were given holdings as large as 30,000 ha or more.[71]

The Council's obvious weakness was its negligible knowledge of the Empire. Most data was gained through reading secondary sources or flying visits undertaken in one of the scouting missions.[72] In official circles, the Council was viewed as an anachronistic institution riddled with internal strife and self-interest. A high Party official frankly remarked: 'Why is it that rather than residing in Rome and making decisions for a country that is unknown and between 4,000 and 5,000 km away it cannot be constituted in IEA itself. Indeed, things are moving in a chaotic way.'[73] Owing to the remote control of the Council, the final endorsement of concessions rested on the central administration in Addis Ababa. From

[69] CD IAO GS 1820, Il problema, p. 2; K. Gandar Dower, *Abyssinian Patchwork* (London, 1949), 235; G. L. Steer, *Sealed and Delivered* (London, 1942), 196.

[70] This limit applied even when the capital was above L.500,000.

[71] CD IAO AOI 1982, DG[A]CL Domande di concessioni agricole in Etiopia, Rm Oct. 1938; Quaranta, *Ethiopia*, 73.

[72] ACS: SPDR 87/W/R/1/LA, Davide Fossa to Osvaldo Sebastiani, AA 10 Sept. 1937; GRA 44/36/2, Lantini to Petretti, Rm 24 Sept. 1937.

[73] ACS SPDR 44/242/R/39, Farinacci to Mussolini, Cremona 25 Dec. 1938.

here, the applications were passed on to the appropriate governor. However, Addis Ababa showed but fleeting interest in the orders from Rome: 'Sometimes the Ministry [of the Colonies] had to wait in vain for months in order to get a reply on some specific cases. Sometimes repeatedly sent urgent messages were allowed to fall into a vacuum.'[74]

In their assessment of applications the governorships took into consideration the availability of domanial land and the feasibility of the programme, a process involving many. This is how one CFA official described the bureaucratic entanglement preceding any land grant:

> The destination of a specific agricultural land for colonization involves not only the direct interest of Directorate of Economic Affairs and through it the Agricultural Inspectorate . . . , but also that of Directorate of Political Affairs concerning the ascertaining of whether there are or are not native interests to be safeguarded; that of the Ministry of Finance, with regard to the expenses involving surveying and diagramming of the land; that of the Quartermaster-General with regard to the execution of preliminary supplies needed for the setting up of the farm; and again that of the Command of the Armed Forces for matters regarding the precautionary safety measures connected with the isolated life of the metropolitan in an environment that is not yet totally peaceful.[75]

As close examination of some unsuccessful applications reveals, there were no nationally valid criteria for which an application was rejected or accepted. Each governorate acted according to its own set of rules. In Härär, the key determinant was the size of the holding, whereas in Amara the location mattered most. So while in Härär most of the applications were refused because they exceeded the maximum ceiling permitted, in Amara the main ground was the unsuitability of the site.[76] Some made the acceptance conditional upon the applicant's prior execution of a particular task, such as an inspection of the site in person.[77]

When the capital invested was less than L.100,000, the application was made through the CFA which, as a Fascist trade union organization, played an important role in the development of IEA agriculture.[78] But such

[74] ACS SPDR 87/W/R/1/LA, Lessona to Fossa, Rm 16 Sept. 1937.

[75] CD IAO AOI 1936, Pini, Considerazióni, p. 4.

[76] CD IAO AOI 1982 DGCL, Domande. The G Am, whose main concern was security, held a position that the concessions should be close to communications, markets and water supplies and be confined to the territories of Wällo Yäjju, Gn, and Lake Tana. In Hr where the key problem was population density, security was not an issue.

[77] In a number of cases such conditions proved effective in putting off applicants lacking seriousness. This was the case with 42 applicants to Agricultural Council who withdrew when summoned to inspect the site. Cf. CD IAO: AOI 1982, DGCL, Domande, p. 5; GS 1820, Il problema, p. 3.

[78] CD IAO AOI 1982, DGCL, Domande.

an organization with powerful vested interests and strong ideological allegiances may not have given an independent and unbiased assessment to an applicant.

Finally, each governorate was empowered to grant small concessions, mainly in the vicinity of urban centres, for kitchen gardens.[79] However, no guidelines were set specifying either the size of land to be granted by these governorates or the department empowered to do so. Lack of clarity in the latter case led to a number of abuses and interdepartmental struggles over claims to the administration of concessions.[80] Equally, failure to set a ceiling encouraged some governors to take a cautious view whereas it induced others to a broad interpretation. The administration in Addis Ababa kept a close watch on the governorships' dealings and the behaviour of liberal governors caused frequent friction.[81] From the central administration's point of view, not only did such an approach become the source of a tacit, albeit unintentional, encouragement for the creation of isolated and scattered farms that were unable to stand on their own feet but also a burdensome addition to the work of the governorates that were hard-pressed by slack financial resources and manpower shortfall. As such it undermined the colonial administration's much cherished initial plan to concentrate commercial enterprise in a few selected areas in order to avoid dispersion of farms and minimize cost.[82]

In the absence of clear guidelines, the governorates pursued a tortuous method that largely destroyed initiatives and discouraged enterprise. For the applicants it was a source of uncertainty and confusion, and several of them were repatriated disgruntled and bankrupt.[83] The practices followed, normally, involved:

[79] CD IAO AOI 1936: Graziani to DG AOI, AA 14 Apr. 1937; Graziani to GG AOI, AA 9 May 1937; Lessona to IA, Rm 14 May 1937.

[80] A typical ex. was Hr where the struggle between Land Office (Ufficio Terriero) and the DCL, presided over by ex-lieutenant-colonel of artillery Malvani on one hand and the Agricultural Inspectorate, chaired by Giovanni Piani, on the other, had developed into serious conflict. Cf. CD IAO AOI 1802: Piani to De Benedictis, Hr 12 July 1939; De Benedictis to Maugini, [AA] 1 Aug. 1939.

[81] CD IAO AOI 1936: Petretti to MAI, AA 25 Sept. 1937; Lessona to GG, Rm 11 Nov. 1937. It refers to the G GS who, broadly interpreting the clause, had begun to grant plots that in the Viceroy's eyes were not small; cf. also CD IAO AOI 1936, Pini, Considerazioni.

[82] CD IAO AOI 1936: Graziani to GG, AA 8 Apr. 1937; Petretti to G Hr. CD IAO AOI 1854, p. 6.

[83] CD IAO GS 1820, Il problema, p. 2; FO 371/22021/2376/40/1, CG (D. F. S. Filliter) to Lord Perth, Naples 7 June 1938; FO371/23380/J1776/296/1, Bird to Halifax, AA 1 Apr. 1939.

1. Technical advice from the Land Office.
2. Planimetric survey of the land.
3. Verification of the domaniality of the land.
4. Verification of the candidate's good moral, civil, and political conduct.
5. Proof of his technical skill.
6. Proof of his financial fitness.
7. Compilation of the charter of concession.
8. Examination and endorsement by the governorate.[84]

The verification of (4), (5), and (6) depended on information supplied from Italy, taking several months. In addition, the completion of stages (1), (2), and (3) paralysed work for months. Contributing to the delay was severe lack of equipment, personnel, and finance. The scattered location of lands requested further compounded the problem. The administration was of the view that it had to think carefully before taking any initiative because it could not afford to keep a technician in a faraway place for many days each time there was an application for a concession, which in many cases was not followed up by the applicant. The situation was further worsened by uncertainties surrounding the definition of a public domain and lack of clear guidance for the composition of contracts. As no clear and authoritative statement existed on it, the decision on domaniality of a particular piece of land was often left to the discretion of the district or subdistrict officer, who behaved arbitrarily. Lands that one officer, defying all evidence to the contrary, declared private property, at later inspection proved to belong to nobody. Equally, fresh surveys had to be conducted and work in progress halted when lands acknowledged by another officer as public domain were subsequently challenged by their Ethiopian owners.[85] Even though there was sympathy among some district officers for the Ethiopian claims, as a rule these became worthy of consideration only when they seemed to have serious political menace or disruption in production. Otherwise, the colonial administration strived to maintain the paramountcy of the settlers' interest. Whenever possible the settler was not to be prevented from obtaining land. In dealing with the Ethiopian claims the usual procedure was to 'induce' them to move off on receipt of compensation. If they refused, their protest was ignored and the administration took measures that it saw fit. One of the standard procedures was to let them to stay on to provide labour under one of

[84] CD IAO: Sc 1801, Attuale, p. 2; GS 1820, Il problema, p. 2.
[85] CD IAO GS 1820, Il problema, p. 4.

several labour contracts normally drawn by the technical officers and ratified by the government.

The composition of the agricultural contracts was a technical officer's nightmare as it required shuttling between the battery of government offices in the colony and liaising with superiors in Italy. Until the end of 1939 no guidelines existed and the agricultural officer had to depend upon his common sense. This was a taxing task for a number of the contracts had to be rewritten several times to the satisfaction of the political authorities who refused endorsement until a particular clause, for example, supply of manpower, was included or removed. 'Different personalities held different views' and much time elapsed before a final draft was thrashed out.[86]

As mentioned earlier, the contracts, had two objectives: to discourage land speculation and the anarchic competition of adventurers; to ensure rapid development by setting out clearly the rights and duties of each concessionaire. In this way the government hoped to nudge the farmers into continuous development of their properties. In the final analysis what these contracts reveal is the colonial administration's underlying lack of confidence in pure economic motives to provide sufficient incentive to concessionaires to develop the land properly. Detailed prohibitions and sanctions corroborate this view. For example, the farmer was specifically forbidden forming partnership with Ethiopians, or lease to them his land in part or *in toto*, without prior approval by the administration. These practices would have encouraged bold speculators. Such restrictions were intended to prevent a repetition of the problems of Eritrean colonization.[87] They also show that the administration thought the spontaneous behaviour of the Italian settlers was not necessarily compatible with the aims of the State.[88]

Built within each contract were two key elements whose main objective was to ensure quick development of the land: annual rent and developmental clauses. Continuous possession of the concession depended upon the discharge of an annual rent. This was computed on the basis of the productivity and the geographical location of the land, bearing two facts in mind—that of minimizing the need for supervision, on one hand, and encouraging the rapid development of the concession, on the other. The underlying philosophy behind rent charge was that it would serve both as a stimulus to the concessionaire to engage himself in the development of

[86] CD IAO GS 1820, Il problema, pp. 3–4; CD IAO AOI 1936, Pini, Considerazioni.
[87] 'Notiziario agricolo commerciale: AOI', *AC* 32/1 (1938), 44.
[88] Ibid; *Azione Coloniale*, 20 July 1939; *Colonie*, 11 July 1939.

the farm motivated by the need to find the necessary revenue that would enable him to discharge his rental obligation, and as a curb on rush on land. Reasonable annual rent was thought to discourage land profiteering and to allow the concessionaire considerable saving to invest on the concession.[89] But despite such intentions, in fact the authorities imposed only a nominal annual rent ranging from L.5 to L.25 per ha. With few exceptions, most of the concessionaires did not pay even this modest amount. The government itself waved the first quinquennium's rent for those cultivating virgin land. By so doing, the administration deprived itself of what it thought of as a potentially effective means of control.[90]

Though varying according to size and organization of the farms, the contracts specified the envisaged land improvement model. Broadly speaking this model was of two types—those designed for profit-oriented medium-sized and large-scale companies and settlement-oriented farms, including medium-sized of the first category. As stated earlier, the first type of concessions were granted on a thirty-year lease, automatically renewable for ninety to ninety-nine years. Such longevity was a stimulus to the big investor. In the settlement-oriented concessions it was the opportunity of ownership that would attract emigrants to IEA. After a trial period, the concessionaire was given the option to purchase as a private property.[91]

Development was built around five- or ten-year plans in which were specified the improvements that should take place as well as the system of cultivation—whether it had to be cash-crop or animal-breeding or a combination of the two. For settlement-oriented concessions, a five-year schedule was fixed: within three months of the contract the concessionaires should take up the farm; within six months delimit the boundary; within a year start development work; by the first year 20 per cent, third year 50 per cent, and fifth year 100 per cent of the agreed programme had to be implemented.[92] Concessions whose holdings were below 50 ha had a considerably reduced timetable. As a rule the total development had to take place within two years from the date of the concession.[93] In

[89] CD IAO AOI 1936: Lessona to GG IA, Disciplinari di concessioni di terreni per uso agricolo, Rm 27 July 1937; Petretti to G Hr. CD IAO Hr 1847; Consulta, 'Avvaloramento', 1570–1; Pesce, 'Disciplina giuridica'.

[90] CD IAO Hr 1847, GH DCL, Verbale della III riunione del Comitato di colonizzazione, n.d. [29 Sept. 1938], p. 2.

[91] MAI, 'Valorizzazione', 210.

[92] DGov Sc 17-7-1939, no. 145: Col quale viene accordata al sign. Tedeschi dott. Aldo una concessione agricola, in *GU* 4/29, 8 Nov. 1939.

[93] DGov Sc: 9-3-1939, no. 223, Concessione accordata al sign. Marini Antonio di un lotto di terreno sito in località Uriel per sfruttamento agricolo, in *GU* 4/15, 2 Aug. 1939; 10-3-1939, no. 224: Concessione accordata al sign. Liri Clemente di un lotto di terreno sito sulla destra della strada AA–Moggio, per sfruttamento agricolo, in *GU* 4/15, 2 Aug. 1939;

profit-oriented concessions, the concessionaire had to develop at least one-third of the farm within the first quinquennium of the concession. The whole programme had to be completed within ten years.[94] The required improvements consisted in clearing land of stones, breaking up the soil, planting hedges, digging wells, and planting trees. The concessions were to be periodically visited by agricultural officers who were to draw up progress reports that would be validated by joint meetings between the officials of the administration and representatives of the farmers. The farmers were allowed to use extensive Ethiopian labour by the way of salary or sharecropping, *compartecipazione*, or other forms of contracts to be ratified by the political authorities.[95]

Common to both types of concession was the requirement that the concessionaire had to man the concession personally and build his principal residence on the site. This was not so much to bind individuals permanently to the soil as to assure direct modern farming of the land and combat abuse by those concessionaires who intended to live in the metropolis 'on the rent of, or subsidies coming from, their concessions'.[96] But the post-1939 contracts made such requirements compulsory only for settlement-oriented concessions whose holders had to cultivate them together with their family.[97] Political expediency demanded only that the holders of profit-oriented concessions be required to reside in IEA and man the farm either personally or employing 'a well-qualified

10-3-1939, no. 225, Concessione, a scopo agricolo, di un lotto sito nei pressi della strada AA–Gm al sign. Carlo Cataldo, in *GU* 4/24, 4 Oct. 1939; 19-9-1939, no. 264, Concessione agricola ai sigg. Barbieri Orlando e Peta Francesco, in *GUGS* 5/25, 9 Mar. 1940; 1-6-1940, no. 88, Concessione al sig. Pettini Cosimo per la durata di anni nove di un lotto di terreno, in *GUGS* 5/25, 19 June 1940.

[94] CD IAO Hr 1813, L'attuale organizzazione, where three types of charters are provided with a fourth temporarily employed until 1939 as annex; Disciplinare tipo per concessione agricola, in CD IAO GS 1822; Disciplinare, in CD IAO Sc 1801. Material on forestry concessions is vast: cf. DG 23-11-1939, no. 1774, in *GUGS* 5/7, 14 Feb. 1940. Extensive material is provided in the Biblio. For material on cotton relating to Cotton Districts see Ch. 7.

[95] The agricultural contracts or charters, known as *disciplinare*, took a certain consistency only towards the end of 1939 and since then, if not in details, in their general principle, they were identical in almost all governorates. A point of significance is that some of them put a special clause forcing the concessionaire to dedicate a certain %age of his land to farming a particular crop or breeding a specific animal suitable to the area, i.e. coffee or sheep in Hr and GS regions. The author is not able to find a facsimile for large-scale concessions except for Cotton Districts and forestry related companies. Equally no copy of charter for the concessions operating in the G Am area is available.

[96] Pesce, 'Disciplina giuridica'.

[97] The exceptions were medium-sized first-category farmers who, like profit-oriented concessionaires, were allowed to reside only in IEA and, in exceptional cases, employ a skilled technical director.

manager'.[98] Doubtless such a policy shift was in response to the inroads of a powerful interest group that had already become part of the rural scene. A brief look at the applicants throws a sufficient light at this background.

To begin with, there was an impressive number of applicants. During the early period of the occupation, requests were mostly for small- and medium-sized concessions. At the end of 1938, 231 requests were made to the Agricultural Council of which 126 were approved, and the CFA received 117.[99] Even though this figure hardly supports the description of 'a numerous army of applicants', requests were considerable,[100] the vast majority directed to the governorate of Shäwa and Härär. Unfortunately, lack of clear data for both of these areas does not allow us to construct a nationwide trend. The Amara and Galla and Sidamo governorates received 652 and 200 applications respectively. Out of these 81 were accepted in Amara and less than 30 in Galla and Sidamo.[101] It is impossible to attribute the high drop-out rate simply to bureaucratic red-tape, as some technical officers suggest.[102] Many applicants may have been speculators deterred by uncertainty of profits.[103]

Even among those settlers who took up concessions, those who possessed the inclination, expertise, capital, and labour to make a success of their venture seem few and far between. The official image of these men as motivated by a passion for farming, rather than sheer economic incentives, and who found in the Empire a territory for developing their talents, needs qualification. Men of substance and talent were dishearteningly few. Like at any frontier, the concessionaires were an atomistic group which included people with different skills, resources, and motivations. There were some lively and colourful characters: adventurers 'searching for a rapid fortune by grabbing whatever comes into their hands';[104] high-ranking government officials and their protégés with no farming experience;[105] individuals who wanted to join in the race for the

[98] DGG 22-4-1939, no. 373: Concessione Agricolo-Pastorale nella zona di Mencherrà (lago Ascianghi) accordata a S. E. Rodolfo Graziani marchese di Neghelli, in *GU* 4/16, 9 Aug. 1939; Pesce, 'Disciplina giuridica'.

[99] CD IAO AOI 1982 DGCL, Domande. [100] CD IAO AOI 1801, Attuale, p. 2.

[101] CD IAO: Am 600, Relazione sull'attività, p. 2; GS 1820, Il problema, p. 1.

[102] CD IAO: Am 600, Relazione sull'attività, pp. 2–3; AOI 1808, Am UA, Relazione, pp. 14–15. [103] CD IAO: GS 1820, Il problema, p. 1.

[104] ASMAI AS III, 2ª Guerra Mandiale PA, Mezzetti to Terruzzi, 13 Aug. 1938; Lessona, *Memorie*, 273–4; Villa Santa *et al.*, *Amedeo*, 206; *CI*, 4 Sept. 1938; Fossa, 'Sane Norme'.

[105] Among this group the most prominent were: Graziani himself who had, in addition to 500 ha in Juba (Somalia), 500 ha at Menchere near Lake Ashanghé (cf. CD IAO AOI 1936, UA Er, Promemoria VI (1938); DG 22-4-1939, no. 373); Nerio Poggi who came to Eritrea with three lorries and two years later, with another partner, owned 1,150 ha,

legendary wealth of the Empire;[106] and lapsed Fascists ordered to make atonement in Africa.[107] From the perspective of promoting agricultural development, this motley could only be viewed as marginal.

The colonial authorities forced several farmers to relinquish their concessions on the grounds of 'lack of technical fitness and financial resources'. In Amara, 15 such cases were reported by the end of 1939 and 22 in the middle of 1940.[108] By the beginning of 1940 11 instances appear in the list of concessionaires of the governorate of Galla and Sidamo.[109] With these the administration dealt severely. But it was powerless in cases related to people with influence and connection within the State or Party,[110] although such people did not directly man their concession. They lived in Addis Ababa or Italy cashing in on the proceeds from their farms worked by Ethiopians.[111]

It can be concluded that scrutiny of the applications, despite the exasperatingly long time it took, was not as vigorous as the authorities claimed and the applicants' motives were not always genuine. The colonial authorities were outraged at being taken advantage of. Expressing such anger, the Duke of Aosta made it unmistakeably clear that land be granted only to authentic farmers. On 5 January 1938 he circulated a letter to governors and military officials, as well as the MAI, demanding that

> land grants should not, I repeat should not, be considered as a reward that can be given to individuals who, notwithstanding their great service [to the country], do not possess the necessary requirements of technical and financial fitness to develop it. Unless we are firm on this point we will be heading for a chain of more or less covert concessions, i.e. land speculations: this must be stopped by all possible means. So, the order of the day is this: land should be given to the farmers.[112]

engaging himself in all sorts of speculations (A. Del Boca, 'Tre generazioni di uomini per domare un oceano di sabbia', *Settimana Illustrata*, 21 Aug. 1956). Cf. also Masotti, *Ricordi*, 103 where he mentions a settler who boasts his friendship with Mussolini and had as a mission to keep other colonists trigger happy and not let them return to Italy.

[106] Del Boca, *Caduta*, 211; Villa Santa *et al.*, *Amedeo*, 206.

[107] A typical ex. was former Party official of Somalia, Marcello Serrazanetti, who was rehabilitated by setting up an agricultural concern in Gm. Tullio Mussolini had a farm of 600 ha near Deré Dawa, with a mission to remind the Italians of the rural origin of the Duce's family (cf. Del Boca, *Caduta*, 211).

[108] CD IAO: Am 600, Relazione sull'attività, p. 3; AOI 1808, G Am UA, Relazione, p. 14.

[109] CD IAO: AOI 1984 Elenco delle aziende; GS 1822, Attività agricole, pp. 6–14.

[110] ASMAI/III, As III P4, Mezzetti to Terruzzi; Masotti, *Ricordi*, 102–3.

[111] These included people such as Graziani, Volpi, and Temistocle Testa, the Prefect of Fiume, and the owner of SIMBA.

[112] CD IAO AOI 1936, Amedeo to MAI, AA 5 Jan. 1938; Villa Santa *et al.* *Amedeo*, 225.

However, nothing would be more erroneous than to conclude that all concessionaires were mere speculators. According to the authorities these formed a tiny minority. The majority were 'honest and upstanding individuals' who struggled as best they could to turn their land into a thriving concern. The enormous gamble that some concessionaires took in massive investment to develop their farm partially supports this claim. But despite eulogistic reports by technicians, the internal correspondence of even some successful farms reveals serious managerial and administrative shortcomings. One such case was that of SIMBA which operated three large concessions in Arsi and Härärgé.[113] The company had a substantial investment in machinery and wheat production at its Arsi farm in 1939 was claimed to amount to almost one-third of the entire contribution made by the settler section. But this was only a partial description of the real picture. Like most settlers' farms, SIMBA repeatedly suffered serious setbacks in production as a result of harsh environment. In 1940, for example, 6,000 ha were sown, yielding 55,000 ql, but just before the harvest the entire wheat fields were suddenly devastated by rust which 'advanced dramatically and deceptively'.[114] And yet, natural disaster was only one aspect of the company's problems. The worst were poor and unrealistic planning and bad management.[115] A secret report of 1940 gives a verdict on what officialdom described as one of the most successful commercial farms:

> The personnel situation [at Asälla] is painful, distressing, and unpleasant . . . Indeed, the best description I can find for what I saw at Asälla is simply tragic: extreme staff shortage, total I could-not-care-less attitude among the workforce, and complete lack of any sort of attachment to the company by all . . . One thing which I had already denounced to the Director is that in my estimate there is 30 per cent inefficiency by farm inspectors—brave men as far as one wishes but capable only of running a family kitchen garden. In terms of administrative efficiency, we are below zero and of this one becomes aware only when assessing the amount of material employed. The vast majority of the workforce is mediocre, destructive [lit. machine-breaker], indifferent, and apathetic.[116]

SIMBA's success rested more upon its virtual monopoly over trade with the Ethiopians rather than its brilliance on agricultural enterprise. The

[113] See Table 19. The concessions predated Italian occupation and two of them—Fadis and Baka—were owned by Emperor Haile Sellassie (cf. ACS ONC 5/A/1, Taticchi to SC, AA 12 Apr. 1938).

[114] CD IAO 1929: SIMBA, Pavirani to Testa, Asella 4 Nov. 1940.

[115] CD IAO AOI 1929, Maugini to Testa, Rm 18 May 1940.

[116] CD IAO AOI 1929, SIMBA, M. Pavirani to Testa, Aselle 7 June 1940.

TABLE 19. SIMBA concessions, 1938

| Place | Area granted (ha) | Areas cultivated (ha) | | |
|---|---|---|---|---|
| | | Directly | Sharecropping | |
| | | | Tenants | Owners |
| Fadis | 10,000 | 616 | 354 | — |
| Asälla | 11,831 | 9,000 | — | — |
| Baka | 30,000 | 250 | 2,800 | 1,500 |

*Note:* The tractor numbers involved for the entire three concessions was 30, mainly of foreign make. In addition the Company had a variety of implements of all kinds including lorries, sowing-machines, and a thresher.

*Source:* CD IAO 1929, SIMBA, Appunto dettato dall'Ecc. Testa, 13 Mar. 1941; CD IAO 1800, Promemoria Per Maugini; 'Vita dell'Impero: Possenti macchine al lavoro', *GV*, 7 Apr. 1940.

company forcefully eliminated any competition by well-established Ethiopian and foreign, mainly Yemeni, merchants from the areas of its extensive concessions. It did so by setting up strategically located trading centres in its catchment areas from where unathorized trading was checked and Ethiopian merchandise—such as cereal, leather, honey—was exchanged with the goods manufactured in Italy, mainly fuel, liquors, clothing, and others.[117]

## EFFECTS OF LAND ALLOCATION POLICY THE CASE OF CFA FARMERS

The MC's offensive to develop Ethiopia agriculturally found a very powerful partner in the CFA, the first pioneers of commercial farming,

[117] CD IAO AOI 1929, Rapporti fra la società [SIMBA] e gli agricoltori proprietari. Almost all Italian farms, including the settlement agencies, obtained exclusive trading rights in their operational areas or 'Zones of influence' but not all of them had been successful in eliminating the local traders or were able to control such trade effectively (cf. DGG 29-9-1937, no. 693, Disciplina del commercio e consumo dei celeali, in *GU* 2/21, 1 Nov. 1937; DGG 1-2-1939, no. 86, Norme per la disciplina del commercio e del consumo deicereali, in *GU* 21/6 (suppl.), 22 Mar. 1939; DGG 20-2-1939, no. 182, Istituzione dell'Azienda Ammassi Cereali del G. Sc, in *GU* 4/16 (suppl.), 22 Mar. 1939.

resembling the White settlers in Kenya. But the CFA's case also stands to demonstrate the frustrating difficulties of commercial farming.[118] At the end of 1936 several CFA farmers, consisting of both agricultural technicians and entrepreneurs, were given by the then MC and their own farmers' union organization the task of exploring possible areas of rapid agricultural enterprise, and told to report on 'the impressions and considerations gained during their research'.[119] Nine of the men were pressurized by the MAI to cut short their tour and engage in wheat production. They were joined later at the end of 1937 by twenty-one other entrepreneurs and their entourage of 114 personnel consisting of farm assistants and skilled technicians. This group brought with them substantial agricultural material. In addition to 1,000 ql of implements, there were 53 high-powered tractors with ploughs, sowing- and harvesting-machines; 40 motor vehicles consisting of workshop vans, small trucks, and motor cars. An account given by one of the second group bears witness to the type of arrangements envisaged:

Towards the end of 1937 the CFA, in agreement with the MAI, held several meetings with various farmers of Upper Italy in Rome at its headquarters of Palazzo Margherita in the presence of representatives from the Ministry and the Fascist Party. The meeting has the scope of convincing the invited farmers to undertake in IEA and with their own means the cultivation of the land. During the session a five-year (1938–39–40–1–2) contract for cereal farming was presented on behalf of the Armed Forces of the Governor-General of Addis Ababa, where a minimum price was set up and all related conditions agreed, such as a pledge on the part of the Armed Forces to anticipate and supply fuel and oil at the foot of the farm for a fixed price and collect the produce from the mouth of the machine inside the concession itself as well as to advance, in addition to the fuel and oil, also all the seed needed. In one of these meetings, some farmers, including myself, asked to go first to Africa to the site of the concession in order to assess the soil and then be in a position to provide the most appropriate equipment. The tendency was then to do everything at once and no time was given to do what instead turned out to be of maximum importance. As a result few went with suitable equipment. The formal promise of the eventual transfer of the concession into ownership and the quinquennial contracts for the farming of cereals were to delude all the farmers into starting, without any delay, and at once taking up the life of a pioneer in IEA.[120]

[118] For additional information on the background of the group see Bono, 'L'opera agricola privata', *GAA*, 13 Nov. 1938, pp. 397–8.

[119] See Ch. 3; ATdR 24/103, Relazione di una Commissione, p. 1; T. Pestellini, 'I territori dell'Impero e la loro valorizzazione agraria', *Georgofili*, 3/3 (1937), 340–60: MAE, Avvaloramento, 345–6; Quaranta, *Ethiopia*, 51–2.

[120] CD IAO AOI 1938: Vincenzo Valducci to Maugini, Fl Jan. 1942, p. 1; 'Notiziario agricolo commerciale AOI', *AC* 32/1 (Jan. 1938), 44.

The first group settled in the immediate surroundings of Addis Ababa before they were accommodated at Addis Aläm, 50 km west along the motorway leading to Näqämti. On this same route 125 km away at Ambo, part of the second group operated. Of those remaining, one was settled at Wänji in Shäwa, where he laid the foundation for future sugar plantation in Ethiopia. The rest were stationed at Čaffa and Boruméda near Däsé, in Wällo. Despite some differences, the vicissitudes of both groups were very similar.[121]

Soon after their arrival an agreement was struck between the first group and the Intendenza Militare, a Quartermaster-General branch entrusted with the promotion of grain production directly or through military commands stationed in Shäwa. The agreements in essence were those mentioned by Valducci. The farmers were to operate in the areas of Moggio and Ambo where the government at the initial stage would grant to each of them at least 500 ha of free land and modern agricultural machinery. It also pledged to advance seeds and purchase the entire produce at a fixed price. However, events developed that soon proved that the government's pledge was untenable. As usual the main stumbling-block was to find lands of public domain with extensive and unbroken holdings. Such miscalculations called for modification of the original plan in the scheme that came to be known as 'zone of influence'. The term was not meant to have any political connotation but only agricultural reference. In order to maximize cultivation, the military chiefs were urged to arrange special agreements between the Italian farmers and the Ethiopians in lands under their respective control. The order set out the type of lands implicated and the modalities of the operations.

> The lands should not have to be exclusively public lands . . . but also vacant private property whose owners declare not to be in a position to cultivate it . . . Consequently, the agricultural activity stemming therefrom, cannot take place on vast organic plots, but is desultory, departing from one centre of expansion, in all lands that can be found in the aforementioned conditions.[122]

The farmers were grouped in four zones that formed their residential and operational headquarters. The division closely met the agricultural conditions of the respective territories. A system that came to be universally known as *compartecipazione*, a bizarre form of sharecropping, developed which had its justification in local agricultural custom. A variety of contracts were signed in the presence of the Italian authorities and local

[121] *Azione Coloniale*, 17 Nov. 1938; *CM*, 17 Dec. 1938.
[122] CD IAO AOI 1930, Graziani, Incremento Culture, AA 11 May 1937, p. 1.

chiefs. In each, the share of produce depended on the contributions afforded by the contracting parties. As a rule, the Ethiopian farmer provided land and carried out all manual work in the field that was to be ploughed by the Italian using modern agricultural equipment. In addition, the Italian advanced the seed and, in some cases, threshed the harvest.

The contracts took five different forms after the seeds supplied by the concessionaire were deducted from the total produce: (*a*) If the concessionaire carried out most of the mechanical work from threshing to ploughing he took two-thirds of the produce; (*b*) three-fifths for any mechanical input which did not involve ploughing; (*c*) one and a half in case he failed to supply the seed; (*d*) and one-fourth if his input was confined to threshing only; (*e*) finally, 50 per cent whenever the concessionaire's input involved seed and threshing; the remainder was shared between the Ethiopian and the Government, taking 40 per cent and 10 per cent respectively:[123]

The scheme evoked the old idea of economic partnership whereby Ethiopian labour and land on one hand, and the capital, expertise, and management of the Italian on the other, entered into a working relationship. Of course, the Ethiopian remained in the role of adjunct to the Italian, assisting and not really controlling production. Reportedly a large number of contracts were signed involving an area of 20,000 ha. Of these only 8,000 ha were actually cultivated and 12,000 ql of seeds distributed by the Italians and, as a result, an increased production was obtained. Pini, a representative of the CFA, enthusiastically describes these developments and their future implications:

> To it was owed the peaceful and spontaneous emergence of very many share-cropping contracts between the metropolitan concessionaires and the natives (who were involved individually) sanctioned as a guarantee by the local chiefs. It gave rise to extensive distribution of seeds by our farmers and set in motion friendly relations which, while, on one hand, permitted the Whites, lonesome in treacherous territories, maximum freedom of movement and action to promote farming activities, on the other, gave hope to effective assertion of favourable relations between native labour and Italian capital and management.[124]

This was a view shared by political circles in both Rome and the Empire, where the experiment was watched with particular interest. For them

[123] CD IAO AOI 1930, Promemoria per il Vice Re, n.d. [1938]. This contract was later modified at the Säbbäta meeting that took place on 7 Dec. 1937 from which the contracts indicated in Table 21 were finally worked out.

[124] CD IAO AOI 1936, Pini, Situazione, p. 1.

too it seemed to offer a classic example of co-operation of Italian capital and management with Ethiopian land and labour. The short-term beneficial effects of such collaboration were not only the sizeable increase in production that was thought to bring the ideal of imperial self-sufficiency within reach, but also a valuable insight into the agricultural resources of the country and an opportunity to test the new machinery. It also had a political and psychological bearing beyond crude economic considerations. It meant a symbolic presence for the State in remote and hostile rural areas. At the same time, the use of mechanical tools would enhance the prestige of the Italian farmer in the face of the Ethiopian familiar only with rudimentary methods of farming. The scheme also aimed to gradually free land from the Ethiopians for colonization and secure a flow of labour into the farms of the Italian settlers.[125] The authorities were led to believe that such a task, though cumbersome, was by no means difficult to achieve. With incentives, such as a salary, Ethiopians eventually would be convinced to relinquish their land and work under Italian guidance.[126]

Initially, the experiment seemed promising. As Pini described,

> The initiative moved, until the period of harvest, to the mutual satisfaction of the parties: for a multitude of the natives who never dedicated themselves with much intensity to fieldwork, it served as stimulus and encouragement; the peaceful work of penetration carried out by the nationals was of such efficacy that it induced even armed groups to give in and set to work; hundreds of hectares that were never put under production were cultivated.[127]

As a result, from the 8,000 ha sown, about 30,000 ql was raised.[128] This modest achievement was overshadowed by a wave of disputes that intensified close to the harvest. All sorts of claimants came out of hiding, claiming a share of the crop. The authorities feared a serious political backlash and in order to pre-empt it they abolished the sharecropping system. Once again Pini, in a graphic description of the plight of CFA farmers, provides an insight into some of the causes:

[125] CD IAO AOI 1936, Lessona to GG, Concessioni di terreno ed utilizzazione degli indigeni che vi esercitano l'agricoltura, Rm 17 Sept. 1937; *Azione Coloniale*, 17 Nov. 1938; Villa Santa *et al.*, *Amedeo*, 223; MAE, 'Valorizzazione', 271–2.

[126] CD IAO AOI 1936, Lessona to GG, Concessioni.

[127] CD IAO AOI 1936, Pini, Considerazioni.

[128] FO371/22020/J371/40/1, Lord Perth to FO, Rm 28 Jan. 1938; Villa Santa *et al.*, *Amedeo*, 223.

With the approach of harvest operations, numerous meddlers; pseudo-landowners who have hardly any evidence to substantiate their hereditary rights, contemptible even by the same local chiefs; and priests, posed to transform an ecclesiastical right to the tithe into claims of ownership, have in the later days swelled the ranks of the claimants and sought to instigate the masses of peaceful sharecropping farmers hoping to find a safe play in the current political circumstances with the hope of driving away these pioneering concessionaires whose activities have so far contributed, and continue to be of great importance, to the increase of production by the natives.[129]

Pini's account gives only a partial picture of events that unfolded in what one district officer lamented as a 'highly undesirable state of things'.[130] The account was criticized by Graziani for its 'inaccuracies and exaggerations'.[131] Poor understanding by the Italians of the complexities of Ethiopian land tenure systems, which allow multiple claimants to one piece of land, was the primary cause of breakdown of the initiative.[132] Added to this was the ambiguity that the concept of *zone of influence* entailed. For many of the concessionaires, the concept opened not only a fertile field for economic activity but also gave them effective control over Ethiopians in their *zone of influence*. As Graziani pointed out, the 'farmers often behaved towards the natives in a thoughtless or bullying manner'.[133] A number of contracts were exacted by use of sheer force, in clear defiance of the government's order that 'It is necessary that land should be occupied only if this was effected with the consensus of the owners to whom may be given in exchange either a rent or a share of the produce as agreed by the contracting parties.'[134]

Mechanical threshing was considered 'an indispensable means for rational and full utilization of the produce in which was spent so much work and sacrifices', but it was not made compulsory. And yet several farmers, lured by the revenues that this offered, forced against their will many of the peasants, including even those outside the zone of influence, to use their machinery despite the official instruction that this should be done only through the propaganda and persuasive work.[135] Their action

[129] CD IAO AOI 1936 Pini, Situazione, p. 2.

[130] CD IAO AOI 1936, Cerulli to MAI, Concessioni provvisorie; Villa Santa *et al.*, *Amedeo*, 223.

[131] CD IAO AOI 1936, Graziani to MAI, Concessionari agricoli del CFA, AA 28-12-1937.

[132] See Ch. 2.

[133] CD IAO AOI 1936, Graziani, Concessionari. Identical view is expressed by General Mezzetti, G of Am, in connection with some of the concessionaires of his area (cf. ASMAI AS III/P4, Mezzetti to Terruzzi, Däsé 13 Aug. 1938).

[134] CD IAO AOI 1930, Graziani, Incremento culture.

[135] CD IAO AOI 1930, Petretti to DS AP, Contratto con agricoltori nazionali, AA 27 Nov. 1937, p. 3.

stirred a lot of discontent among the Ethiopian farmers. This situation was further aggravated by the licence granted to each farmer to have an armed police control to protect his farm and the right to collect tributes from the Ethiopians on behalf of the government in his zones of influence. Owing to such extra-economic powers, those acquiring farms ended up as *imperium in imperio*: they punished reluctant elements in the population and nominated new village chiefs, sometimes removing even those appointed by the central authority. Others engaged themselves in illicit speculative activities and signed numerous agreements promising exorbitant rent knowing very well that they had neither the financial resources nor technical capability. Instances where the concessionaire lived in a nearby city from where orders were passed to the Ethiopian farmers to cultivate the land were not unknown,[135] reminiscent of the feudal lords of the *ancien régime*.

Other contributing factors to the setback of the initiative, mentioned by Pini himself, were less of the Italian farmers' own making. Tractors, oxen, and vehicles promised by the government either arrived too late or were not strong enough to cope with the terrain. Largely owing to this, many of the farmers were unable to maintain their part of the agreement and farming was carried out largely by Ethiopians using traditional methods. Problems were further compounded by the modest yield of newly introduced crops of the Italian variety for which the Ethiopians held the Italians to be largely responsible. Because of this, the Ethiopians strongly objected to sharing their produce, unmoved by the barrage of threats and abuses by the Italian farmers. Endless disputes and conflicts supplanted the initial phase of friendly co-operation. Some Italians received arrest warrants charging breach of contract, malpractice, and gross-misconduct against the Ethiopians. Some farmers were attacked by a gang armed with spears, knives, and staves. Their sense as a master race in the eyes of the Ethiopians and prospects for a more solid co-operation in years ahead were soon shattered.[136]

The situation was tense, and the subdistrict officers feared that unless swiftly addressed, discontented Ethiopians would turn to the nationalist cause, providing the patriotic movement with a greater flow of recruits. It

[135] Sbacchi, 'Italian Colonialism', 370–2; ASMAI AS III/P4, Mezzetti to Terruzzi.

[136] CD IAO AOI 1936: Pini, Situazione, pp. 2–3; id., Promemoria, AA 2 June 1938. Pini points out that the Ethiopians owed 6,770 ql to CFA members and interprets it as being an 'irreparable loss of the produce with significant economic damage for the concessionaires and the development in the minds of the native of the belief that he could get away with impunity from the obligation that he had freely taken, which was ratified by the government, and appropriate for himself what rightfully belongs to the White man'.

was against this background that, on 7 December 1937, a meeting was convened by the then Addis Ababa governorate at Säbbäta, about 25 km from Addis Ababa on the Addis Ababa–Jimma motorway. It was attended by the farmers and the local officers from the respective zone of influence. In the major review that followed, the controversial zone of influence with all the attributes attendant upon the term, including police force and tax collection, was abolished.[137] With one exception, the existing sharecropping contracts either were revoked or modified. The Italian farmers were persuaded to settle for a share from the preceding year's harvest that was drastically 'reduced to a bare minimum'. Yet even this was strongly resisted by a substantial number of Ethiopians, with the authorities either conniving with their resistance or impotent to intervene.[138]

Two brothers, Sostene and Carlo Boidi, were exempted. Unlike their fellow-countrymen, the Boidi brothers seem to have earned a high standing among the Ethiopians with the result that they commanded the respect even of hard-line Fascists. The key to their success was their business ethics, entirely alien to Fascist thinking: they combined thrift and hard work with respect for local tradition and friendly treatment of the Ethiopians. By a communiqué of 19 April 1938, their activities were officially sanctioned. This document defined their areas of operation and set out in great detail the different type of sharecropping contracts they were allowed to establish. Their work went ahead smoothly, and within a few years they were able to conclude 3,000 contracts covering an area of more than 2,000 ha.[139]

Even though the government pursued the policy of 'avoiding, as a matter of general principle, sharecropping contracts with the Ethiopians' and grants of land for direct cultivation, it was not able to devise a better

[137] CD IAO AOI 1930, Amedeo di Savoia to GG Commando Forze Armate per Intendenza AOI, Contratti con agricoltori nazionali, AA 28 Jan. 1938. In this letter not only does he declare the term 'zone of influence' abolished but also removes all extra-economic activities that were attributed to the farmers; mechanical threshing becomes optional again to Ethiopian farmers even though the respective political authorities were called to give support for the publicity work of the farmers among the Ethiopians for its use. Villa Santa *et al.*, *Amedeo*, 223.

[138] On 23 Feb. 1938 the government ordered the computation of the quantity owned by the Ethiopians and the prompt settlement of it; yet as it appears from Pini's complaints (CD IAO AOI 1936 Pini, Promemoria, AA 2 June 1938) no action was taken by early June 1938.

[139] CD IAO AOI 1936: Cerulli to MAI, Concessioni provvisorie; Cerulli to G GS and Commando Settore Occidentale [Ambò], Contratti di compartecipazione con gli indigeni, AA 19 Apr. 1938; Cerulli to MAI, Contratti di compartecipazione con gli indigeni, AA 22 June 1938. ACS GRA 43/34/26, Renato Trevisani [Delegato centrale per l'AOI della CFI], Relazione sulla missione svolta in AOI, Apr.–May 1938, p. 37; ACS ONC 3/B/1, Taticchi, Relazione mensile to SC, AA 14 Sept. 1939; Villa Santa *et al.*, *Amedeo*, 240.

method of handling the problem of land and labour shortages. Later a modified version of sharecropping was introduced in Shäwa. Despite the Ethiopians' resistance to its restoration, the contracts of the Boidi brothers with the local population remained the prototype for this new scheme.[140]

However, the sudden policy shift outraged most of the 'pioneer column', who denounced it as betrayal and cowardliness. In their view the government was simply capitulating 'to the most puerile complaints by Ethiopian intriguers' and to malevolent advice from fellow countrymen 'who never even deigned to follow closely the life of their farms'. They pointed out the political dangers in allowing the Ethiopians to win an 'outstanding concession'.[141] Indeed, the government's action had serious moral and material repercussions upon the farmers. A memorandum written on 13 May 1938 gives a brief but clear picture of their feeling when it graphically describes their situation as

> too mortifying for it reduced to total and absolute inertia and economic passivity those very ones who were the first and most courageous agricultural entrepreneurs to reach the colony and sacrifice their time, health, and capital with the sole hope of developing the conquered land. Mortifying and equally humiliating because it established conditions of inferiority, that were less than dignifying, *vis-à-vis* the natives, who took every care not to maintain their (freely assumed) commitments towards the concessionaires who found themselves powerless and without any prestige.[142]

Turned from the confident empire-builders of the previous year into rather desperate and cantankerous entrepreneurs, the farmers became an economic drain. Political authorities were deluged by letters from the CFA headquarters crying for 'an authoritative intervention before the first initiative of colonization was irreparably compromised'.[143] The CFA officials castigated the government's behaviour in Addis Ababa towards 'their patient work of penetration' as lethargic at its best and hostile at its worst.[144] The government in Addis Ababa insisted any action would only create disturbances without having the desired effect. Orders from Rome combined with internal pressure led the colonial administration to agree to a 'package deal' with the 'intent of helping the national farmers who, in the first year of their activity, for a number of complex circumstances

[140] CD IAO AOI 1936, Cerulli to MAI, Concessioni provvisorie.

[141] CD IAO AOI 1936, Pini, Situazione, pp. 2–3.

[142] CD IAO AOI 1936, Promemoria: situazione dei concessionari del I Gruppo, 13 May 1938, pp. 2–3. The document deals with the whole range of problems encountered by the first and second CFA groups.

[143] CD IAO AOI 1936, Pini, Situazione, p. 4.

[144] Ibid., p. 3.

were unable to obtain just remuneration from their work'.[145] In light of this, early agreements were substantially reviewed. Payment for seeds due at the end of 1937 was frozen; fuel repayment was suspended up to February 1940 and amortization of the agricultural machinery, originally expected to be over by the end of 1939, was made to start by early 1940.[146]

According to the agreement reached between the CFA and the MAI, twenty-one farmers of the second CFA group were to be provided with 1,000 ha each in Ambo and Däsé. It was mistakenly believed that in these areas there was at least 15,000–16,000 ha of public land.[147] The available land was of poor quality and attempts to direct their activities towards more fertile zones were frustrated by absence of land free from Ethiopian claims. The Ethiopians were reluctant to move off the land even where exchange was arranged. Lack of security, combined with high fuel and labour costs, made the life of the farmers even more uncomfortable.[148] Two repatriated, the rest pressed for the revision of the existing agreements and were able to win a number of concessions. The guaranteed price of wheat was brought from L.200 per ql for an average yield of 8 ql per ha to L.305 per ha whenever the production unit per ha was 4 ql. The price decreased in proportion to the increase of yield per ha. In addition, there was a land reclamation fund of L.250 for each hectare reclaimed.[149]

As Table 20 shows, in early 1940 the farmers' plight had slightly improved but the areas they were farming were of far more modest size than they had been originally promised. Rather than contributing to imperial self-sufficiency in foodstuffs, the concessionaires remained totally dependent on state subvention. None succeeded in cultivating their concession. Out of about 4,000 ha in Däsé, only about 1,500 ha were farmed. Of the six farmers, two had not commenced farming. The situation was by no means different in Shäwa, as the cases of Scagliarini & Sons and A. Bisacchi show (see Tables 15 and 16).

These two CFA cases illustrate the disappointment common to many land-hunters and the colonial administration's incompetence and lack of direction. Their enterprise, which had begun in a mood of determined

[145] CD IAO AOI 1936, Cerulli to MAI, Concessioni provvisorie.

[146] Ibid.

[147] CD IAO AOI 1936: Muzzaini [President of CFA], to MAI IA, Rm 28 Sept. 1937; Produzione; CFA accordi; Lessona to CFA, Trasferimento.

[148] CD IAO AOI 1936: Rossi to Bonfondi, Promemoria.

[149] CD IAO AOI 1936: Cerulli to MAI, Concessioni provvisorie; Promemoria: Avviamento alle imprese agricole dei concessionari del II gruppo, 13 May 1938. CD IAO Am 600, Relazione sull'attività; Villa Santa *et al.*, *Amedeo*, 240–1; Sbacchi, 'Italian Colonialism', 380–3.

TABLE 20. Landholdings and tractors of the CFA, 1940

| Concessionaire | Place | Date of concession | No. of tractors | Areas of concession (ha) | %age |
|---|---|---|---|---|---|
| Bros. S. & C. Boidi | Addis Aläm | 1937 | 11 | 900 | 5.49 |
| Borgnino Carlo | Addis Aläm | 1937 | 3 | 320 | 1.95 |
| Gentile Francesco | Addis Aläm | 1937 | 2 | 350 | 2.14 |
| Morello Orazio | Addis Aläm | 1937 | 4 | 320 | 4.88 |
| W. & G. Raisi | Addis Aläm | 1937 | 5 | 430 | 1.95 |
| Scotto Ernesto | Addis Aläm | 1937 | 4 | 260 | 2.62 |
| Bisacchi Ugo | Ambo | 1938 | 5 | 1,450 | 1.58 |
| Busotti Pasquale | Ambo | 1938 | 3 | 680 | 8.85 |
| Greppi Giuseppe | Ambo | 1938 | 5 | 840 | 4.15 |
| Iotti E. & C. | Ambo | 1938 | 3 | 750 | 5.17 |
| Madella Beniamino | Ambo | 1938 | 5 | 1,400 | 4.57 |
| Mattioli Cesare | Wänji | 1938 | 3 | 700 | 8.55 |
| Raddi & Osti Quintilliano | Ambo | 1938 | 3 | 650 | 3.97 |
| Scagliarini Roberto | Ambo | 1938 | 5 | 1,500 | 9.16 |
| Toschi L. & Bros. | Ambo | 1938 | 2 | 490 | 2.99 |
| Valducci Vincenzo | Ambo | 1938 | 4 | 1,450 | 8.85 |
| Atti Enzio | Däsé | 1938 | * | 290 | 1.77 |
| Bagnoli Roberto | Däsé | 1938 | 2 | 664 | 4.05 |
| Martini Gino | Däsé | 1938 | * | 500 | 3.05 |
| Pavia Giuseppe | Däsé | 1938 | 2 | 1,100 | 6.72 |
| Rampolla | Däsé | 1938 | 0 | 140 | 6.72 |
| Roggero Beniamino | Däsé | 1938 | 4 | 1,100 | 0.85 |
| TOTAL | | | 85 | 15,684 | |

*Note:* * Indicates that figure is not available.

*Source:* CD IAO: AOI 1808 Am UA, Relazione; AOI 1802, [AOI Concessioni]; Sc 1801, Attuale; Hr 1813,

optimism, had become a more sober affair, beset by doubts and anxieties. Yet CFA's experience was not an isolated case, but the reality for many concessionaires.[150]

## THE ORGANIZATION OF LABOUR

The viability of commercial farming largely depended on the ability to tap and exploit cheap indigenous labour. As an official of the MAI stated: 'It is clear that it is futile to grant land for agricultural purposes unless it is possible to secure for the enterprises the labour at the scale required and on the most favourable terms of salary.'[151]

Contrary to all expectations, labour recruitment proved far from easy and throughout the occupation the companies experienced a serious labour shortage. Aware that labour was a critical factor for the success of the settlers' infant agriculture, the Agricultural Council had earlier called on the government to devise proper ways and means to settle the issue. It wrote that

> rejecting a priori, for obvious reasons, any idea of compulsory labour by the natives, it is necessary to think of pursuing a policy that allows an influx into the White farms of numerous natives disposed to work for a low salary, in the same style practised in all Tropical African countries, which is indispensable to putting the enterprises on a sound economic footing. These problems cannot be solved without the continued good office of competent organs of the colonial administration and a comprehensive understanding of the needs of colonization.[152]

Similar pleas were made by the concessionaires themselves. Much to the dismay of the interventionists, the colonial administration's response was cool at its best and inefficient at its worst. At a theoretical level, a number of measures had been worked out since early days to counter the problem, but in practice, they failed. The consternation of the concessionaires increased. When viewing the fact that the majority of the settlers expected that as members of the ruling race they had a right to the services of the subject people, this was understandable. So, many of them regarded the use of constraint as the only effective method for overcoming the 'inertia' of the Ethiopians. But the government was

[150] CD IAO Sc 1328, Ciccarone, Relazione sulla missione; and 1329, id, Relazione.
[151] CD IAO AOI 1926, Angelo de Rubeis Promemoria per S.E. il Sottosegretario: Mano d'opera indigena per le aziende agrarie nelle nuove terre dell'Impero, Rm 12 May 1938.
[152] Consulta, 'Avvaloramento', 1571.

concerned about the political implications such a policy of forced labour might cause, and was reluctant to openly sanction compulsory recruitment as the farmers demanded. It equally understood that unless a conscious effort was made to meet the demands of the struggling settler farms, its vital interest would be seriously compromised. A statement by Nasi sets out the government's dilemma:

It is necessary for the agricultural concessionaires to bear in mind that the government cannot and will not deal with the 'Black slave trade' in order to supply the native labour they require. Indeed, the government can and must help in every possible way in the recruitment of labour required by private farms. But this and 'the slave trade', which is being demanded, are poles apart.[153]

Although labour proved an intractable problem, it was resolved to some degree, in connivance with the administration, by pressures and brutalities that had all the characteristics of forced labour elsewhere in colonial Africa. This makes the declarations of the Italian authorities against such measures appear hollow. At the time of Nasi's pronouncement, two dominant patterns had emerged as the standard forms of statutory exchange of labour: waged or *salariato*, and sharecropping or *compartecipazione*. Both were widely used with the tacit endorsement, and at times open encouragement, of the colonial administration. The two systems were modelled on labour codes prevailing in other Italian colonies, especially Somalia and Eritrea.

Waged labour was recruited by Ethiopian chiefs who, as in other parts of colonial Africa, were also the main instruments for raising forced labour for public road works, military and paramilitary services, and the cultivation of certain cash crops.[154] In substance if not in detail, the pattern of recruitment was identical throughout the Empire. After a relatively careful census of the population, each chief was responsible for supplying a specified number of workers from the able-bodied, including women and children,[155] under his control. The amount of labour to be provided

[153] 'Il mio credo: Raccolta di circolari del Generale Nasi edita dal governo dello Sc', in L. Goglia and F. Grassi, *Il colonialismo italiano da Adua all'Impero* (Rm, 1981), 395.

[154] CD IAO: GS 1820, Il problema, p. 10; CD IAO AOI 1929 SIMBA, Norme per la mano d'opera aziendale. Labour problems relating to Cotton Districts are extensively discussed in Ch. 7.

[155] Use of child labour was widespread and was widely reported. Thus Mrs Fanin, one of the Fascist admirers among the British aristocracy, states that during her visit in 1938 she saw 'quite a number of children employed with the road gangs' (F0371/22030/310). The Italians justified the employment of children on educational grounds (ACS GRA 44/36 1 Archimede Mischi to GG, *Autonomia nel campo alimentare,* Moggio 15 May 1937).

fluctuated, with its peaks at high agricultural season when its shortage was most intensely felt.[156]

However, as the chiefs and their subjects tried to resist the companies' wishes, labour recruitment proved extremely difficult. Initially, companies like SIMBA used the pre-Italian Amhara governors who successfully delivered the required labour, imposing on recalcitrants severe physical and pecuniary penalties—measures described in the company's records as 'Amhara's effective ascendancy over the subdued people'. But with the emergence of the police force and the pursuit of a policy of de-Abyssinization, most of these governors were stripped of their traditional power, and, replaced by warrant chiefs. Cases in which the latter were elected by their ethnic group on the proposal of company officials were not uncommon; but several of these chiefs lacked roots in tradition and, as a result, were often rejected by the population, making their efficiency in obtaining the required labour very low.

Data on chiefs openly conniving with their subjects against the company are extremely rare,[157] but, as a rule, gaining whole-hearted support from the chiefs needed long negotiations involving hard work, great energy, and additional financial inducements on top of a token monthly payment as part of their service.[158] But resort also was made to punitive measures, coercing the chiefs into active collaboration with the company officials. A typical example was SIMBA, which abolished the chiefs' monthly salary and introduced a remunerative system that was proportional to the size of workforce supplied. The system began to prove successful. But even then the companies continued to be confronted with the problem of a reluctant and generally inadequate workforce. Well-to-do peasants were often able to avoid any work by paying their way off to the chiefs.[159] The recruits induced to work had to be organized into labour gangs (*squadra di lavoratori*) and closely supervised, often by armed guards. Otherwise they simply stopped work or fled.[160]

The Ethiopian chief was only an auxiliary instrument. The real power rested in the *residente* and *commissario* who were the direct contact between the company and the chiefs. Any power the chiefs had was

[156] CD IAO AOI 1848, Fonti to G Hr, Concessioni, pp. 4–5; P. Bono, 'Opera agricola'.

[157] A typical case is that of the local *cadis* at SIMBA's farms who, unlike the Amhara leaders, refused to use force against their Muslim co-religionists unwilling to work for SIMBA (cf. CD IAO AOI 1929, SIMBA, Norme).

[158] CD IAO AOI 1929, SIMBA: Norme; Generalità e programmi di massima, 27 Feb. 1942, pp. 12, 16.

[159] CD IAO AOI 1929, SIMBA, Norme.

[160] A. Chiuderi, 'Agricoltura indigena nel GS ed i mezzi per farla progredire', *AC* 36/10 (1942), 272; Masotti, *Ricordi*, 138–9; see also Ch. 7.

delegated from these officers. The success in labour recruitment, to a certain degree, was very much a matter of how committed these officers were to the civilizing mission.[161] Some officers did their best in collaborating with the management of the companies by talking to the chiefs and encouraging them to be loyal agents; explaining the significance of a particular crop to the metropolitan economy; and punishing and dismissing incompetent and uncooperative chiefs. But such measures offered no lasting remedy to the problem: puppet chiefs proved themselves to be in no better a position than their predecessors and, as a result, they in turn succumbed to the same fate of dismissal or punishment, or both.[162]

In several areas the companies were allowed to recruit their own labour and, as long as this ran smoothly, the administration saw no reason to intervene. But the policy gave rise to ample abuses. Where the labour needs of struggling farms became acute, forced labour became an inevitable option. Thus SIMBA, for example, in its Villa Baka concession organized the inhabitants of the surrounding villages into eight 100-strong labour gangs who provided their service by turns.[163] Each day before the start of work, the company checked each worker on a register. The absentees were immediately reported to the local police who searched out truants. At the beginning of the week, each worker was given a ticket on which each evening a company official recorded the quantity and the quality of work done. A ticket had to be completed within a week, and those failing to do so were required to work the following week; otherwise no payment was made.[164]

The labour-gang or ticket system worked with some satisfaction, but at the expense of the peasants' own subsistence farming. The situation called for the intervention of the central administration who, recognizing the danger of unrest and disruption in agricultural production, sought to offer the Ethiopians some protection. At the end of 1938 the Viceroy's government ordered the companies to disengage and send home all peasants during the sowing and harvesting season.[165] This measure had a devastating effect on the companies as they experienced a marked reduction in the supply of labour specially at a peak period of their activities. A compromise of sorts was reached whereby various *residenti* of the region

[161] CD IAO GS 1820, Il problema, p. 9.

[162] CD IAO AOI 1929, SIMBA, Norme.

[163] Later modifications set the number at 120, while the term was shortened from eight to six weeks (cf. ibid.). In GS it lasted a month (cf. Chiuderi, 'Agricoltura').

[164] CD IAO AOI 1929, Pagamento mano d'opera.

[165] Chiuderi, 'Agricoltura', 272.

combined forces to supply a labour quota by turns. Although the system interfered less in Ethiopian subsistence farming, as a remedy for labour shortage it proved inadequate. Despite strict security, most of the workforce just melted away in the course of the journey. Those reaching the destination were subjected to a tough regime. Even then it was possible for incidents similar to those described by one officer in relation to the labour gang of Jubdo mines in Wälläga to occur:

> If by any chance a leopard or any night animal had a nasty idea of strolling around close to their camps, it offered the best opportunity for total confusion whereby everybody returned to his homeland, where under normal circumstances a visit by a leopard would have hardly raised much emotion once they were locked inside their dwellings. In such a case it became necessary to start the recruitment afresh amid general annoyance—in the first place that of the *residente*.[166]

As the Italians were quite aware, one of the reasons for resistance was the realization by the Ethiopians of unprecedented opportunities opened for them by rival interest groups pleading for their labour.[167] But this was only part of the explanation. The crux was that, unlike the working classes of Europe, the Italians were dealing with people who owned the means of their subsistence. For them money income was not a necessity but a luxury and this made their bargaining position a very strong one. Therefore, as one agricultural technician complained, the Ethiopians were reluctant to work in farms unless offered what he described as 'absurd wages equivalent to those they earned from the construction and building companies'.[168] But owing to the need to render their enterprise profitable and keep their produce competitive in the international market, the farms were not in a position to compete with current market wages. The concessionaries were of the view that a rise in the price of labour would create a serious crisis to the point of putting a large proportion of their farms out of business.[169]

[166] Masotti, *Ricordi*, 138–9. Assessing the severity used Masotti wrote that 'I have to confess that I would never have liked to be the *residente* of Jubdo where a Genoese colleague only with his notorious harshness succeeded in keeping them together.'

[167] Such awareness is also highlighted by Renato Trevisani, an official of CFI: 'Despite all this [= attempt to control wages of the natives], the natives are already well aware that their services represent much sought-after merchandise, as the fact that the success or failure of an employer's business depends on the profit margin that the native considers to be worth for his work stands to prove' [ACS GRA 43/34/26, R. Trevisani, Relazione, p. 40; ACS ONC 4/33, Benigno Fagotti, Relazione, Bishoftú 12 Apr. 1938].

[168] CD IAO AOI GS 1820, Il problema, pp. 9–10.

[169] Fossa, *Lavoro*, 238–9, 253; M. Rava, 'Diario di un secondo viaggio nell'ovest Etiopico', *Nuova Antologia*, 405 (1939), 135, 153–4; F. Santagata, *L'Harar* (Milan, 1940), 208; N. Bonfatti, 'Tutela razziale del lavoro in AOI', *REAI* 26 (1938), 1407.

Wages in road-building ran from L.4 to L.6 *per diem*, and construction workers in Addis Ababa, Härär, and other urban centres earned L.7 per diem and those with greater skills from L.10 to L.20.[170] In contrast, an agricultural labourer's wages ranged from L.1 to a maximum of L.3 per diem depending on age, sex, conditions of work, and in a few cases climate. The normal pay was L.3 for men; L.2.50 for women; and L.2, L.0.50, and L.1 for children depending on performance and type of work.[171] The payment of these wages in lire provided little inducement.[172] As a result agriculture, alongside the mining industry, was left at the mercy of an unreliable Ethiopian workforce which knew that its labour was a much sought-after asset. The effects were as devastating as in the early period of occupation when

> many natives have abandoned the European estates; to make them stay the others have been obliged to increase agricultural wages; where coffee is collected from wild plants many natives who until last year were not paid, or else received very small wages and rarely, have also abandoned the old coffee work or have demanded a higher wage. In the small independent farms the natives have in some cases preferred to leave the coffee for building work where more favourable opportunities are offered.[173]

The intense competition for labour resulted in considerable friction between various groups of Italian employers, notably between the farmers and the construction companies. The government was pressed to standardize wages throughout the country and as result a number of decrees were enacted to this effect. But the government possessed neither the political will nor the ability to carry out its decisions effectively.[174] Orders prohibiting the employment of Ethiopians for non-agricultural work in

[170] R. Pankhurst, 'Italian and "Native" Labour', 62–5.

[171] CD IAO: AOI 1929, SIMBA, Pagamento mano d'opera; AOI 1848, Fonti to G Hr, Concessioni, pp. 4–5. Most common way of payment was part in kind and part in cash. Items paid were barley, *durrah, čat*, firewood, or *dergo*—food ration (cf. Santagata, *L'Harar*, 210; ACS GRA 43/36/26, Trevisani, Relazione, 40–1).

[172] ACS GRA: 46/41/19, Cusmano and Corvo, Situazione economica, AA 1 Apr. 1938; 43/34/26, Trevisani, Relazione, 64–72.

[173] Prinzi, 'Manodopera', 1716.

[174] Salary standarization was initially attempted by the G GS immediately after the occupation invoking severe penalties against the transgressors of both the salary limit as well as payment in thalers; yet this as well as the efforts of other governorates failed to obtain their objectives. (Cf. Pankhurst, 'Italian and "Native" Labour', 60–3; D Gov Sc 9-3-1937, no. 76, Fissazione della retribuzione da corrispondersi ai manovali indigeni, in *GU* 2/3, 1 Apr. 1937; D Gov Am, Determinazione dei salari massimi per opera e lavoratori indigeni, in *BUGA* 2/11–12, Nov.–Dec. 1937 which introduced the law with the explicit view of penalizing 'employers' abuses and unjustifiable cornering of labour'.)

the vicinity of concessions were also unsuccessful in arresting 'the exodus from the fields'.[175]

Although attractive in many ways, sharecropping too did not prove to be a lasting solution. The system took two forms depending on whether the land was domanial or under private ownership. In domanial lands, the Ethiopian resident was considered as a tenant and the concessionaire as a landlord. In such circumstances, sharecropping took the form of labour rent whereby, for the privilege of living on the settler's estate and receiving land for the cultivation of foodstuff, the Ethiopian paid a labour rent to the Italian landlord. As a rule, the small- and medium-sized concessions tended to settle the Ethiopians at the periphery of their farms, forming villages or isolated homesteads that served as reservoirs of labour.[176] The practice on large farms was to allocate land in lieu of wages, and the Ethiopian tenant either appropriated the entire produce for himself or shared it with the concessionaire according to a pre-agreed clause. The tenant was obliged to supply his own and his family's labour, freely in the first case, or against some nominal payment in cash or in kind or both in the second.[177] Abuses by the concessionaires were common, as in Wällo where government officials and individuals with influence and connection demanded 40 per cent of the produce without making any financial contribution to the production process.[178]

The sharecropping system was quite deliberately and consciously fostered by the government which encouraged the Italian farmers to keep Ethiopian cultivators within their concessions.[179] The contract aimed to

[175] Santagata, *L'Harar*, 210–11; ACS GRA 43/34/26, Trevisani, Relazione, 39–40; CD IAO Misc. 1038, Alberto Pollera, Europei e indigeni nella valorizzazione ed economia dell'Impero, As 7 Oct. 1938, p. 18.

[176] The need for a native reserve, or *villagio indigeno*, modelled on that of Kenya and South Africa was advocated by many Italians. The Duke of Aosta, under whose regime racially segregated development policy was introduced, saw the advantages of dividing the entire country along racial and ethnic lines, further expounding it. (Cf. CD IAO Misc. 1038, Pollera, Europei e indigeni, pp. 2–5; CD IAO AOI 1936, Graziani to MAI, Concessioni terriere; ATdR 24/104, G. R. Giglioli, Rapporto al MAI in relazione al viaggio nell'Unione del Sud Africa, Fl 5 Aug. 1937, p. 31; Prinzi, 'Manodopera', 1718–19; Villa Santa *et al.*, *Amedeo*, 214–15.)

[177] Santagata, *L'Harar*; CD IAO AOI 1936, Graziani, Concessioni, AA 19 Nov. 1937.

[178] CD IAO AOI 1936, Pini, Considerazioni.

[179] Most post-1939 concession charters contain a clause on the following model: 'The concessionaire should stipulate with the native workers within the concession area salary or sharecropping contracts . . . and should be clear that they [the natives] cannot be displaced without the approval of the local political authorities and for reasons of well-proven incapacity and indiscipline on the part of the [native] cultivators or the inability of the farm to effectively absorb their manpower.' (CD IAO Sc 1801.) For the other governorships cf. CD IAO: Hr 1812, annexes nos. 4, 5, 6; GS 1822, Disciplinare Tipo per Concessione Agricola.

bind the Ethiopians to the land of the company where they were 'gainfully' employed by establishing them in what was euphemistically described as 'cultivating settlements'.

An interesting ramification of this was the emergence of labour colonies. The companies were unwilling to accept delays or shortfalls in production on their farms; to prevent such an eventuality, particularly where the local workforce was scant, labour was imported from elsewhere. This experiment, chiefly employed by the SIA Arussi in its ex-Belgian concession,[180] directed by Jacopo Gasparini, the former governor of Eritrea whose name was associated with the Täsänäy State Cotton Farm, was gradually gaining ground.[181] After a number of unsuccessful attempts to solve the labour problem, the SIA Arussi brought in Härärgé *näftäña* who had been confined to penal colonies.[182] Later on, the Sidamo migrants from surrounding areas joined them. The scheme aimed to gather each ethnic group into villages with its own chiefs, religious institutions, and water supply. After agreement was reached with the chiefs, a series of sharecropping contracts were established: the Ethiopian settler was given a piece of land for his own use with seeds and agricultural tools, and entitled to keep a few poultry and cattle. He was also provided with a hut, cash loans, medical assistance, and a well for irrigation. In return he worked the company's land at a rate which varied according to the yield produced or inputs provided by the parties. Sources available fail to give details as to the magnitude of this programme. Yet we know that the SIA Arussi by about 1940 had settled two groups of Shäwan Amharas from the JeJega areas. On the whole, this was considered one of the best long-term strategies to overcome labour shortages, particularly in areas where large companies operated. Yet the system was in its infancy when the Ethiopian War of Independence broke out, and it would be speculative to contemplate whether it could have offered a lasting remedy.[183] But a similar experiment in Somalia, where such a system was extensively used

[180] Formed in 1913 and known as Société Belge des plantations d'Abyssinie, the Belgian concession was one of the largest modern concessions predating the Italian occupation. Largely based on dry farming, it extended over 2,000 ha. Although coffee was the main cash crop, a variety of fruits were grown. At the time of occupation it had an estimated labour force of 2,000 and *c.*1,700,000 coffee plants. (Cf. ASMAI 54/22, L'attività economica del Belgio e le aziende agricole nell'Etiopia sud orientale, AA n.d. [4 Feb. 1927]; F. Santagata, 'Il sistema economico degli Arussi', *REAI* 27/9 Sept. (1939), 1078–9; Cerulli, 'Colonizzazione', 74–5.)

[181] See above, Ch. 1.

[182] The Italian officials claimed that they settled them on humanitarian grounds as they needed urgent accommodation owing to the fact that, following the advent of the Italian conquest, their Somali subjects refused to work for them.

[183] Cerulli, 'Colonizzazione', 74.

and for a long period, gave poor results. On the other hand, we know that there was already marked reluctance on the part of free Ethiopians to leave their homes and subsistence farming to work, virtually as slaves, for foreign masters.[184]

Where land was under private ownership, the Ethiopian landlord allotted a certain percentage of his land to cultivate an industrial crop of interest to the company who purchased the produce at an agreed price.[185] The most common procedure was to divide the harvest on the basis of input provided by the Italian concessionaire (see Table 21). Whenever possible, the pre-existing tenancy arrangements between the Ethiopians were maintained. In such cases the landlord, i.e. a *gult*-holder, was the signatory who, however, listed his tenants who continued to pay him their customary duties.[186]

Sharecropping contracts were notarized by local authorities in the presence of the chiefs. They had to be fully explained to the Ethiopian and signed by both parties. The size of plots varied from 0.5 to 3 ha, and the cultivator worked, in theory at any rate, if not always in practice, under the supervision of the concessionaire who distributed seeds and sponsored research.[187]

As we saw earlier, ideally, the sharecropping system was a partnership between the Italian entrepreneur and the Ethiopian farmer and avowedly served a double purpose: to uplift the material welfare of the Ethiopian by organizing rationally his agricultural enterprise and to provide the concessionaire with a secure source of production. Yet in practice, it was a mechanism to incorporate the Ethiopian population into the world market at an arbitrary level of remuneration. In the process, the Ethiopian peasant not only became the basis of exploitation, rooted in traditional economic structure, but also surrendered his independence to become the servant of business interests. A relatively independent community passed into the control of a remote financial magnate. Sharecropping was attractive to the Italians for its functional versatility: primarily, it could claim roots deep in tradition and local agricultural practices;[188] in addition to coping with labour deficiency, it aided expansion of production keeping costs far below the market level. Finally, it helped to maintain the racial

[184] CD IAO Misc. 1038, Pollera, Europei e indigeni, p. 13; see also Ch. 7.

[185] CD IAO 1929, SIMBA: Rapporti esistenti fra la Società ed i coloni, n.d.; Santagata, *L'Harar*, 210.

[186] CD IAO 1929, SIMBA: Rapporti. [187] Ibid.

[188] See Ch. 2, on the traditional Ethiopian tenancy system; CD IAO 794, Ciocca, Elementi; CD IAO Sc 3025, Moreschini, Principali contratti.

prestige of the Italian concessionaire as the manager and supervisor of the farm while the Ethiopian carried out all the demeaning work.[189]

Originally, the sharecropping system was intended as a stop-gap until Ethiopians were forced on to the labour market and the land problem was definitely settled. Instead it was becoming a permanent feature of the rural economy, and gradually taking over from salaried labour.[190] Reports attest that, with the exception of Shäwa, in many of the governorships sharecropping was the backbone of the activities of many concessions. In SIMBA's case, for example, of 970 ha of its Fadis Farm, 354 ha were sharecropped; equally out of a total of 5,724 ha at Villa Baka, 5,403 ha was conducted under the same system in which 1,030 Ethiopian tenants and owners were engaged.[191]

The Ethiopians' response to the programme was on the whole unenthusiastic. At SIMBA's farm of Villa Baka, for example, the indigenous population reacted adversely while the Amharas surprisingly showed a remarkable degree of co-operation. As former lords and, understandably, Italians' key political adversaries, the Amharas had much to gain by co-operating with the new overlords. To overcome the resistance of the Ethiopians, political pressure was blended with punitive measures and economic incentives. Intense propaganda activity was carried out by the agricultural technicians who employed, as SIMBA's report points out, 'at once patience, adequate rewards and fine exemplary measures'—the latter a genteel expression for physical punishment.[192] Recalcitrant individuals were penalized and those co-operating rewarded. Thus the Amharas who co-operated with SIMBA were allowed to maintain their former status of landlords or *gultäña*.[193] It was claimed that where their co-operation was successfully enlisted, the activities of chiefs and notables proved most effective. In such cases some companies, such as SIMBA, strengthened their position by restoring, in agreement with the political authorities, part of their former powers with a view to enhancing their authority over their subjects.

Resistance took various forms. Often little or no care was given to the

[189] Sbacchi, 'Italian Colonialism', 367; Saverio Caroselli, 'Aspetti economici dell'avvaloramento agrario nell'Impero', *AC* 35/2 (1941), 53.

[190] Most post-1939 concession charters encourage the concessionaire to sharecropping contracts. Cf. CD IAO: GS 1822, Attività agricole; Hr 1812, Concessioni per trasferimento di proprietà, 1813, L'attuale organizzazione; Sc 1801, Attuale.

[191] CD IAO AOI 1929, SIMBA, Generalità, pp. 13–14; Bodini to DG Soc. An. SIMBA, AA 10 Jan. 1941; Relazione su Fadis al 31 Dec. 1940.

[192] CD IAO 1929, SIMBA, Generalità, p. 16.

[193] CD IAO 1929, SIMBA, Agricoltura indigena.

company's unprofitable crop. Unprofitability gave rise to the most common phenomenon of moonlighting whereby the Ethiopians misappropriated the crop by night to sell it for a remunerative price at an unofficial market: 'Cases of joint land management with the natives are not rare. But the system has not been brilliant. Where he was not closely supervised and well integrated, the native tended often to breach the trust laid upon him as an associate by ably stealing the common produce.'[194] Insufficient resources, combined with the problems of operating in an often inaccessible and vast area, made effective control difficult. Yet the concessionaires, in their drive to control production and fight fraud, continuously improved their methods of surveillance, devising new and more stringent measures.

A typical case was the system followed by SIMBA at its Villa Baka Farm where it fostered the production of crops that had little local consumption. In addition, it set up an *ad hoc* commission, composed of members representing the interests of both contracting parties. Ethiopian interests were represented by two chiefs and the company by a valuator and registrar. Before the harvest period, the Commission toured each farm, assessed the total produce and the company's share, and recorded them in two copies—one for the Ethiopian and one for the company—bearing the signatures of each member of the Commission. When depositing at the company's storage, the Ethiopian took with him his copy and was issued with a receipt. The company kept a register where all payments and evaluations were carefully recorded.[195] Most of the crop was paid for in kind, with the exception of coffee and *čat*, a valuable narcotic cash crop extensively cultivated in part of the south. For these two, payment was usually in cash—in the region of L.1 if the plant was irrigated and L.0.5 if dry.

The Ethiopian could challenge the Commission's judgement by appealing directly to the company administration. But once the assessment was accepted, any failure to comply brought serious penalties. In the cases of coffee and *čat*, the normal procedure was to seize the crop and confiscate an equivalent amount to the monies due. Despite the company's boast of efficiency in production control and elimination of fraud, this system was the source of interminable disputes and subject of many amendments.[196] Rome was wary of its long-term repercussions. In a letter written to the Director-General of the Company, Mario Dini, Maugini's

[194] G. Piani, 'L'agricoltura indigena nel G dell'Hr ed i mezzi per farla progredire', *AC* 33/7 (1939), 408.

[195] CD IAO 1929, SIMBA, Rapporti.

[196] Ibid.

TABLE 21. Sharecropping contracts between Italians and Ethiopians

| Type of contract | Settler | | | Ethiopian | | |
|---|---|---|---|---|---|---|
| | Input | | Share of produce[a] | Input | | Share of produce[a] |
| | Capital | Labour | | Capital | Labour | |
| 1. | 6 harrows<br>6 sickles | Ploughing<br>Threshing | 44% | Land<br>Seeds | Sowing<br>Others[b] | 56% |
| 2. | 2 harrows<br>6 sickles | Ploughing<br>Threshing | 33% | Land<br>Seeds | Sowing<br>Others[b] | 67% |
| 3. | 2 harrows<br>6 sickles | Threshing | 17% | Land<br>Seeds | Ploughing<br>Sowing<br>Others[b] | 83% |
| 4. | 2 harrows<br>6 sickles | | 5% | Land<br>Seeds | Ploughing<br>Sowing<br>Threshing<br>Others[b] | 95% |
| 5. | Seeds | | Seeds with 22% interest | Land<br>Seeds | Ploughing<br>Sowing<br>Threshing<br>Others[b] | 100% less seeds and interest |

[a] This depended on input level.
[b] Includes manuring, harrowing, weeding, harvesting, binding, and transporting to the threshing floor. See App. 3.

*Source:* CD IAO 1936, Cerulli to MAI, Contratti di compartecipazione.

verdict was: 'The company's task should not be confined to *exaction*. This certainly is one of the worst solutions and it will not be too long before the consequences manifest themselves.'[197]

The picture that emerges is that, contrary to the expectations of the colonial authorities, the policy of achieving autarky through commercial farms was a disastrous failure. Rather than contributing to economic self-sufficiency, the farms remained a drain on the slender resources of the State who had to ensure their survival through generous subsidy and fiscal concessions.[198] As both the cultivated areas and the yield gradually decreased, initial enthusiasm subsided. In 1940 an estimated area of about 20,000 ha was cultivated with production of a mere 130,000 ql. The estimate for 1941 was lower than this but 200,000 ql were expected.[199] But the optimism faded away when climatic change ruined most of the crop.

In official circles, the reason for failure was sought in the conventional explanation of the distrust of Italian capital to move into a new environment, particularly at a time of turbulent international conditions compounded by uncertain local markets. There is little doubt that these were significant contributing factors. But they were not the sole reasons. The policies for commercial farming rested on grandly conceived but impracticable ideas.[200] They were not preceded by any feasibility study nor did they seem to have balanced the impact that various interests might have during implementation. Their pace of development was determined by the interplay of three major forces: the technical knowledge of the farmers, and availability of land and labour.

Despite eulogistic reports in the contemporary media, the settlers showed a low level of experience and technical expertise. Attempts to find land free from Ethiopian claims had little success. The continuous intervention of the political authorities, aided by a package of coercive measures, never succeeded in remedying the chronic labour shortage. The situation was further compounded by the lack of a coherent set of policies, administrative disorganization, personality clashes, bureaucratic entanglement, and inadequacy of economic and manpower resources. Under the best of circumstances, the Italians would have had to be patient as the official agricultural development strategy matured. But the brief period of the occupation simply was not long enough for their plans to come to fruition.

[197] CD IAO 1929, Maugini to Mario Dini, Fl 28 Jan. 1940.
[198] Del Boca, *Caduta*, 212; D Gov Sc 13-2-1940, no. 201, Concessione *GUGS* 5/25, 19 June 1940. [199] CD IAO 1774, Pini, Attività; Villa Santa, *et al.*, *Amedeo*, 252.
[200] Villa Santa *et al.*, *Amedeo*, 221–2.

TABLE 22. Commercial and demographic concessions, 1941

| Governorate | No. of companies | Concession areas (in ha) | | |
|---|---|---|---|---|
| | | Granted | Cultivated | Balance |
| Amara | 107 | 4,162 | 2,300 | –1,862 |
| GS | 112 | 21,381 | 6,500 | –14,881 |
| Härär | 124 | 65,878 | 11,700 | –54,178 |
| Shäwa | 229 | 33,194 | 15,600 | –1,794 |
| Eritrea | 126 | 23,343 | 12,000 | –11,343 |
| Somalia | 129 | 28,780 | 25,500 | –3,280 |
| TOTAL | 829 | 176,737 | 73,600 | –103,137 |

*Note:* The data is given by a CFA official, Pini, and, compared with those supplied by the agricultural offices (see Tables 11–13) is much more inflated.

*Source:* CD IAO AOI 1774, Pini, Attivitá. (Totals as in the original.)

In spite of this failure, the government did not relinquish its commitment to settler farming. Yet it was compelled to make a major reappraisal of its economic strategy. Greater attention was now paid to those who argued that Ethiopians could, given due incentives, produce certain crops more cheaply and even eminently than the settlers.[201] The result of such reassessment was a positive, though timid, government gesture of encouraging Ethiopian farming as a way of maintaining revenue in view of achieving agricultural self-sufficiency and bolstering the fiscal base of the colonial state and subsidizing the survival of settler farming.

[201] Piani, 'L'agricoltura indigena', 406; CD IAO 1926, Terruzzi to Duce, 27 Apr. 1938.

TABLE 23. Settler agricultural production and annual consumption needs, 1941

| Type | Amara | GS | Härär | Shäwa | Eritrea | Somalia | Total production | Annual need | Balance |
|---|---|---|---|---|---|---|---|---|---|
| Wheat | 10,000 | 2,000 | 25,000 | 28,000 | 5,000 | — | 70,000 | 1,010,000 | –1,009,930 |
| *Téf* | 3,000 | 100 | 400 | 4,500 | 5,000 | — | 13,000 | 979,500 | –966,500 |
| Barley | 2,500 | — | 2,500 | 3,000 | 5,000 | — | 11,000 | 305,650 | –294,650 |
| Durrah | — | 2,800 | 6,000 | 3,000 | 8,000 | — | 17,000 | 735,000 | –718,000 |
| Corn | 50 | 50 | 5,000 | 12,000 | 150 | 70,000 | 90,000 | 506,000 | –416,000 |
| Oilseeds | 150 | — | 4,200 | 6,000 | 1,050 | 6,000 | 17,000 | 493,400 | –476,400 |
| Coffee[a] | — | — | — | — | — | — | 6,500 | 26,375 | –19,875 |
| Dry vegetable | 100 | 200 | 50 | 1,250 | 200 | — | 1,700 | 466,000 | –464,300 |
| Potato and fruit[b] | 500 | 2,300 | 2,000 | 810 | 2,000 | 500,000 | 507,600 | 193,000 | +314,600 |
| Cotton Fibre | — | 3,300 | 150 | 800 | 500 | 7,000 | 11,750 | * | — |
| Castor Fruit | — | 100 | 5,000 | 100 | — | 1,000 | 6,200 | — | — |
| Poultry | 500 | 500 | 500 | 2,000 | 5,300 | 200 | 9,000 | 3,060,000 | –3,051,000 |
| Beef | 25 | 10 | 10 | 10 | 25 | — | 80 | 154,566 | –154,486 |
| Pork | 150 | 50 | 200 | 800 | 300 | — | 1,500 | 6,840 | –5,340 |
| Lamb | 80 | 20 | 50 | 50 | 100 | 100 | 400 | 253,450 | –253,050 |

*Notes:* Cereals measured in ql and animals by number.

* indicates that no figure is available.

[a] Estimated at 2,619,000 plants and 0.25 kg per plant.

[b] Total fruit 502,110 ql of which: GS (100), Härär (1,000), Shäwa (10), Eritrea (1,000), Somalia (500,000).

*Source:* CD IAO AOI 1774, Pini, Attività. (Totals as in the original.)

# 7
# *Modernizing Ethiopian Peasant Agriculture?*

## THE DIMENSIONS OF CHANGE

Until demographic and commercial colonization efforts patently failed and the first flush of Fascist enthusiasm died away, modernization of indigenous agriculture was a mythology included in the speeches of higher authorities for propaganda purposes. Vague thoughts of uplifting the peasantry, at least with a view to preventing the spectre of social problems, were not absent. They were expressed by a few visionaries in the proper political garb of the Fascist administration, and were accepted as indispensable declarations showing that it was part of Italy's civilizing mission.

The paramount concern of the Italians was to obtain enough land for their own settlers. Although the Italian authorities avowedly claimed to respect 'legitimate' Ethiopian land claims, the issue of reconciling the imperatives of the metropolitan colonization with the interests of the Ethiopians remained unresolved.[1] Colonization dictated that settlement be directed to areas where climatic conditions, the cost of manpower, and the security of capital were most favourable to making the enterprise profitable and large-scale production feasible. But these were also areas that were densely populated and the settlers had to compete for agricultural resources with the Ethiopians who were almost exclusively dependent on land for their subsistence.

For the Italian authorities, dealing with indigenous agriculture entailed many complicated economic and political factors to which there were no clear-cut solutions: massive displacement of the Ethiopians from the areas destined for settler agriculture might incite rebellion and open warfare, or cause migration to the urban centres where they might form a socially and politically restless proletariat. On the other hand, the

[1] Brotto, *Regime*, 24.

Ethiopians' labour was essential to develop the colonists' farms.[2] Also the Ethiopians, by their sheer numbers, were regarded as a large market for metropolitan goods and their produce was gradually perceived as indispensable in achieving the goal of imperial self-sufficiency in foodstuffs. But as the Ethiopians were believed to have far fewer wants and could therefore sell at a much cheaper price than the Italian settlers, there was also a constant fear that competition from their low-cost produce might undermine any progress on the settlers' farms.[3]

Nevertheless, the Italians maintained that the occupation of Ethiopia delivered the Ethiopian peasantry from bondage and backwardness. Such a view was echoed by the chief intellectual architect of Italian colonization in Ethiopia, Maugini: 'A new era of peace and state of well-being is opened to the masses of IEA as many of the present miseries that these people are experiencing will be once and for all eliminated.'[4]

From the official point of view, the list of such miseries included archaic farming methods, absence of well-developed markets, and rudimentary socio-economic infrastructure which acted as bottlenecks to improving the peasant economy.[5] But it was in the land tenure system as manifested in the *gäbbar* system that the Italians saw the chief constraints paralysing Ethiopian agriculture:

> the *gäbbar*, a tax-paying peasant, subject to the tyrannical rule of an Amhara landlord, with no title to land which belonged to the Emperor or his feudal lords or churches, burdened by tithe imposed by law and exorbitant taxes imposed by the lesser chiefs, robbed by raiders, compelled to maintain in his house the soldiers of the Emperor and those of the provincial governors, does not obviously feel any incentive to improve and increase production, nor, even if willing, will he be in a position to carry out any initiatives in this respect; on the other hand, nor does such initiative come from the temporary *gult*-holders or, call them whatever you like, feudal lords who, not having the security of permanent tenure, have no other interest than maximum possible exploitation, in collaboration with the wretched population.[6]

From the start the Italians had hailed themselves as the harbingers of a new economic order that was likely to break the precarious cycle of the

[2] *GI*, 9 Mar. 1939; G. Mondaini, 'I problemi del lavoro nell'Impero', *REAI* 25/5 (1937), 748–9.

[3] Giglio, *Colonizzazione*, 53–8; id., 'Rapporti', 1561–5.

[4] Maugini, 'Primi orizzonti', 20.

[5] Ibid.; G. Pistolese, 'Problemi dell'Impero', *EI* (July 1938).

[6] ASMAI, B/245, Italo Papini, La produzione dell'Impero, p. 29. This view of proven inaccuracy, though amply refuted by, and contradictory to, the Italians' own findings, remained part of official orthodoxy. Cf. Ch. 2.

existence of Ethiopian masses. But the need to involve the Ethiopian peasantry in a more productive agriculture was felt largely after the belated realization that some crops could be better grown by the Ethiopians rather than by settlers. The year 1938, aptly called a 'year of reappraisal and restructuring' or *anno di assestamento*, was a turning-point, if not a watershed, in Italian official thinking. Public pronouncements were repeatedly made emphasizing that the object of the Italian policy must also be to harness the productive potential of the Ethiopians in order to assure them a better life by raising their purchasing power and improving their standard of living. Of course, these were self-interested measures, for improved lifestyle of the colonized people would bring a good reputation to Italian rule, while increased production would boost imperial autarky in foodstuff. The essential colonial economic outlook remained unchanged, even though measures were taken whose expressed chief purpose was to create a favourable economic climate, improve the production techniques, and develop economic infrastructures. The Italians believed in the long-term feasibility of their scheme. The intervention of Ethiopian Independence disrupted the programmes. Even though to refute their claims would be a long and arduous task, what was most clear from the outset was the fact that the scheme consisted of tentative and sporadic ventures, undertaken at the lowest possible cost and with little or no overall planning. The best that one can do is to describe these initiatives and the Ethiopian reaction to them.

## THE CHALLENGES OF RESTRUCTURING THE *GÄBBAR* SYSTEM

The reform of the *gult* system was interpreted as necessary for the creation of an economic environment conducive to healthy agricultural development, for *gult* was seen as the basis of Ethiopia's underdevelopment and a source of exploitation of peasant production by the former rulers. Condemnation of the *gult* system had been a lurid theme in Italian propaganda leading up to the invasion, and was thought of as a valuable political weapon to drive a wedge between classes within Ethiopia. In areas where draconian measures were pursued, its abolition led to an indiscriminate declaration of all *gult* tenures as public domain. Stiff resistance followed, forcing the Italians towards 1939 to distinguish between gult of lordship or *gult di signoria* and hereditary gult or *gult di diritto terriero*. The first was recognized as public domain and the second as private property.[7]

The Italians congratulated themselves that with the abolition of the *gult* system they erased the exploitation characteristic of rural life that had formed the basis of the Abyssinian Empire. However, needless to say, such claims did not square with reality. Their policy towards *gult* tenures was dictated more by political considerations and the need to expand public domain for colonization purposes than genuine concern for agricultural development.

Under Graziani, conversion of *gult* into state domain was implemented, particularly in regions recently conquered by the Abyssinians, as part of the so called de-Abyssinization or *de-Amhärization* policy. By stripping the Amharas and their allies of their wealth and depriving them of their economic lifeline, he planned to accelerate pacification and Italian domination over the territory. In his view, successful elimination of the Amharas from their position of power would enhance Italy's colonial prestige as a master race in the eyes of the Abyssinian's former subjects, such as the Galla and other ethnic groups.[8]

On the other hand, loyal chiefs, both of Amhara and other ethnic groups, were allowed to keep their *gult* and the privileges associated with it intact, or were rewarded with tracts of public domain from where they continued to levy tributes as in former times. Some chiefs, like Ras Häylu of Gojjam, who was made a Commander of the Order of the Colonial Star for the part he played in the pacification process, tried to appropriate whatever they could lay their hands on, sometimes using false claims. In places like Kämbata, the indigenous leaders who were quick to confiscate, with the tacit approval of Italian authorities, the *gult* of some rebel Amhara rulers, were forced to reinstate them once the latter submitted.[9]

The exemption of Church *gult* lands from state domain most belies Italian claims to have abolished *gult* tenures. Since the beginning of invasion, the Italians understood the need for the Church's co-operation in order to secure control of the country. The Church was also viewed as a stronghold of Ethiopian nationalism. Using various methods, including terroristic tactics, the indiscriminate massacre of leading religious

[7] ASMAI PB/6, Missione Studi, Promemoria Per S.E. il G., Gn 27 Sept. 1939.

[8] ACS GRA 38/33/3: Graziani to MC, 20 Jan. 1937; Graziani to Commissariati Regionali Dembidollo and Irgalem, 17 Jan. 1937.

[9] DGG 17-7-1936, No. 84, Restituzione beni a Ras Hailu, in *GU* 1/8 [suppl.], 26 Apr. 1936; ACS GRA 40/33/5: Bocca to GG AOI, AA 27 July 1937; Capo Gabinetto to Belly, AA 24 July 1937; Lessona to GG, AA 19 Dec. 1937. ACS GRA 40/33/9: Graziani to GG, AA 7 Oct. 1937; Moreno to GG, Gm 18 Oct. 1937.

personalities, the total dismantling of the administrative structure, and severance with the outside link, the Italians succeeded in manipulating and weakening the EOC. But a *modus vivendi* was reached only when the Italians agreed to suspend for a decennium the law transforming the Church *gult* properties into public domain in return for the Church's co-operation.[10] Graziani, who saw Church property as the only valid 'powerful weapon to bend the church to our will', explained to Mussolini:

> Given the great significance of the issue and taking into account that in the final analysis the transformation of *gult* into public domain boils down to a question of pure formality and that the influence of the Church in this country is of much greater relevance, the best policy is to postpone for a period of ten years the nationalization of *gult* properties and leave to the Church the total disposal. . . . This will have an immediate effect in allaying the mind of the clergy and winning its support, thus sparing the government from any financial cost. As a result [our] principle remains uncompromised while the task of pacification will be made much easier.[11]

Of course, there were also economic considerations that presented the maintenance of status quo of EOC lands as a wise alternative, a point supported by a case study conducted by a dynamic and liberal *commissario* of the Commissariat of Bishoftu, Attilio Scaglione. Scaglione, who carried out a survey of 98 churches with a religious population of 1,311 and owning 1,137.75 *gasha* or *c.*45,480 ha, warned that any abolition of EOC lands would bring unbearable costs to the State treasury.[12]

The Italian decision to allow loyal *gult*-holders to maintain former privileges made their policy unpopular. But grievances over the arbitrary

[10] ACS GRA 46/41/15, Autocephalia della Chiesa Copta Abissina', p. 52. This document, consisting of 74 pages, contains most of Graziani's correspondence with Rome concerning the EOC. The author finds the tactics and intrigues used to sever EOC links with the Alexandrian See the most revealing because it contradicts the most commonly held view esp. as it relates to the conduct of the metropolitan *Abuna* Qérlos. With the reorganization of the EOC, the higher clergy, like other State employees, became relatively well-paid salaried officials in the service of the State (cf. DGov 11-3-1938, no. 134918, Assegnazione ai due vescovi copti del territorio dell'Am di L.6000 mensili, in *BUGA* 2/2–3, Feb.–Mar. 1938; Sbacchi, 'Italian Colonialism, 256, 258. For more information on Italian policy towards the EOC, see Haile M. Larebo, 'The EOC and Politics in the Twentieth Century', in *Northeast African Studies*, 10/1 (1988), 1–5.

[11] ACS GRA 46/41/15, pp. 53–4. In settlement areas while other categories of lands were confiscated for use of the Italians, church lands remained normally unaffected. The clergy also had been one of the main obstacles against the advancement of the settlers' interest. In places like Qalitti in Bishoftu, the Ethiopians successfully resisted sharecropping with ONC because of the opposition of the local clergy. (Cf. ACS ONC 5/A/1, Taticchi to SC, AA 3 Dec. 1939.)

[12] CD IAO 791, Commissariato di G. di Biscioftù, La chiesa Etiopica, pp. i–x.

method in which the *gult* was transformed into public domain and massive evacuation of the original occupants to accommodate the settlers, combined with gratuitous harsh treatment of the population by the government officials and the colonists, made most of the settlement areas highly combustible. As no land was surveyed for allotment, rather than searching for lands that were indisputably *gult*, the general procedure was to occupy a colonizable land on the assumption that it was *gult* and then invite any claimants to ownership to substantiate their claims. It was a procedure that gave rise to ample abuse. The Ethiopians had to present documents from the previous Ethiopian regime's land registry, or other records such as receipts for tax payments, all of which were not easy to procure. The situation helped only the land-hunters who, on the advice of the authorities roamed from one governorship to another in search of domanial lands, and 'picked the first green valley that befell their eyes'.[13] Often occupation by the Ethiopians was seen as a sign of fertile land, and many settlers deliberately selected land that was in an exceptionally high state of cultivation, shunning away from abandoned lands.

Contrary to the Italian claims, the public domains, where settlements took place, were not merely wastelands, lands with no other master than the State, or uncultivated lands whose occupation would have inconvenienced no one.[14] Even the *gult* belonging to rebels and their sympathizers or royal families were not lands empty of people. They were lands occupied by peasantry who for generations had ensured their productivity. Official claims that evacuation was peaceful are belied by the confidential correspondence.[15] Referring specifically to Härär where 'there was massive colonization by expropriating the natives from their lands and deporting them somewhere else', and comparing it with the earlier misguided land policy of Eritrea, the Secretary of Italian Africa, Terruzzi, warned Mussolini of widespread and simmering discontent that, unless remedied quickly, would inevitably open the floodgate of 'an eternal cycle of violence'.[16]

[13] Mondaini, 'Tradizione', 115; Villa Santa *et al.*, *Amedeo*, 226. CD IAO AOI: 1936, Pini, Considerazioni; 1926, Terruzzi to Duce, Rm 27 Apr. 1938.

[14] Such claims were even made by General Guglielmo Nasi, the G of Hr, where such massive displacement took place. Comparing Italy's 'generous and progressive' land-alienation policy with the precedents of his misconceived Abyssinian conquest, he wrote: 'The key principle followed by the Italian government [in land alienation] was—unlike the Amhara conquerors who expropriated cultivated lands—to occupy and develop uncultivated lands.' Cf. Nasi, Opera, 18.

[15] Mondaini, 'Tradizione', 115.

[16] 'From various parts but particularly from Hr, to which, because of the major tranquillity that the region enjoys, applications are directed, concerns are being expressed to me against the way the present colonization is carried out, i.e. in vast areas expropriating the

Nor was the discontent confined to the cultivators of *gult* lands. As we saw earlier, evacuation also affected private owners adjacent to *gult* lands to which, despite safeguards, their plots were amalgamated to make the fragmented colonizable lands economically viable. In exchange the owners were given lands elsewhere in areas that were assumed to be of little or no relevance for the settlers. Of course, many Ethiopians, especially the *čisäña* or tenants, were 'persuaded' to stay on to provide labour. The scale and terms of such employment varied. But as a rule, in such instances, labour contracts were drawn in which the Ethiopians were often allowed to cultivate a small tract of land, variable in size but often not exceeding 5,000 sq m. In return, they undertook to work for the settler either at call or for stipulated proportion of the year. The settler had the power to dismiss them at will.[17] In terms of law, the contracts had effectively transformed the Ethiopians into mere squatters with no security of tenure other than their labour contracts. The settler held considerable power over them to the extent that he could move them at will or change the terms of agreement under which they held their allotments. Only in 1939 did the government made spasmodic attempts to give the Ethiopians some protection by demanding the colonist to resettle or provide them with an alternative occupation if he decided to make them redundant. However, such clauses applied only if the expiration of the contract was amicable; otherwise, the Ethiopians were left at the mercy of the settler who retained the power to dismiss them or revoke their contract at any time for disciplinary breach or insubordination.[18]

The Italians were aware that land confiscation was a highly delicate matter to the Ethiopians who, immaterial of their ethnic origin and whether the land was *gult* or *rest*, felt very strongly about their possessions.[19] The Italians boasted that in most cases the Ethiopians were compensated fairly and that lands appropriated for colonization were few and far between. Strictly speaking it is true that the Ethiopians received compensation.

natives of their lands and deporting them somewhere else. Undoubtedly such a method can cause serious political repercussions. The revolt of Bahta Hagos in 1894 in Eritrea was caused exactly by the contemporary mistaken land policy and even currently one of the major charges directed against us by the rebels—by both Amhara and Galla—is that of expropriating for the White colonization their lands to which the natives, immaterial of their race, feel a particular attachment.' (CD IAO AOI 1926, Terruzzi to Mussolini, Rm 27 Apr. 1938.)

[17] ACS ONC 7: Contratto fra gli indigeni di Guntutà e l'ONC, 1 Mar. 1939; Nasi to ONC, Contratto di salariato fra colonizzatori ed indigeni, AA 20 May 1939. See App. 1–3.

[18] ACS ONC 7, Nasi to ONC, AA 20 May 1939.

[19] CD IAO AOI 1926, Terruzzi to Duce, Rm 27 Apr. 1938.

But the two parties had totally different concepts of land values and usage, and therefore what counted as a fair price for the Italians was not so to the Ethiopians. Nor were the Ethiopians consulted on the price. The owners were summoned and told how much their land was worth after the land was surveyed, often by unqualified people. Theoretically, the Ethiopians could refuse to move without adequate compensation. But an interview with an individual, whose land at Bishoftu was appropriated in this way, reveals that the Ethiopians were likely to lose their land whether they agreed to it or not. Thus, most accepted the decision as a *fait accompli*. But the payments were slow. Thus, in Addis Ababa between 1937 and 1938 property worth about L.10,000,000 was confiscated for colonization and new urban planning, but less than L.2,000,000 was paid.[20]

The Ethiopians were paid much less than the market value for their land. In Addis Ababa, for example, land taken from Ethiopians was parcelled out into small plots and sold at between 15 per cent and 60 per cent higher than the price of compensation. The precise scale of such confiscations is not easy to establish. Yet cases suggest that the Italians, particularly at an early stage, did not hesitate to take any land thought essential for the viability of settler farms. By the end of 1939 lands given in this way to the settlers totalled 186,000 ha, not including lands made public domain but not yet allocated.[21]

The Cusmano and Corvo Commission in its secret report to the Duke of Aosta deplored the displacement policy as politically dangerous and exploitative:

> Expropriation even when effected following the criteria of generously rewarding the owners will result in transforming the mass of best peasants, as is the case particularly with the Galla, into a mob of have-nots, who, ultimately by *force majeure*, will end up by dedicating themselves to banditry. . . . One should not forget that the bonds that tie a native are very few. . . . What can one expect then when they are removed from fields which they cultivated for centuries and which represent for them, together with their herds, an integral part of their life?[22]

At least until 1939 the final decision on the status of *gult* and its transformation into domanial land rested on the exclusive judgement of often unqualified local authorities, who often behaved arbitrarily, aided

[20] Sbacchi, 'Italian Colonialism', 403.

[21] Mondaini, 'Tradizione', 368; MAI, 'Valorizzazione', 208–10.

[22] ACS GRA 46/41/19, Cusmano and Corvo, Situazione economica, AA 1 Apr. 1938, pp. 11–12.

by local chiefs. The latter were open to extortion, and cases where land concession was bartered with the colonists for influence and speculative purposes at the expense of the peasantry were not uncommon.[23] Most officers not only often ignored or condoned such practice but themselves issued many illegal permits often in connivance with the local Fascist Party or the colonization office. General Mezzetti, the Governor of Amara, attributed the rebellion of his governorship largely to such unauthorized dealings:

> What is most serious and which is a particular cause of the disorder is that abusive seizures have been carried out by the higher representatives of the local Fascists as well as a member of the Fascist Federation of Gondär. The latter has occupied the vast planes of Grado—without any knowledge of the Governor at Gondär—and there he had set up a farm where seventeen metropolitan workers are employed and has built a plant worth about L.250,000 . . . A vast region that was demarcated as domanial and where also are settled a number of colonists, happened to be the property of our loyal leader who has at his disposal a very skilled lawyer and he is going to assert his rights by legal means. I hope to come to terms with him.[24]

Like Governor Mezzetti, officers sympathetic to the Ethiopians felt incompetent to protect them against sharks or influential colonists who illegally seized large tracts of land. One such case was that of a certain official who had a vast concession on the outskirts of Däsé. He taxed the peasantry heavily and pocketed all the revenue, including the tithe due to the state. The peasants reported this to the local *residente* each time, carrying, in accordance with Ethiopian tradition, a pair of yokes on their shoulder as a symbol of their oppression. A tense relationship developed between the colonist and the *residente* whose authority was curtailed by the fact that the abusive concessionaire boasted a special relationship with Mussolini.[25]

In other instances, different *residenti* pursued policies that were glaringly contradictory. Thus, for example, where one *residente* declared all his subdistrict *gult* land, the neighbouring officer claimed that the land under his jurisdiction was prevalently *rest*. On other occasions, the distinction between *gult* and *rest* lands was merely technical, which the officers agreed to ignore in the overriding interests of colonization. The agricultural office of the governorship of Galla and Sidamo gave a typical

[23] Masotti, *Ricordi*, 102.

[24] ASMAI AS, III., 2a Guerra Mondiale, p. 4, Mezzetti to Terruzzi, Däsé 13 Aug. 1938; Del Boca, *Caduta*, 198–9.

[25] Masotti, *Ricordi*, 102–3.

example. To the South-West of Jimma, 500 ha located 5 km from the centre of the city was acknowledged as private property, and yet the inhabitants were removed and the land confiscated; equally land lying at 6 km South of Jimma was recognized as domanial and subsequently dispossessed, even though the presence of *tukul* and thriving coffee plants suggested otherwise.[26]

These arbitrary land policies had the cumulative effect of alienating most of the Ethiopian population, including those who were reportedly indifferent or sympathetic to the Italian invasion.[27] In Amara opposition was led by the middle-ranking nobility who took to arms and, with land alienation as the rallying cry, commanded wide popular support, and succeeded in confining the Italian presence to a few garrisoned urban centres.[28] With the exception of Shäwa, armed resistance in settlement areas was sporadic and, as in Arsi, easily crushed. But the general condition in the countryside was a state of simmering rebellion and mounting discontent, while the cost of promoting the misguided policies drained the treasury.

However, as it would be erroneous to assert that colonized lands were lands free from the Ethiopian claims or of secondary interest for them, it would be equally misguided to conclude that colonization was established only on the best lands and solely by means of violence or arbitrary dispossession. These two extreme positions can be juxtaposed. Cases of peaceful agricultural development and even of amicable land transactions were not uncommon. Illustrative examples are those instances of Italians who leased land from the Ethiopians or worked as their tenants on mutually agreed terms of contract or replenished their cattle stock through direct purchase and without the interference of the political authorities or agency officials. In Shäwa a conspicuous number of Italians ran garden-farming as tenants for the Ethiopian landlords. It seems that many of the latter, taking advantage of the opportunities opened before them, had attempted to make their land lucrative by squeezing their tenants

[26] CD IAO GS 1820, Il problema, p. 4.

[27] Pankhurst, 'Economic Verdict', 81.

[28] ACS GRA 40/33/5, Graziani to Ministro Africa, AA 4 Dec. 1937 where he reports that the rebels mentioned land alienation as one of the motives in their plea for the defection of Ras Häylu: 'Who will not be offended and whose heart will not boil with rage when seeing that his country is oppressed, his *rest* and property is confiscated.' A similar message was sent to the Gojjamese people by Shäwan patriots: 'You know that it [Italy] had betrayed the Ethiopian people, for having summoned to the capital [AA] all the notables and elders from all regions, it told them: "You will have no *rest*, no property, no cattle, no children. From now onwards everything belongs to me." ' (Cf. ACS GRA 36/31/12, Zeude Asfaw, Blatta Tacle, Mesfin Silesci to Notabili ed Anziani del Goggiam, 22 Oct. 1937.)

as much as they could. In doing so they had drawn against themselves the criticism of the technical officers who called for an authoritative intervention.[29]

Towards 1939 the government restrained itself from randomly creating domanial lands. Unless political and economic realities dictated otherwise, only lands that were indisputably *gult* would be transformed into public domain. Certainly, the government, by now alarmed by the failure of settler agriculture and the growing assertion of Ethiopian resistance, was wary of disruption in production and possible political complications.

Earlier the government had reimposed *asrat*, which had been abolished ostensibly as a humanitarian gesture—to show the Ethiopians that the Italian government was a benevolent father who, unlike the exiled Ethiopian regime, did not intend to burden them with heavy taxes. But in reality, the government lacked human and economic resources to collect it. *Asrat*, imposed on the existing Ethiopian principle—on products of the soil—and from which the Italian farmers were exempt for a decennium, was paid in kind. This was a clear departure from the policy of Emperor Haile Sellassie who, as we saw earlier, favoured payment in cash.[30] Although a systematic fiscal system was not yet fully developed, the new taxes were neither less burdensome nor less complex than those they superseded.[31]

Attempts to raise locally sufficient foodstuff led to the setting-up of strategically located Grain Purchasing Centres, manned by the Italians with the aid of Ethiopian grain agents who bought the grain from the indigene at fixed prices that were below the price both of the market and the grain produced by the Italians.[32] The effect of this combined with

[29] CD IAO Sc 1801, Attuale. Another typical ex. is that of Calà Giuseppe who personally purchased numerous cattle from the Ethiopians and leased more than 100 ha of land from an Ethiopian called Qäňazmach Tullu Tadesse without the knowledge of the director of the Holätta Farm (ACS ONC 8/A, Calà Giuseppe [Sebastiano] to Direzione Central ONC, Randazzo 12 Oct. 1949).

[30] GG 27-9-1937, no. 748, Ripristino in tutti i territori dell'AOI dei tributi erariali tradizionali, in *GU* 2/21, 1 Nov. 1937; Circolare 25-11-1937, no. 163647, Riscossione della decima, in *GU* 2/23, 1 Dec. 1937.

[31] In addition to *äsrat*, there were taxes on livestock, house, income; taxes on servants, improvements, consumption of electricity, and even taxes on taxes in order to defray the cost of collection.

[32] DGG 8-12-1936, Istituzione di un comitato centrale per regolare e coordinare l'azione dei comitati locali per la vigila dei prezzi sui mercati dell'AOI, *BUGA* 2/1–2, Jan.–Feb. 1937; DGG 29-9-1937, no. 693, Disciplina del commercio e consumo dei cereali, in *GU* 2/21, 1 Nov. 1937; D Vice GG AOI, 18-10-1937, no. 726, Costituzione del Comitato Alimentare Centrale, in *BUGA* 2/9–10, Oct. 1937. No Italian source at the author's disposal gives the price paid by the Italian authorities for the grains produced by the Ethiopians. Information supplied by a foreigner well familiar with the area on a few important items in Arsi region suggests the Italians were paid three times more for the same goods produced by the

sudden imposition of traditional taxes on agriculture and forceful introduction of *lira* currency was to remove incentive to produce and drive the population to inactivity. Reports suggest that sowings became limited almost to the minimum necessary for personal consumption and less was available for marketing. Even when the products were available sale remained unattractive. The severe measures used to reduce the population to subjugation and the loss of stock, which often was the result of confiscation by the Italian cattle-thievers, carried out with tacit support by the government, and many punitive expeditions left many farmers without farm animals and in appalling economic condition. This caused serious food shortage, particularly of wheat, especially in early 1938.[33] Between 1934 and 1941 a substantial number of able-bodied Ethiopians had abandoned their lands—to which most returned only after liberation—for other more lucrative pursuits or to join the army—first the Ethiopian and then that of the Italian or resistance movement. Even in regions which, like Wälläga, were less affected by the war and by the government's devastating scorched-earth policy pursued particularly under Graziani, the state of war had caused impoverishment as a result of complete neglect of agriculture.[34]

These policies make the declarations of the Italians void. Rather than creating an environment conducive to the development of Ethiopian agriculture, what stands out perhaps is that their extraordinary, almost gratuitous, failure to win the confidence and co-operation of the Ethiopians and handling land issues combined with arbitrary dealings with local customs, such as changing market days,[35] had the effect of intensifying rural tension.

## DEMOGRAPHIC VILLAGES

A number of incentives were offered to integrate Ethiopians into the new agricultural system and programmes created as propaganda and to increase local production. Any increase in the productivity of indigenous

Ethiopians. So, whereas the government purchased for L.250 a quintal of Italian-produced wheat or barley, it paid for the Ethiopians L.80 if wheat, and L.60 if barley (FO371/22021/J2363/40/1, CG Bird, Situation in Arussi, AA 24 May 1938; ACS ONC AB, Specchio del grano e cereali versati dai colonizzatori: produzione esercizio 1940–41, n.d. [2 Apr. 1941]).

[33] FO371/23380/J1776/296/1, Bird to Halifax, AA 1 Apr. 1939; *äsrat* for the year totalled only 30,000 qls (cf. ACS GRA 43/34/26, Trevisani, Relazione).

[34] ASMAI 13/145, GG AOI CSFA, *Monografia del Uollega* (AA, 1938), 144.

[35] FO378/22021/J2363/40/1, CG Bird, Situation in Arussi, AA 24 May 1938.

agriculture would decrease the financial burdens of Italy and bring the Empire towards self-sufficiency. The Ethiopian farmers who were selected to take part in these programmes were eligible for land, subsidies, and tax advantages. As a rule, ploughing and threshing was done mechanically by the government, who also provided seeds.[36] There were two main types of settlements: the first directed to the accommodation of liberated slaves; the second to provide a workforce for settler agriculture. The workforce consisted of former *ṫisäña*, who were displaced by the Italian colonists, and the former landlords, the Amhara *näfṫäña* of the South, released from confinement following the Duke of Aosta's policy of 'clemency and attraction'.[37] The exact number of villages constructed by the Italians is uncertain, but it was not more than a dozen. Ideally, these villages were supposed to parallel the settlements for the metropolitan colonists. Each Ethiopian settler was given a small plot, a farmhouse, an ox, and tools and supplies. A centre equipped with a school, a marketplace, a café, and a health centre completed the picture. The farms were expected to grow indigenous grains, with an emphasis on wheat.[38]

The settlement of former slaves was in the territory of Soddo residency, in the Lake Čamo zone along the road leading from Dallé to Soddo —which used to be the heart of the greatest slave-markets of the region— 50 km from Ṫembaro, Boloso, and Sorä.[39] The land belonged to a well-known Abyssinian slave merchant, Fitawurari Wubshät, killed for his trade in human traffic. The village was known as Villaggio Vittorio Bottego, in memory of a ruthless explorer who lost his life at Dago-Roba, Wälläga, in one of his military-cum-scientific expeditions, organized by the Italian government and Società Geografica Italiana.[40] By 1938 700 people had been settled in a well-located and -watered 400 ha. There was a plan to expand the scheme in case the size proved insufficient to the settlers' needs. The experiment was aimed to solve the problem of training liberated slaves and teaching them to become useful members of the community.[41]

Other Ethiopian settlements were scattered everywhere and largely consisted of Amharas released from various internment centres. Härär

[36] Piani, 'Agricoltura indigena', 408; CD IAO Ex AOI 1097, GG AOI IA, Relazione sull'attività svolta dall'IA durante il 1939 in AOI, AA Gennaio 1940; CD IAO Misc. 1038, Pollera, Europei e indigeni, p. 13.

[37] FO371/22021/J1214/40/1, CG Bird to PSSFA, AA 16 Feb. 1938.

[38] Quaranta, *Ethiopia*, 31.

[39] FO371/22021/J2728/40/1, Lord Perth to Visc. Halifax, Rm 9 July 1938.

[40] Del Boca, *Dall'unità*, 746–9.

[41] FO371/22021/J2728/40/1, Perth to Halifax, Rm 9 July 1938; Quaranta, *Ethiopia*, 30–1.

Governorate had four.[42] At its Babile's Amhara Settlement Centre, sixty families were settled, almost all of them deported from Jejega area to supplement SIA's labour shortage. Several hundred released from Dananä concentration camp were resettled with the government providing them with lands, animals, L.1,000 to build a hut, and seeds. SIMBA resettled 120 destitute Amhara families from Ankobär in one of its Ärsi farms under the same terms that SIA employed.[43] In the governorship of Galla and Sidamo a number of Amharas held in Filtu concentration camp were settled in domanial lands, giving one-fourth of their crops to the government.[44] Six hundred and eighty-two families displaced by EDR were accommodated within a distance of 10 km from the agency's settlement area; on average each family was provided with 1.02 ha and L.250 as compensation for their lost property and start-up capital. There were settlements at Gudär and Mäkanisa at the outskirts of Addis Ababa.[45]

By the time of the restoration of independence, the programme of demographic villages had not gone beyond the drawing board. Yet the Italians claimed that in the case of the ex-slave settlement in Soddo, the results so far achieved were encouraging.[46] With the exception of the Native Settlement Centres at Mäkanisa and Gudär, many of which were planned in each governorate, the rest were not centrally planned. They were isolated undertakings by the respective governorate to overcome local problems. What is more, the true purpose of their establishment was to serve either simply as a cheap workforce depot for the nearby labour-starved settler farms, or as a penal colony for political suspects, or both as was the case at Mäkanisa Farm.[47] In the case of the former slave settlement, it had an additional social function. As one technician remarked, it aimed to bring into the productive system those 'agriculturally inactive' sections of the population.[48]

Information on the demographic villages as whole is scanty. Only the ex-slave settlement of Soddo and the Mäkanisa Farm provide the best data. Villaggio Vittorio Bottego's settlement was used as a showcase. A letter published from Jimma emphasized that the village was proof that

[42] They were located at Jejega, Siré, Asälla, and Goba and set up during 1938 and 1939 agricultural campaigns (cf. Piani, 'Agricoltura indigena', 408).

[43] Sbacchi, 'Italian Colonialism', 272.

[44] Ibid. 271.

[45] IAC, *Main Features of Italy's Action in Ethiopia 1936–1941* (Fl, 1946), plate 60; Sbacchi, 'Italian Colonialism', 272.

[46] Quaranta, *Ethiopia*, 31; FO371/22021/J2728/40/1, Perth to Halifax, 9 July 1938.

[47] Sbacchi, 'Italian Colonialism', 272.

[48] G. Pistolese, 'Le attività agricole dell'AOI', *EI* (Feb. 1939).

Italy's commitment to the abolition of slavery was not mere 'humanitarian rhetoric'.[49] The correspondent admitted that the Italian

> civilization, based as it is upon a solid historical and social tradition, has no need of being wrapped in sentimentalism in order to be generous towards the vanquished and the weak. Populations which had never before understood that work is a human law from which no one is exempt, are now beginning to realize the civic necessity of individual work. The success of the first Village of Liberty proves it in a very clear manner.[50]

The Italians, however, insisted that the initiative had more than publicity value. It was one of the many strategies employed by the government to ease the problems created as a result of the proclamation ordering the abolition of slavery, issued simultaneously with the first advance of Italian troops.[51] To highlight the magnitude of the problem, the Italians claimed that from the governorship of Galla and Sidamo alone (the main slave-market), about 400,000, and from Härär 20,000, slaves had been freed. Some of these liberated slaves returned to their original home, and a few ventured into the field of petty trade. But the vast majority were left with no means of subsistence. In the eyes of the Italians, the use of the slave-village scheme was only one of several practical responses to overcome this problem. As part of such combined effort a large number were enrolled in the native militia and in labour-gangs, while a fair proportion were left with their old masters, working as share-croppers under new terms of contract drawn and controlled by the local *residenti*. The initiative of the governorate of Galla and Sidamo involved only a small fraction. However, reports suggest that the scheme was being followed elsewhere especially at Jubdo, in Wälläga, where the use of ex-slaves was planned in the labour-starved mines.[52]

But the slaves showed little enthusiasm, which technicians attributed as owing to indolence born of age-long servitude. The district officers combined physical punishments and persuasion in their attempt to overcome this and foster a sense of independence.[53]

The Mäkanisa Farm consisted of about thirty-two plots, where an equal number of families were accommodated. The farm had a demonstration and trial camp. The Ethiopian settlers were almost exclusively

[49] *GI*, 3 July 1938; FO371/22021/J2728/40/1, Perth to Halifax, Rm 9 July 1938.
[50] FO371/22021/J2728/40/1, Perth to Halifax, Rm 9 July 1938. [51] Ibid.
[52] Ibid.; M. Moreno, 'La politica indigena italiana in AOI', *AAI* 5/1 (1942), 64–77.
[53] FO371/22021/J2728/40/1, Perth to Halifax, Rm 9 July 1938; Quaranta, *Ethiopia*, 31.

the *ťisäña*[54] families that were forcibly dislodged to make a room for the Italian settlers of the nearby pilot demographic colonization farm consisting of Milizia della Prima Centuria Agricola.[55]

These experiments were in their infancy. However, contrary to the authorities' claim, the results were not always encouraging. In places like Siré (Härär), for example, the programme was a disastrous failure.[56]

## AGRICULTURAL CAMPAIGNS

Among the institutions that closely affected the indigenous agricultural system were the agricultural offices established in 1937–8. Each agricultural office had a department dedicated to native agriculture, with the task of studying indigenous farming and the application of measures whose purpose would be the increase of Ethiopian agricultural output. Some of these were equipped with research stations dealing with plant physiology and pathology and had their own model farms, where indigenous and foreign varieties of crops, such as Mentana and Kenya B1, were grown. Samples of local crops were sent for testing in the laboratories of the RIA AI in Florence.[57]

The offices used various techniques whose stated purpose was the expansion of Ethiopian production. Open-air schools in agriculture and visits to orientation farms were organized.[58] Where there was a public press, articles of particular interest to peasants were published.[59] In the long term, the campaigns aimed at raising agricultural output and lifting the standard of living of the rural peasantry. In the short term, the chief concern was to produce food-crops needed for the continuously growing

[54] It is difficult to assess the terms and the scale of these settlements. But the ideal model appeared to be the contract between ONC and the inhabitants of Guntuta (see App. 1).

[55] Known also as Prima Centuria Agricola di Pre-colonizzazione (cf. Ch. 4, n. 146, CD IAO Ex AOI 1097, Relazione sull'attività).

[56] Piani, 'Agricoltura indigena', 408.

[57] Quaranta, *Ethiopia*, 29.

[58] G. Piani, 'Attività autarchiche nell'Hararino', *Autarchia Alimentare* (June 1938); *IOM*, 5 Apr. 1939; *Piccolo*, 19 Apr. 1940.

[59] Glancing at articles published in the local weekly *Yäqésar Mängest Mälektäña* and the periodical *Yä-Roma Berhan* one finds articles such as: 'Fertilize the Fields'; 'Noxious Insects to Agriculture'; 'The Locusts'; 'From the Moment of Harvest Think of Sowing Wheat'; 'How You Should Improve the Food of your Cattle'; 'Report in Time the Appearance of Locusts'; 'Plant Trees'; 'Increase the Cultivable Area of your Land'; 'Prize Competition for Winning Native Farmers'; 'The Family Gardens'; 'Roads, Commercial Development, and Agricultural Machinery' (cf. CD IAO: Sc 1801, Attuale; Ex AOI 1097, Relazione sull'attività).

Italian population. Key difficulties were in the field of experimental research and organization of production.

Research demanded prior collection of data on the existing crops of the area, their relevance to the economy of the Empire, their ability to adapt to various seasonal conditions, and their resistance to parasite attack; in addition appropriate soil and altitude, and exact time for sowing had to be established. Once the type of crop to be promoted was identified, the technicians decided on the category of farmers to be targeted. As a matter of principle, the choice of crop for the Ethiopians had to preclude any potential or actual risk of competition, or interference, with metropolitan production. The way in which the decision to promote wheat production was reached in the governorate of Galla and Sidamo illustrates this point. Wheat formed a respectable, though minor, part of the local diet. As in the rest of Ethiopia, its cultivation started from an altitude as low as 1,700 m above sea level and went up to 3,000 m; it was most productive between 2,200–2,400 m. The Italian settlers in Galla and Sidamo were numerically few and most of them were settled under 2,000 m, an altitude below the ideal conditions for wheat production. Under such circumstances an appeal to indigenous agriculture was the most appropriate option.[60] Most of the Ethiopians of the area were skilled farmers. What was more, the climatic and soil condition of the territory allowed the Italian settlers to dedicate themselves more profitably to other crops, such as coffee, tobacco, oilseeds, vegetables, and fruits, that could offer a much higher profit-margin.[61]

However, despite careful planning the risk from Ethiopian competition was not always avoidable. Indeed, Ethiopian competition was one of the most frequent and indignant complaints of the settler agriculture. These complaints seem to be not without substance. At the end of 1938 as low-priced vegetables from Härär made their way to Addis Ababa, the ONC's hitherto-profitable vegetable trade sunk deep into the red. During this same period Jimma market became saturated with what the technicians called 'useless potatoes produced and sold for a ridiculous price' by the natives. Although a minor episode, it shook the confidence of the colonists, unable to sell their own produce for over a fortnight. Commenting

[60] A. Chiuderi, 'Alcune considerazioni su una campagna granaria nel GS', *Georgofili*, 7/7 (1941): 117; for the rest of the territories this same rule applies. Cf. R. Ciferri, 'Frumenti e granicoltura indigena in Etiopia', *Georgifili*, 6/5 (1939), 233–42; G. Mazzoni, 'Prima impostazione del problema dei frumenti nelle terre alte dell'AOI', *Georgifili*, 7/7 (1941), 99–105; CD IAO RGGS UA 1818, Attività, 6–7.

[61] Chiuderi, 'Alcune considerazioni', 119.

on the case, one agricultural officer remarked, it was a frivolous event but 'sufficient as a symptom of a situation which will not fail to manifest itself in a more decisive manner in the future'.[62]

Organizationally, agricultural campaigns were the result of the co-ordinated efforts of political pressure and technical assistance on the one hand, and the co-operation of the Ethiopians and their chiefs, on the other. The success partly depended on the harmonious working of all these groups. The political authorities in their respective governorate issued awards and guaranteed the purchase of the crop at harvest. All chiefs and local farmers were summoned to participate. With the aid of the technicians from the agricultural office, the *residente* or the *commissario* gathered the chiefs and the farmers under his jurisdiction; he explained the government's plan to increase production and communicated the facilities provided by the government to the farmers for this purpose.

The facilities varied according to the circumstances and the needs of the farmers, but, as a rule, destitute peasants were furnished with seeds from the government granary with the obligation to refund it at harvest; most of these consisted of *asrat* collected earlier from the peasantry;[63] for those applicants with lands inadequate to their working capacity, domanial lands were made available on a temporary basis; agricultural tools, such as improved harrows and ploughs, were distributed freely. Areas showing the best results were provided with threshing machines, an item much appreciated by the Ethiopians. The technicians of the agricultural office provided assistance throughout sowing, cultivation, and harvest.[64] The peasants could choose the plot to be reserved for the crop and in this they demonstrated, as a rule, considerable ability and skill. To the surprise of the Italian technicians, they knew the quality of the soil, and which soils were more receptive to what grains, and at which altitude and in which season each type and strains thrived best. The Italian officers, who earlier used to describe native agricultural knowledge and farming methods as primitive and worthless, began to praise and even copy it.[65] Financial incentives and medals were distributed by the technicians as prizes for the most successful farmers and chiefs. In Amara area, such awards were reserved for wheat production, in Shäwa for wheat and kitchen-garden,

[62] CD IAO GS 1820, Il problema, p. 13.

[63] Ibid.; Chiuderi, 'Alcune considerazioni', 118; G. B. Lusignani [Capo dell'UA Am], 'Iniziative cerealicole dell'UA Am', *Autarchia Alimentare* (June 1938), 37; *Colonie*, 28 June 1938; 'Iniziative agricole nel territorio di Borana e Hararino', *EI* (July 1938); *CE*, 25 Aug. 1939.

[64] *CE*, 28 June 1938; *Piccolo*, 1 July 1938, and [13–14] Aug. 1938.

[65] ACS ONC: 3/A/1. Balconi to Taticchi, 27 June 1937; 3/B, Di Crollalanza to Taticchi, Rm 17 Dec. 1938.

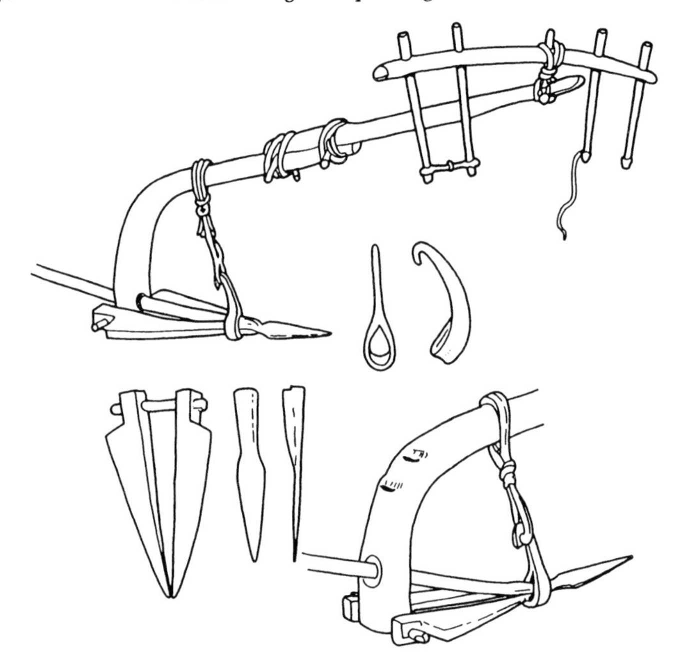

FIG. 4. Planned agricultural tools
(*a*) Traditional Ethiopian plough
(*Source:* IAC, *Italy's Action in Ethiopia*, pl. 63.)

and in Härär for coffee. The amount of prizes was between L.400 and L.1,000 and was given to a single farmer or village.[66]

The local chiefs had the task of distribution of seeds and surveillance of the farmers' plots during sowing and harvest; any abuses on the part of the peasantry were reported by them to the authorities. As a reward, the chiefs were given the right to collect a certain percentage, normally

[66] CD IAO EX AOI Gn 600, Relazione sull'attività; DRegG, 21-6-1937, A.XV, Concorsi a premio per la coltivazione del grano e delle patate agli agricoltori indigeni, in *BUGA* 2/3–4, May–June 1937; DGov 18-7-1938, no. 46, Concorsi a premio da assegnarsi agli agricoltori indigeni che avranno ottenuto nel prossimo raccolto una maggiore produzione unitaria di grano ed altri cereali, in *BUGA* 3/15, 1 Oct. 1938; DGG 25-5-1938, no. 651, Concorsi a premio tra i coltivatori indigeni residenti nei territori del G AA, nei Commandi Settori di Ambò, Debra Brehan e del Commissariato di G Adama, in *GU* 3/15, 1 Aug. 1938; CD IAO Hr 1813, G Hr DCL IA, Provvedimenti a favore degli indigeni; *Azione coloniale*, 21 June 1938; 1 Sept. 1938; *Colonie*, 11 Nov. 1938; 31 Apr. 1939; *IOM*, 20 Sept. 1938; *La Stampa*, 11 Aug. 1938; FO371/23380/J1776/296/1, Bird to Halifax, AA 1 Apr. 1939; FO371/23380/J4710/296/1, ACG Gibbs to PSSFA, AA 1 Nov. 1939.

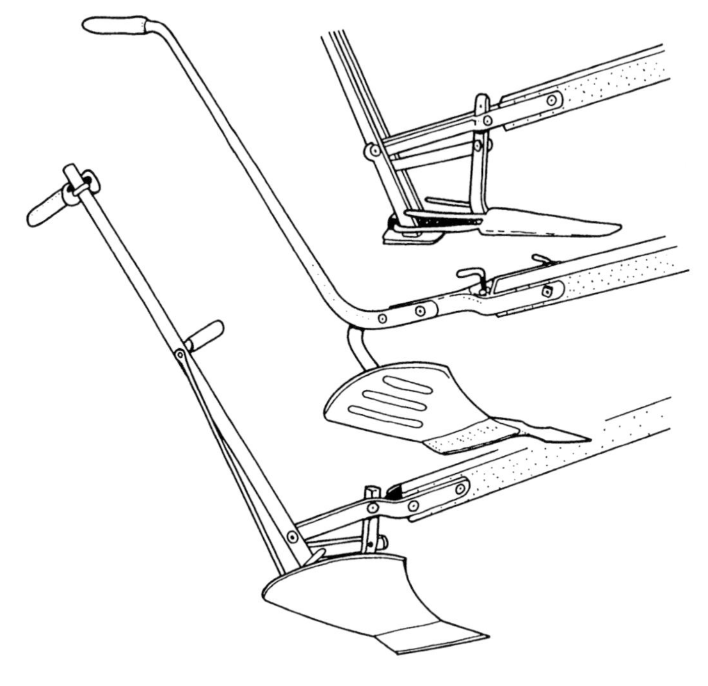

(*b*) Improved plough

between 2 and 3 per cent of the produce from farmers under their jurisdiction, assessed in cash. In Amara, chiefs of successful districts were paid substantial cash, between L.500 and L.1,000 per person.[67]

Participation of peasants in the campaigns took place through a written agreement by the chief or the farmer which stipulated the quantity that each subscriber agreed to sow or have sown. Those failing to sow the seeds taken from the government or the quantity specified in the written declaration were fined.[68]

In some areas agricultural campaigns started as early as mid-1937, their origin closely linked with Mussolini's demand that the Empire should reach economic autarky by mid-1938. With some exceptions, the technicians' reports suggest that participation was relatively widespread; yet the outcome of the campaigns did not always match the expectation of the planners.

[67] DRegG 21-6-1937, A. XV; DGov 25-5-1938 no. 651, in *GU* 3/15, 1 Aug. 1938.
[68] Chiuderi, 'Alcune considerazioni', 117–18.

Friction between interventionist political authorities, particularly the Fascist Party members, and self-opinionated technical officers was a constant problem. During the early campaigns, the agricultural office, consulted only indirectly, gave lukewarm support. At other times the campaign suffered from inadequate preparation by technicians, as was the case with the 1939 campaign in Galla and Sidamo: seeds were insufficient, proper storage was lacking, and sacks scarce. At other times the technicians failed to carry out properly their auxiliary support, such as setting up orientation farms, supply of simple farm machinery and selected seeds. In areas like the Amara governorate where Italian hegemony was not fully asserted, the campaign was arrested by lack of co-operation by, or straight hostility from, the chiefs, and the distribution of seeds and technical assistance remained impossible.[69] Propaganda was also drastically curtailed by fuel restriction that came into force as an austerity measure in September 1939.[70]

For the technicians, each campaign provided them with experience that allowed them to improve their subsequent action on the basis of lessons learnt. But their efforts were hampered by shortage of funds, the vast areas they had to cover, and language barriers which made conversing with the peasantry—who were by nature unresponsive to change—difficult. It was little wonder that even those devoted men did not have much effect on existing agricultural practices. Their reports dwell also at length on the many technical mistakes that undermined the confidence of the Ethiopian farmers in them. They pointed to the introduction of new varieties of seeds which, distributed after hasty trials, showed poor resistance to parasites; this was the case with *mentana*, *cappelli*, and other Italian varieties that were widely used, particularly in Shäwa and Härär. In 1939, a year of violent rains, they were completely destroyed by smut.[71]

Perhaps even more demoralizing was the technicians' continuous battle against bad weather, draught, marauding insects, and locusts. In 1939 the locusts destroyed many thriving fields in Wälläga and northwest areas of the Galla and Sidamo territory; there were environmental limitations, the overcoming of which demanded extended time and long acquaintance with the area. They pointed out insufficient atmospheric precipitations

[69] *CE*, 28 Aug. 1938; Lusignani, 'Iniziative'.

[70] CD IAO Ex AOI 1097, Relazione sull'attività.

[71] Castellani, 'Problemi fitopatologici dell'Impero: Osservazioni ed orientamenti', *Georgofili*, 6/5 (1939), 554; E. Castellani, 'Ruggini e granicoltura nell'Africa tropicale montana', *Georgofili*, 7/8 (1942), 127.

that arrested the full development and maturation of the winter crops and prevented spring sowing.[72]

## COMMERCIALIZING PEASANT AGRICULTURE THE CASE OF COTTON

Attempts to develop an integrated Roman world economy led to the policy of dividing the Empire into sectors, to be granted to Italian concessionaires who would have the right to buy and market certain items of Ethiopian produce. The scheme was planned to cover every aspect of the Empire's agricultural life. Owing to the abrupt and sudden end of the occupation, none of the schemes were translated into action except that for cotton production.

Given the significance of cotton to the Italian metropolitan industry, great attention was given to the possibility of growing cotton. Even though cotton was grown in southern Italy, Eritrea, and Somalia, the needs of Italian industry were not met.[73] The policy of self-sufficiency following international trade sanctions led to a rapid increase in domestic production matched with a small rise in the colonies, as appears from Table 25. The cotton produced in Italy and the colonies amounted to just 5.5 per cent of the total needs of the domestic industry. With the sharp decrease in supply (Table 24), the sanctions exposed the vulnerability of dependence on foreign imports. It was estimated that Italy needed to cultivate a minimum of 900,000 ha with an average production of 525 kg of raw cotton per ha and the Italians were confident that Ethiopia could start to supply this amount annually within a short period.[74]

It is against such increasing domestic economic pressure that expansion of cotton production in the Empire has to be assessed. Increased dependence on export and the crisis that followed the sanction led to the

[72] Chiuderi, 'Alcune considerazioni', 118; *CE*, 3 Feb. 1939; *CI*, 24 Aug. 1938; *CM*, 17 Dec. 1938; 'I cereali della zona di Macallé', *II*, Dec. 1938; *Colonie*, 23 Dec. 1938; Lusignani, 'Iniziative'.

[73] FO371/22030/310, Bird to Halifax, AA 1 Apr. 1939. What had been done in Eritrea was very little, while in Somaliland there were flourishing cotton plantations in the restricted regions watered by the Juba River along the banks of the Wabi Shibälé, where two crops were taken annually, planting taking place in Apr.–June and again in Sept.–Oct. There was a government-owned ginnery at Vittorio d'Africa, a few miles inland from Merca.

[74] G. Mangano, 'La politica del cotone', *Georgofili*, 6/5 (1939), 270; 'La cotonicultura in Etiopia' *AC* 31/5 (May 1937), 202.

Royal Decree of 7 October 1937, which saw the creation of a government concern, the ECAI, an affiliate of the ICI,[75] itself an association of all the Italian cotton industrialists for co-ordination of cotton production. The agency had a capital of L.25,000,000; of this, half was contributed by the State and half by the ICI. The agency's publicly stated goal was 'to develop, regulate and expand cotton farming in Italian Africa', including Libya.[76] As expanded later, such a task involved:

1. Research the cotton resources of the Empire and laying out schemes of development.
2. Regulating all cotton production in the Empire, administering and co-ordinating sales.
3. Advising the concessionaires.
4. Promoting production, plant protection, provision of insecticides, and disease control.[77]

Once approved by the MAI, the decisions of the agency had statutory power and were binding upon all those engaged in activities related to the agency's field of action. The agency was run by a board of directors, consisting of a president and ten other members appointed by the joint decree of MAI and the Ministry of Finance.

An overview of the composition of the Board reflects the presence of interlocking interests of which the agency was an expression. Two of the members were nominated by MAI, three by the ICI. The rest were representatives from the Ministry of Agriculture and Forestry, Ministry of Corporations, Governorate-General of both IEA and Libya, and Textile Products Corporations: each of these appointed one member. The Board was headed by Angelo de Rubeis, the Director-General of Colonization and Labour of MAI; Guido Mangano, a specialist in cotoniculture, acted as its General Manager. Despite such an impressive composition, as a decision-making body the Board excluded among others the indigenous representatives whose presence was vital to the successful cultivation of cotton. This was the weakest point of the scheme.

The agency operated under the joint supervision of the MAI and the Ministry of Finance and had Rome as its headquarters, making the agency

[75] The task of the Institute with regard to cotton was: (1) Information gathering; (2) Market regulation; (3) Manufacturing regulation; (4) Co-ordination of raw cotton supply; (5) Enforcement of liquidation of bankrupt firms, establishing new ones, and arbitration of labour dispute. (See Quaranta, *Ethiopia*, 59.)

[76] RDL 7-10-1937, no. 2513, Costituzione dell'ECAI, in *GURI*, 19 Feb. 1938; Mangano, 'Programma di massima di attività dell'ECAI', *AC* 32/9 (Sept. 1938), 44.

[77] Mangano, 'Programma', 414.

TABLE 24. Italy's cotton import, 1929–1938 (in ql)

| Year | USA | India | Egypt | Others | Total |
|---|---|---|---|---|---|
| 1929 | 1,662,369 | 480,539 | 239,763 | 62,110 | 2,444,831 |
| 1930 | 1,328,439 | 478,845 | 185,437 | 54,675 | 2,047,396 |
| 1931 | 1,047,009 | 365,024 | 215,717 | 76,907 | 1,704,657 |
| 1932 | 1,465,651 | 146,068 | 230,652 | 50,630 | 1,902,001 |
| 1933 | 1,649,526 | 263,132 | 255,367 | 29,739 | 2,197,764 |
| 1934 | 1,153,101 | 320,276 | 335,446 | 64,192 | 1,874,015 |
| 1935 | 868,975 | 264,351 | 290,974 | 62,835 | 1,487,335 |
| 1936 | 735,123 | 68,372 | 138,226 | 71,869 | 1,014,040 |
| 1937 | 982,847 | 177,079 | 299,207 | 205,102 | 1,664,235 |
| 1938 | 951,509 | 136,196 | 263,520 | 210,497 | 1,570,722 |

*Source:* G. Mangano, 'La politica del cotone', *Georgofili*, 6/5 (1939), 268. (Totals as in the original.)

an organizationally amorphous body with no independent existence of its own. Inside the colonies its presence was limited to a few experimental stations at Alamata, Adama, and Abala. The results achieved in these three stations were insignificant. Only in the Abala were they fruitful.[78] For important decisions the agency largely depended on the reports of *ad hoc* commissions carried out during their short but expensive trips to the colonies. The commissions charged exorbitant consultancy fees and, as a result, were disliked by the colonial administration.[79]

After preliminary studies, the government was faced with two options: that of setting up large-scale irrigated farming along the lines used at Gezira in Sudan, Täsänäy in Eritrea, and Lower Juba in Somalia; or rely on mobilizing existing indigenous farming through propaganda and assistance, as was the case in some African colonies, such as Uganda and the Belgian Congo and, to a small degree, Eritrea and Somalia. From the government's point of view, the first method was unremunerative; it required extensive work involving building dams and canalization, and

[78] Villa Santa *et al.*, *Amedeo*, 249. CD IAO: AOI 1830, Pesarini Alfredo to A. Maugini, Relazione distretti Gobbo-Uoldia, e Lekemti, n.d., p. 3; AOI 1826, Achille Pajella, Brevi notizie ed osservazioni sul Distretto Cotoniero del Lago Margherita, Rm 1 Mar. 1943, p. 10.

[79] ACS GRA 46/41/19, Cusmano and Corvo to Amedeo d'Aosta, AA 4 June 1938.

TABLE 25. Cotton production in the metropolis and colonies, 1932–1938 (in ql)

| Year | Italy | Colonies | Total |
|---|---|---|---|
| 1932 | 2,600 | 11,000 | 13,600 |
| 1933 | 9,500 | 9,000 | 18,500 |
| 1934 | 8,000 | 10,000 | 18,000 |
| 1935 | 18,600 | 10,400 | 29,000 |
| 1936 | 38,600 | 6,000 | 44,600 |
| 1937 | 42,000 | 6,000 | 48,000 |
| 1938 | 82,000 | 6,400 | 98,400 |

*Source:* G. Mangano, 'Politica', 270. (Totals as in the original.)

clearing and levelling of lands. The government therefore favoured the second alternative as the best paying method.[80]

Once the technicalities on method of exploitation were over, affairs began to operate on two major problems. How to organize the production geographically, and to whom to give its management? The first issue was settled with the creation of Cotton Districts in 1938. As set out in the Royal Decree of January 1938, these were areas mapped out as specially favourable for cotton growing and within which the administration, through the ECAI, aimed 'to increase, in preference to other crops, cotton production, using the free co-operation of the native populations, understanding that the production of food-crops and animal breeding, indispensable to the life of these same populations, is respected.'[81]

The authority of creating such cotton districts rested with the MAI which also had the task of granting concessions. This same Ministry defined the boundaries, approved the contracts and granted special privileges to the concessionaires. The colonial administration in IEA, on the other hand, issued rules concerning cotton farming and measures regulating import of seeds and protecting them from insects and other diseases, while the daily control *in situ* rested with the agricultural offices and the

[80] CD IAO AOI 1106, IAO, Rapporto schematico sulle attività cotonicole svolte dall'Italia in Etiopia, F1 11 July 1949, pp. 1–2; Vicinelli, 'L'Istituto del "Distretto Cotoniero" nell'AOI', *AC* 33/1 (Jan. 1939), 34; MAE, *Avvaloramento*, 392.

[81] RD 7-1-1938, no. 443, Istituzione di distretti cotonieri nell'AOI, in *GURI*, 10 May 1938 and *GU* 3/13, 1 July 1938.

TABLE 26. Cotton District areas

| Area | Area (ha) | Altitude (m) | Rainfall (cm) | Minimum temperature during the growing season (°C) |
|---|---|---|---|---|
| Awash | 1,000,000 | 1,944.95 | 80 | 17.82 |
| Baro | 4,000,000 | 520.33 | 110 | 29.7 |
| Mätämma | 1,400,000 | 760.36 | 90 | 24.75 |
| Näqämti | 1,700,000 | 1,801.02 | 130 | 19.8 |
| Qobbo | 900,000 | 1,400.87 | 110 | 21.78 |
| Soddo | 3,200,000 | 1,500.90 | 119.8 | 21.78 |
| Ṭana | 1,400,000 | 1,500.90 | 119.8 | 23.76 |
| TOTAL | 13,600,000 | | | |

*Source:* DM 16-5-1938, IDC: Auash; Cobbò; Metemma-Dongur, in *GU* 3/17, 1 Sept. 1938; DM 1-12-1938, IDC: Lechenti and Soddu, in *GU* 4/12, 16 June 1939; Ethiopian Ministry of Agriculture, quoted by G. Nicholson, 'A General Survey of Cotton Production in Ethiopia and Eritrea', in *Empire of Cotton Growing Review*, 32 (Jan. 1955), 4.

local administration.[82] Within four months of the formation of ECAI, three Cotton Districts were created. By 1940 a total of nine Cotton Districts had been mapped out as suitable for cotton farming, including those already existing in the two old colonies, Täsänäy in Eritrea, and Juba in Somalia. The districts, scattered throughout the Empire, involved vast areas of land varying in altitude, size and rainfall (See the Map shown in Fig. 5, p. 283, and Table 26).

1. Awash Valley Cotton District centred on Adama, a town on the upper reaches of Awash River, situated at about 65 miles south-east of Addis Ababa on the Djibuti railway.
2. Baro Cotton District, lying along the Baro river, centred on Gambéla in Galla and Sidamo.
3. Lake Ṭana Cotton District included various localities on the shores of Lake Ṭana and centred on Gorgora, 42 miles away from Gondär.
4. Mätämma Dongur Cotton District, situated along the Sudan frontier to the west of Gondär, centred on Mätämma.

[82] *GURI*, 10 May 1938.

5. [L]Näqämti Cotton District, lying along the Sudan Frontier in the Wälläga province, centred on Sayo.
6. Qobbo Cotton District centred on Alamaṭa, a town situated 370 miles north of Addis Ababa, on the main highway to Asmära in Eritrea.
7. Soddo Cotton District, 190 miles distant from Addis Ababa, centred on the banks of Lake Čamo, in Galla Sidamo.

There were other Cotton Districts which, like Bako-Burgi Cotton District South-west of Soddo, never became operational. In the old colonies, Eritrea and Somalia, the cotton production was restructured along the lines of the new policy. Täsänäy State Farm became Täsänäy Cotton District and was given to SIA; while in Somalia there was Lower Juba Cotton District run by SAICES.[83]

As to the second problem, the government was to take the commercial farms in partnership, planning to give the organization and management of each Cotton District to private companies. But the government was torn between the needs of safeguarding its authority over the conduct of the enterprise and that of guaranteeing the profitability of the scheme. With bad bargains the commercial farms might acquire too much power and end up being an *imperium in imperio*. Equally, too much stringency on the part of the government was bound to generate little interest in the scheme. It was an understandable fear of compromising wider national objectives with limited commercial interests. The government saw as the most important factors in this regard the timescale and the size of the concession as well as the financial incentives. The result was that a compromise was struck. The concessions would operate for a limited period of twenty-five years, with an option for a further twenty-five-year extension. The government had the right to terminate the concession earlier if it considered the management unsatisfactory for political or economic reasons. In case of early termination, the companies would retain the ownership of their industrial plants and the equipment as well as the land they stood on, together with the eventual cotton stock and other mobile property. But clauses empowering the government to purchase them at a price assessed by a three-body commission composed of representatives of the Governor-General, the ECAI, and the company concerned, show

[83] DM 16-5-1938, IDC: Auash; Cobbò; Metemma-Dongur, in *GU* 3/17, 1 Sept. 1938; DM 1-12-1938, IDC: Lechenti la cui organizzazione e la gestione vengono affidate alla ICAI; Soddu la cui organizzazione e la gestione vengono affidate alla ICAI, in *GU* 4/12, 16 June 1939; CD IAO 1106, Rapporto schematico, p. 3; DM 1-12-1938, IDC Basso Giuba, in *GU* 4/4, 16 Feb. 1939; *GURI* 105, 10 June 1938; *BULMAI* 6, June 1938; FO371/22030/310, Bird to Halifax, AA 1 Apr. 1939; 'Notiziario agricolo commerciale AOI', *AC* (Mar. 1938), 235; Vicinelli, 'Istituto' 41–3; Villa Santa *et al.*, *Amedeo*, 246–7; Quaranta, *Ethiopia*, 62–3; MAE, *Avvaloramento*, 392–7; Nicholson, 'General Survey', 10–14.

the government's anxiety to safeguard its own position. The agreement left the government practically as a strict landlord endeavouring to maintain almost a feudal relationship between itself and the company.[84]

As regards the size of the concessions, the government made sure that the Cotton Districts were extensive enough to make the scheme an attractive enterprise. Districts were scattered all over the Empire, on average stretching over half a million hectares of land (Table 26).[85] Theoretically, the granting of a concession to a company did not entail any ownership of land, except for that allotted for its processing plants and experimental fields, and the building of administrative and residential centres. The government upheld its right to carry out public and reclamation works, as well as granting of lands to third parties.[86] Equally, the existing rights of the Ethiopian cultivators and the methods of land transaction were not to be affected. And yet the companies successfully asserted their claims to State lands within their districts to which they had preferential rights. Included among these were lands where Ethiopian rights were partial. Otherwise, what the company really had within each Cotton District was an exclusive right of exploitation consisting in commercial monopoly over trade of cotton produced by the Ethiopians to the exclusion of other competitors. Except in a few cases, even the settlers' farms, willing to cultivate cotton within such districts, were no exception to this rule;[87] these cultivated cotton only after having obtained clearance from the company beforehand, and strictly abided by the rules prevailing inside each district.[88]

In its bid to attract private capital the government pursued an open-door policy. Any company with proven financial and managerial ability and technical knowledge was welcomed. Undertakings were made to create conditions that would ensure the long-term profitability of cotton production.[89] As an additional incentive, the government offered extensive economic inducements without being too stringent on what the

[84] RD 7-1-1938, no. 443.

[85] CD IAO 1106, Rapporto schematico, p. 3.

[86] 'Notiziario agrario commerciale: AOI', *AC* 32/9 (1938), 477. These rights are clearly stated in a clause in art. 7 of the contracts.

[87] As it appears from art. 7 of the contracts, a special arrangement was to be made for the agricultural farms whose work within a particular Cotton District predates that of the concessionaire company.

[88] Vicinelli, 'Istituto', 37.

[89] Angelo de Rubeis summed up the measures to be adopted to this effect inside the Cotton Districts. Accordingly, the government would: (1) involve the largest possible number of cotton outgrowers; (2) enforce measures necessary to promote cotton production; (3) establish a fair purchase price; (4) create conditions and infrastructure that would make cotton production more attractive; (5) ensure that the administrative personnel, *residente* and *commissario*, were knowledgeable in economic and political matters. (Cf. CD IAO AOI 1936, Angelo de Rubeis, Promemoria per S.E. il Sottosegretario: Produzione cotoniera a mezzo degli indigeni, Rm 12 May 1938.)

companies should do[90] and a ten-year clemency period during which the companies as well as their subsidiaries were exempt from all taxes except for a token payment of registration and commission fees.[91]

Despite such favourable terms, the metropolitan capital remained unimpressed. Only two companies showed interest: the ICAI and the Ethiopian National Cotton Company (Cotetio). Both of them were of recent foundation, thus with limited operational experience. Cotetio was formed in July 1936 by the Italian industrialists through ICI and had a capital of L.14,000,000 which by 1940 was raised to L.50,000,000. ICAI was formed in 1938 by Milanese industrialists. Its capital was L.1,000,000. Both companies operated from Milanese headquarters. The Näqämti and Soddo Cotton Districts were granted to the ICAI,[92] while Cotetio retained a monopoly over the rest.[93] How much of their capital these two companies transferred to their respective concessions is unknown. But what records make clear is that they functioned on a shoestring budget and their technical know-how proved inadequate. Owing to such financial limitations, ICAI was forced to give up its Näqämti concession to Cotetio, who thus exercised near monopoly of cotton production.[94]

[90] The incentives granted by the administration consisted mainly of: (1) Promoting cotton production among the Ethiopians through the State machine; purchasing raw cotton, regulating the local markets and ensuring stable labour. (2) Remittance of inputs up to their destination. (3) Granting gratuitously lands required for administration, plants, and residence. (4) Facilitating foreign exchange for the purchase of inputs (such as seeds, machinery) from overseas. (5) Providing and maintaining feed roads. (6) Granting fiscal exemptions as contemplated by the law. (7) Concession of favourable freight tariffs on railway, motorway, and waterway transport. (8) Grant of all facilities that in the future would aim to benefit new concessionaires of Cotton Districts.

[91] The main fiscal concessions were: (1) Exemption from direct income tax on industrial revenues; (2) Exemption from taxes on lands and buildings; (3) Reduction of commission fees to quarter of the normal rate; (4) Exemption from customs duties on imported seeds, agricultural machinery and vehicles as well as their spare parts; (5) Exemption from stamp duty and exchange, administrative fees connected with the concession including contracts related to building constructions, supplies and insurance and all acts proving the fusion and increase of capital. Cf. RD 7-1-1938, no. 443; DIM 18-5-1938: Distretti cotonieri, agevolazioni in materia fiscale e doganale, in *GU* 3/18, 16 Sept. 1938; DIM 1-12-1939: Esenzioni e riduzioni fiscali a favore dell'ICAI, in *GU* 4/13, 1 July 1939; DM 31-5-1940: Concessione alla Cotetio di alcune esenzioni e riduzioni fiscali e doganali sui tributi coloniali, in *GU* 5/35, 31 May 1940.

[92] DM 1-12-1938: IDC Soddu; DM 1-12-1938: IDC Lechenti; 'Notiziario agricolo commerciale; AOI', *AC* 32/5 (1938), 235.

[93] DM 16-5-1938: IDC Cobbò; DM 16-5-1938: IDC Auash; DM 16-5-1938: IDC Metemma-Dongur. The exception was Baro (Cotton District which was run by Compagnia Cotoniera del Baro (COTOBARO), a subsidiary of Cotetio; 'Notiziario agricolo commerciale: AOI', *AC* 32/3 (1938), 139–40.

[94] DM 31-5-1940: Revoca della concessione del distretto cotoniero di Lekemti alla ICAI e passaggio dello stesso alla Cotetio, in *GUGS* 5/35 (suppl.), 31 Aug. 1940.

The obligations laid upon the concessionaires were minimal and vague.[95] One clear example is clause (d) of article 2 which demanded the companies obtain 'as much production as is possible to procure within the first two years through intense propaganda'. Even the description of the Cotton Districts defined as areas destined for cotton production 'in preference to other crops, . . . provided that the production of staple food-crops and animal husbandry' required for the life of the Ethiopian population is maintained, lent itself to differing interpretations. The decree did not specify how such requirements were to be assessed; equally, it said nothing about the nature of these same requirements: were they to be applied exclusively for consumption or extended also for sale?

Ambiguities such as these were the source of occasional conflicts. A typical example was the campaigns conducted by Cotetio in 1938 which led to a showdown between the company and the authorities in Addis Ababa. On balance the relationship between the companies and the district political officers, though not always idyllic, was good, sustained by common racial sentiments and a deep sense of civilizing mission. But the situation with the higher authorities, particularly the regime in Addis Ababa, was quite different. Perhaps resentful of being excluded from negotiations in the metropolitan capital between the MAI and the companies, from the outset Addis Ababa displayed a fleeting interest in the Cotton Districts, further exacerbated by an ill-defined relationship between the companies and the colonial administration within the IEA.

The 1938 campaign revealed that the companies' commercial interests did not necessarily tally with national objectives. The companies' main aim was to secure within a short time at least part of its operating revenue and recoup investment on processing plants. By broadly interpreting the ambiguous clauses, Cotetio not only had forced many peasants to cultivate cotton at the expense of staple-food production but also ventured into areas where the possibilities of cotton farming were reportedly poor. The government feared that ventures into unpromising lands would undermine Ethiopian farmers' confidence in the scheme. It also viewed overemphasis on cotton as compromising its efforts towards ensuring self-sufficiency in the production of foodstuffs. The Duke of Aosta protested against the company's conduct, but was rebuffed as unwarrantedly interfering into the company's internal affairs. The dispute was settled only after an appeal to the MAI who empowered the colonial

[95] The obligations were standard as were the inducements. In addition to their respective Charters, see also CD IAO 1106, Rapporto schematico, p. 2.

administration to set the target of food-crop production in each Cotton District.[96]

Ostensibly, the Cotton Districts were not under the direct management of the concessionaire company. Such a role belonged to the Ethiopian farmers. The company was to provide them with extension services, advance cash, and foodstuffs. These were deducted from the cotton price at sale when paying the growers. The farmers were expected to continue to live as they had in the past, the only difference being that in addition to growing the usual cereals and cattle herding, they were now encouraged to undertake cotton farming as well and sell the crop to the company at a mutually agreed price. Such an undertaking relied much on the willingness and capacity of the indigenous population to participate in the scheme.[97]

The Italian authorities did not doubt that such co-operation would materialize. Given the many obvious advantages that the scheme offered, they expected the Ethiopian would welcome it. The enterprise was not something superimposed from outside, because the social system to which the Ethiopian farmer was accustomed remained undisturbed. He cultivated cotton on his own property using family labour. What was more, cotton cultivation was an ancient practice and had considerable importance in the country's social and economic life. Accounts on the eve of Italian occupation reveal that cotton was the most popular of commodities on sale in the Addis Ababa market.[98] Of course, most of this cotton was reportedly of poor quality, consisting of either spontaneously grown wild plants, or plants cultivated, rather carelessly, by families in a little patch within their compound walls or intercropped. Nevertheless, being exclusively rain-grown, it had the advantage that it required no capital outlay or technical innovation. It was on this cotton that an old and widespread family-based local cotton craft industry had thrived. Such knowledge of the crop was invaluable, and the Italian administration understandably thought the promotion of cotton for export was building upon a locally produced commodity. More importantly, the guaranteed price, combined with intense propaganda, was thought to convince the

[96] Villa Santa, *et al.*, *Amedeo*, 247; Cerulli, 'Colonizzazione', 67; art. 2(c) of IDC of Soddo and Läqämti.

[97] Villa Santa, *et al.*, *Amedeo*, 247; Quaranta, *Ethiopia*, 60; CD IAO 1106, Rapporto schematico, p. 2.

[98] L. D. Farago, *Abyssinia on the Eve* (London, 1935), 44; R. Di Lauro, 'Panorama politico-economico dei GS', *REAI* 26/7 (July 1937), 1084; A Pajella, 'Contributo allo studio dell'economica agraria nel territorio dei GS', *AC* 32/4 (1938), 179; UA Am, 'Aspetti generali e zootecnici del Lago Tana', *AC* 32/6 (1938), 264.

Ethiopian farmer that by growing cotton he was making a good financial investment with opportunities to raise his standard of living.[99]

But contrary to the authorities' optimistic forecasts, the picture that emerges from the few existing authoritative reports was that of strong Ethiopian resistance, rudely shattering Italian expectations. Resistance was by no means co-ordinated and uniform, but widespread and consistent enough to be the most crippling constraint on both the expansion of cotton and the enthusiasm of the concessionaire companies.[100] Administrative and technical difficulties were added to further complicate these problems.

Administratively, each Cotton District was divided by the companies into blocks of various sizes, known as *zona*. In each *zona* two forms of cotton cultivation were followed. Private fields consisted of a number of small plots set aside by each taxpayer and cultivated exclusively by family labour. In most cases this land belonged to the cultivator and in a few cases it involved State land given to landless peasants. The fields were scattered and their supervision was administratively exacting. Collective fields consisted either of State lands granted to the company's use or a suitable plot chosen by the peasants themselves. Most of these fields were strategically located in a few selected areas and employed unpaid communal labour recruited from surrounding villages. These two types of fields within a block ranged from a handful to over one thousand. At Dugunà block in Soddo Cotton District for example, private fields only numbered 900. On the other hand, some collective farms had the potential of engaging more than 1,000 day-labourers, as was the case with the 600-ha Motoarato field at Bonaya, Näqämti Cotton District.[101]

The technical and administrative control and the agricultural efficiency of these was vested in the block inspector (*capo zona*) who was responsible to the district chief (*capo distretto*). The block inspector had an extremely tight schedule: he had to help, and above all control, the Ethiopian farmers—a hard and demanding task. This involved a regular tour to the private fields instructing the Ethiopian farmer to farm well, informing him about the sowing programme, keeping him up to date with his work, purchasing cotton by weighing it, and paying cultivators. He maintained constant contact with Ethiopian chiefs upon whose politics of collaboration

[99] Quaranta, *Ethiopia*, 59–60; CD IAO 1826, Pajella, Brevi notizie, pp. 3–4; CD IAO 1830, p. 4; A. Chiuderi, 'Agricoltura indigena', 6; CD IAO 1106, Rapporo schematico, p. 2.

[100] CD IAO 1106, Rapporto schematico, p. 4.

[101] CD IAO AOI 1826, Pajella, Brevi notizie, pp. 3–4; CD IAO 1830, Pesarini to Maugini p. 4; Chiuderi, 'Agricoltura indigena', 273.

he depended for supply of labour to collective fields which absorbed most of his daily energies. He ran a small meteorological observatory from where he despatched the monthly data to District headquarters. In the execution of such tasks he was aided by his assistants who supervised part of the crops. In each Cotton District there were a handful of civil and mechanical engineers for the construction of buildings, technicians for the manning of machinery and, in Soddo District, an administrator who was in charge of a plethora of tasks ranging from accounting, workshop, and carpentry, to construction and correspondence with the headquarters in Addis Ababa. In a well-staffed Cotton District like that of Soddo, the personnel was between ten and fifteen strong. But it was the block inspector and district chiefs who were responsible for the agricultural efficiency of the project. Their actions were backed by the local political authorities who criss-crossed the districts during key periods in the agricultural calendar to enforce cotton growing.[102]

Legally, the district chiefs should have had at least one field inspector for each 1,000 ha and one Ethiopian chief for every 300 ha under cultivation.[103] But in practice this was often not the case. Instances where field inspectors were in charge of as many as 1,500 ha were common.[104] The issue, however, was not only that of inadequacy of personnel, but also the level of energy and competence of those working. As appears from a secret report by Alfredo Pesarini, a man with twenty years' experience in cotton fields in Egypt, lack of job-satisfaction combined with poor wages and bad working conditions meant the technical officers showed little enthusiasm for their work and many anxiously awaited the termination of the contract, when they would be repatriated or engaged in more lucrative activities within the Empire.[105] Management came and went on

[102] CD IAO 1826, Pajella, Brevi notizie, p. 2.

[103] A clause in art. 2 of the contracts asserts that 'the district chief employed by the company must have at his disposal for the purpose of supervision of, and propaganda amid, the natives at least one assistant per 1,000 ha, and one native-chief cultivator per 300 ha.

[104] CD IAO AOI 1826, Pajella, Brevi notizie, p. 5.

[105] The list of salaries available is only for the Soddo Cotton District personnel whose monthly, is given as follows: Chief District L.6,000; Administrator L.5,000; Assistant Chief District L.3,300; *Capo Zona* L.3,000; Assistant *capo zona* L.2,000–2,500; Mechanic L.1,500–2,000; Carpenter L.1,500–2,500; Builder L.1,500–2,000; Driver L.1,500. Each month L.300 from the workers', between L.500 and L.550 from the personnel of the district headquarters, and L.800 from the technicians' salaries was deducted to cover meals' cost. The company paid the full wages of the workers' servants, and part of that of the personnel while the technicians paid for their own. With the exclusion of the technicians posted in AA, for the rest of the personnel the employment was on a yearly basis (cf. CD IAO AOI 1826, pp. 9–10).

average within a year. Such a rapid turnover of officials threatened the continuity of carefully studied programmes.[106] This situation was further aggravated by the fact that generally they did not speak the local language and no systematic approach to cotton cultivation had been worked out. Work continued in an *ad hoc* and haphazard fashion.[107]

This, however, does not mean that cotton production was left to chance. On the contrary, the companies conducted 'cotton campaigns' in their respective Cotton Districts. As a pamphlet issued in Amharic by Cotetio in June 1938[108] reveals, the campaigns were orders instructing the peasantry on production techniques and conservation. This involved proper ploughing, row planting, proper care and maintenance, weeding and picking, as well as selling exclusively to the company. The peasants were ordered to use only new seeds distributed by the company and to destroy local ones.[109] The district chiefs saw that such orders were translated into action. Directives were passed to the block officers, allocating them the minimum amount of land that they should aim to get cultivated in their respective zone.[110]

From 1938 until the end of Italian rule the companies carried out three campaigns. Even though in 1938 the Italians claimed that Ethiopia 'can grow good American cotton to the extent of about 900,00 ql yearly, which would be almost sufficient to supply the Italian mills,'[111] when compared to the size of the territory covered by the Cotton Districts, the results achieved were modest.[112] Progress was made only in Adama, Alamaťa, and above all Soddo where the companies succeeded in setting up experimental fields, laboratories, and ginneries. Some of the premises and ginneries were found to be still in a good state in 1955.[113] The type of cotton grown was American Upland consisting of several varieties—Acala, Delfos, Stoneville 5, Cliett, Bagley, Rogers 3, Cocker Wilt 100, Lankart Watson. Towards the end of occupation, U4 strains from Barberton Experimental Station were introduced. The best results were obtained only with varieties Acala and U4. Yields ranged between 150 and 200 lb of lint per acre.

[106] Ibid., pp. 9–10; CD IAO 1830, Pesarini Maugini, pp. 3–4.
[107] CD IAO 1830, Pesarini Maugini, pp. 3–6. [108] See App. 5.
[109] CD IAO AOI 1936, Cotetio, Norme elementari per la coltura del cotone AA n.d. [date given in pencil indicates June 1938]. See App. 4 and 5.
[110] CD IAO 1826, Pajella, Brevi notizie, p. 5; 'La produzione del cotone nell'Impero', *IC* 15 Apr. 1939.
[111] 'Cotton Growing in Ethiopia', *International Cotton Bulletin*, 16/62 (1938), 210.
[112] CD IAO 1106, Rapporto schematico, p. 5. See Tables 26, 27, and 28.
[113] Nicholson 'General Survey', 10–12; Maugini, 'Appunti', 408.

TABLE 27. Cotton production in 1938–1939 campaign

| Place | Cultivated areas (ha) | Production raw (ql) | Production ginned (ql) |
|---|---|---|---|
| Awash | 834 | 260 | 80 |
| Gambéla | 10 | 30 | — |
| Näqämti | 80 | 40 | — |
| Qobbo | 550 | 320 | 100 |
| Soddo | 1,100 | 1,800 | 600 |
| Ťana | 60 | 30 | — |
| Täsänäy | 1,800 | 4,225 | 1,450 |
| Somala | 6,184 | 15,121 | 5,041 |
| TOTAL | 10,618 | 21,826 | 7,291 |

*Source:* 'Notiziario agrario commerciale', *AC*, 34/11 (Nov. 1940), 482.

In some sections of Soddo district, where the best results were obtained, there were reports of 750 lb per acre with Acala variety.[114] Despite these mixed results it became increasingly clear that the cotton production was built on a very precarious foundation.

Given the limited resources at their disposal, the block officers found the quota allocated to them unrealistic. Those who attempted to meet the target at any cost were faced with disastrous consequences. A typical case was the collective camp of Motoarato of Bonaya in Läqämti Cotton District where in 1940, against expert advice to limit cultivation to 200 ha, agreement was made with tractor-drivers to cultivate 600 ha. The avaricious contractors ploughed through the night but with such carelessness and haste that the seeds failed to germinate. The incident only served to undermine the confidence of the labourers in the technical experts. When instructed to sow the field again, 400 labourers, whom the authorities had recruited with difficulty, vanished. Convincing them to return proved an impossible task.[115]

The peasants were reluctant to produce the goods required by the companies. The exception was in part of Qobbo Cotton District where

[114] Maugini, 'Appunti', 408; CD IAO 1106, Rapporto schematico, p. 4; 'Notiziario agricolo commerciale': AOI', *AC* 32/9 (1938), 477.

[115] CD IAO 1830, Pesarini to Maugini, p. 5.

TABLE 28. Scale of unginned cotton production, 1938–1941

| Year | Cultivated areas (ha) | Production (ql) |
|---|---|---|
| 1938–9 | 2,534 | 2,156 |
| 1939–40 | 6,740 | 7,070 |
| 1940–1 | 7,100 | 10,000 |
| TOTAL | 16,374 | 19,226 |

*Source:* 'Notiziario agrario commerciale', 482; Villa Santa *et al.*, *Amedeo*, 248.

the technical officers, aided by a strong backing of the local *residente*, were able to secure the co-operation of the chiefs and their subjects.[116] Otherwise, what the technicians' reports clearly illustrate is lack of interest of the peasantry in the scheme. Resistance was not open or organized but subtle, local, day-to-day, and passive, similar to peasants' reactions elsewhere in colonial Africa or in American slave plantations.[117] An official from Soddo Cotton District complained that the peasants expressed great enthusiasm for each proposition put forward by the company, but they signed agreements without having the slightest intention of carrying them out. During their inspections, the field officers discovered that only the worst land was allocated to cotton or, where there was no continuous weekly inspection by the company officials, no care was given to cotton fields. The large number of private fields existing in any one block, the need for a continuous presence in collective fields, and the variety of tasks to be performed in any one day by the field inspectors, made any attempt at rigorous inspection almost impossible.[118]

Confronted with such difficulties, the companies shifted their emphasis from reliance on private fields to collective fields. Administratively, the strategy seemed to offer better control of the production of cotton and labour. But for its opponents, this method was only a tool of 'lazy' technicians who wanted to sacrifice productivity to administrative expediency, surrendering to the Ethiopians' demands. They pointed out that the collective fields had unfailingly proved to be unproductive largely

[116] Ibid., p. 2.

[117] Documents say that 'open opposition was not uncommon' but do not describe the nature of it (cf. CD IAO 1106, Rapporto schematico, p. 4).

[118] CD IAO AOI 1826, Pajella, Brevi notizie, pp. 2–3.

because the Ethiopians reserved the best lands for their own food-crops.[119] In practice, the shift from private to collective fields neither eased the administrative problems nor improved production efficiency, but aggravated existing difficulties.

Collective fields seemed to offer more satisfactory use of labour, which was unpaid and easily recruited. But in practice passive resistance left the companies impotent, dependent on the co-operation of the chiefs. Yet not all chiefs were willing to collaborate and the companies had to make frequent recourse to the use of force. Some chiefs were replaced and others fined. In some places stringent measures were adopted allocating to each chief a labour quota proportional to the number of people under his jurisdiction. But, despite their spurious claims to authority, even willing chiefs found difficulty in persuading the people they 'ruled' to work in collective fields. Sometimes the labour force they despatched was far below the requirements of the company. In the case of Motoarato Camp in Näqämti, the chiefs were able to supply only 460 workers per day against 1,000 stipulated. Once in the fields, this same labour proved recalcitrant and used every pretext to avoid work or to suddenly disappear and walk back home.[120] Reading through the available data one often comes across laments like this officer's:

> To cultivate a very modest acreage enormous effort was put. The chiefs promised manpower and totally failed to despatch it. Each labourer sought to work as little as possible; labour was not the same every day and because of this, for example, after one has ultimately succeeded in convincing the labourers that the seed should not be planted deeper than one or two centimetres, the following day most of them do not turn up and so everything has to be explained afresh to the newcomer.[121]

As with all agricultural undertakings, demand for labour fluctuated seasonally according to the area. At peak season, cotton in areas like Näqämti and Baro Cotton Districts demanded a large workforce and competed for scarce manpower. Near Baro Cotton District, Jubdo had a well-established mining industry belonging to Italian national Alberto Prasso, which was taken over by the Italian government.[122] By the mid-1930s it employed a workforce of about 2,000—perhaps the largest of any

[119] CD IAO: AOI 1830, Pesarini to Maugini, p. 7; AOI 1106, Rapporto schematico, p. 4.

[120] CD IAO: AOI 1826, Rapporto schematico, p. 4; AOI 1826, Pajella, Brevi notizie, p. 7; Villa Santa *et al.*, *Amedeo*, 249–50; Masotti, *Ricordi*, 138–9.

[121] CD IAO AOI: 1826, Pajella, Brevi notizie, pp. 3–6; 1830, Pesarini to Maugini; p. 7.

[122] Masotti, *Ricordi*, 137.

economic sector. Like the Cotton Districts, these mines depended on corvée labour whose recruitment was identical to that of the Cotton Districts in collective farms. Every two months each *residenza* provided a certain number of people who were 'persuaded' to work in the mines in exchange for very small pay and equally small tax exemption. The pay was normally in kind, such as food; contrary to what some sources allege, cash payment was an exception.[123]

The chiefs who recruited labour were not beyond reproach. As in collective cotton farms, it was largely the financially underprivileged who ended up in the mines while the well-to-do often gained exemption by illicit means. Earlier, the companies were dismayed when they discovered that most of their recruits happened to be, as one official described it, the old, crippled, and half-blind. A medical test was introduced whereby each labourer underwent a relatively rigorous examination. Escorted by a dozen uniformed officers, the workforce trekked on foot to the workplace. Despite strict security, many absconded during the journey.[124] Those who reached the mines were subjected to an extremely tough regime of control and, like in Somalia, the measures adopted against the dodgers were so brutal as to be 'indistinguishable from slavery'.[125]

Mining was not the only competitor for labour. The demand from public work such as road-building, settlers' farms, and subsistence agriculture was equally acute. As labour became increasingly difficult to obtain tension developed and the government in Addis Ababa had to intervene. In Wälläga Cotton District an additional clause had to be incorporated cautioning the company not to upset the local labour market.[126] In Soddo the intervention was designed to protect subsistence agriculture and balance labour demands between the company and public work. Resistance in this District had often strained the fragile relationship between the *residente* and company officials. The block inspector at Umbo, Achille Pajella, gives an illuminating example. The Ethiopians were given an order by the residente to repair a road. To evade work in

[123] Ibid. 138. [124] Ibid. 138–9.

[125] Ibid.; Margery Perham to the Editors, *The Times*, 7 July 1948; D. Mack Smith, *Mussolini's Empire*, 108–9.

[126] DM 31-1-1940: IDC Uollega, in *GUGS* 5/29, 17 July 1940 where in art. 2 clause (b) is incorporated a demand for the company 'not to divert, for the purpose of cotton farming, the manpower necessary to new cultivations which the administration of the AOI wishes to expand in the areas of the Cotton District, to the existing cultivations within and which the administration wants to increase, and to the mineral enterprises which are already operating in the District or will be set up inside the District or its bordering zones; see also Chiuderi, 'Agricoltura indigena', 272–3.

the collective cotton farms, the peasants presented the residente's order to the company, demanding ten to twenty days absence. As it was an order from the political authority the block inspector had no option except to let them go. Afterwards the peasants informed the *residente* that they were prevented by the company's technician from carrying out the road repair and went to attend to their own plots. When he realized that the work was not completed within the scheduled time, the *residente* wrote angry letters of complaint demanding the company discharge the Ethiopians immediately so that they could attend to road repairs. The *residente* was not convinced by the company's explanation, and in order to avoid misunderstandings the latter had to resort to actions such as taking the *residente* around the empty cotton camps, then to the road construction site, so that he could draw his own conclusion. At the end both the company and the government had to give in to the peasants by reviving as a compromise a defunct labour practice. Accordingly, an agreement stipulated between the company and the district chiefs, and sanctioned by the *residente* and *commissario*, timetabled six working days per week—three to cotton cultivation, and three to the peasant's own work.[127]

Important questions arising from such a compromise are why was the peasantry so unresponsive to incentives to involve it in cotton production and what were the reasons for cotton fields to be unattractive? In the collective fields, the workers were provided with food while at work to spare the need to work on their own fields. But the work was carried out under duress and without pay. Moreover, as one technician quite perceptively lamented, the cotton field 'does not belong to the peasant and so he does not care about it'.[128]

The official explanation for the peasant resistance was a conventional stereotyping of naturally idle and thriftless Ethiopian peasantry with a deep-seated uneconomic attitude and apathy as regards change.[129] One official of the Soddo Cotton District wrote:

> The Wällamo has the treacherous character of a person who tries to evade in any possible way any form of imposition. He obeyed his Abyssinian master for he

[127] CD IAO AOI 1826, Pajella, Brevi notizie, pp. 7–8. The stipulation simply reinstated one of the pre-Italian labour practices governing the *hudad* land, practically abolished under the reign of Emperor Haile Sellassie. According to this defunct practice, two days per week were allocated for work on the *hudad* of the local official and three days if the area had a *hudad* belonging to both government and local official. (Cf. C. D. McClellan, *State Transformation and National Integration: Gedeo and the Ethiopian Empire, 1895–1935* (East Lansing, Mich., 1988), 70.)

[128] CD IAO: AOI 1830, Pesarini to Maugini, p. 7; AOI 1106, Rapporto schematico, p. 4.

[129] CD IAO AOI 1106, Rapporto schematico, p. 4.

commanded him with whips, an instrument which we Italians do not feel at ease to use. To each proposal of work he always replies affirmatively but he rarely or never keeps his word. He prefers the sweet idleness to the hard work of cotton-farming, alternating herding his cattle seated under a shade of a tree and giving a little care to his small plot.[130]

But rather than lazy and apathetic, the peasants appear hard-working and calculating. When they were searched out in their villages, they were not caught sitting idle, feasting, or drinking but attending to their own food production.[131] It is a colonial bias to equate hard-working and progressive peasantry with those working to the interest of the settlers. In this respect the Italian official thinking was no exception.

The peasantry was acutely aware of the disadvantages associated with cultivating cotton. First, cotton was primarily an export crop, vulnerable to the bewildering fluctuations of the world market and lacking the flexibility characteristic of subsistence crops. It could at no time be used as part of the diet. It was planted during the long rains and harvested in December–January. These were critical months for the subsistence crops in which manpower was already severely stretched. Unlike cotton, these crops serve as staple food in domestic consumption. Their surplus could be exchanged for items such as salt and clothing, and in years of disastrously low yield could be held back as means of famine relief.[132] Cotton cultivation competed with these well-tried and flexible subsistence crops and could not be done without necessarily disrupting their production or cutting back labour from them. Moreover, unlike the local variety, the cotton distributed by the Italians showed minimum resistance to parasite, extreme susceptibility to changes in weather conditions, and needed abundant rains.[133] This made the crop extremely unreliable.

Secondly, cotton was labour intensive and all crops had to be sown afresh each time on laboriously prepared ground. Before harvest the cotton plants demanded periodic thinning and weeding—time-consuming jobs.[134] By contrast, subsistence crops were less demanding of labour, except perhaps at harvest time. Under the circumstances the peasants were not prepared to compromise their own best plots or waste their

[130] CD IAO AOI 1826, Pajella to Maugini, pp. 4–5. [131] Ibid. p. 8.

[132] J. Tosh, 'Lango Agriculture during the Early Colonial Period: Land and Labour in a Cash-Crop Economy', *JAH* 19/3 (1978), 428; M. Vaughan, 'Food Production and Family Labour in Southern Malawi: The Shire Highlands and Upper Shire Valley in the Early Colonial Period', *JAH* 23/3 (1982), 362–3.

[133] CD IAO AOI 1826, Pajella, Brevi notizie, pp. 1–2.

[134] CD IAO AOI 1106, Rapporto schematico, p. 4.

labour in order to maintain an unpredictable government-imposed export crop. So, like African growers elsewhere, the Ethiopian peasantry continued to insist on allocating the best lands for their staple food crops.[135] Added to these drawbacks was the lack of economic incentives for cotton production. The price paid for cotton never fully compensated for the loss of labour which otherwise would have gone into food production. The price of foodstuffs had risen tremendously since the conquest and the Ethiopians had began to earn high revenue from it. Thus when the company (ICAI) put before the peasants that cotton was a profitable crop they did not find the idea very convincing, for they easily inferred that of the profit that can be made from one and the same plot used for cultivating cotton or corn, the earning from the first one is very modest. In fact from 2,000 sq m of cotton field an average of 40 kg of cotton is harvested from which by selling to the company the Ethiopian peasants could make only L.80, whereas the cultivation of corn in this same plot could earn about 150 kg which in 1940 was sold at the Soddo market for L.140.[136]

For cotton to be attractive to the peasant in any significant way, it had to be equally, if not more, profitable than food crops. But the political authorities, to protect the companies from a capricious world market to which cotton was subjected, adopted from the outset the policy of keeping the price paid to the Ethiopian peasant ridiculously low.[137]

> Such a price should start from the premisses that cotton should be produced at a very low cost so that it would offset the heavy transport and freight expenses and provide the metropolitan industry with fine raw cotton at economic prices. Therefore, the native should be paid little for his cotton particularly at the early stage since the native can never understand any later price less than what he earlier enjoyed.[138]

In Soddo, where ICAI purchased fine-quality cotton for L.2 per kg, the same quantity of low-quality cotton of local variety was sold for over

[135] For comparison in other part of Africa see A. O. Anjorin, 'European Attempts to Develop Cotton Cultivation in West Africa, 1850–1910', *Odú*, 3/1 (1960), 13; H. Slater, 'Land, Labour and Capital in Natal: The Natal Land and Colonisation Company, 1860–1948', *JAH* 16/2 (1975), 257–83; Tosh, 'Lango Agriculture', 425–7; R. Palmer, 'Working Conditions and Worker Responses on Nyasaland Tea Estates, 1930–1953', *JAH* 27 (1986), 109–11, 116–18; T. Bassett, 'The Development of Cotton in Northern Ivory Coast, 1910–1965', *JAH* 29 (1988), 272–4.

[136] CD IAO AOI 1826, Pajella, Brevi notizie, p. 5.

[137] CD IAO AOI 1106, Rapporto schematico, p. 5.

[138] CD IAO AOI 1936, Angelo de Rubeis, Promemoria per S.E. Il Sottosegretario: Produzione cotoniera a mezzo degli indigeni, Rm 12 May 1938.

L.10 in the local market.[139] As a producer of fine-quality cotton the peasant could have obtained a higher price, but the monopoly by the companies prevented previous free market practices.

The most tangible effect was a sudden revival of local black markets and an outbreak of cotton theft 'whereby the farmer harvested cotton at night and hid it in order to sell it in one of the remote markets'.[140] The political authorities seemed to be aware from the start, of the obstacles facing export cotton from a well-established local industry. In order to export cotton, the local market must be saturated. To do so a clause in the charter of the companies demanded the companies to leave at the disposal of the Ethiopians cotton 'sufficient enough to the customarily made local craft'.[141] Whether the companies complied with this obligation is uncertain. But the existence of such a parallel market formed a formidable challenge to the monopolistic tendencies of the companies and became the haven for those dissatisfied with their merciless exploitation. Theft was widely reported, especially in the private fields. Anybody caught was subjected to harsh punishment.

As efforts to eradicate petty theft proved ineffective, the political authorities concentrated their fight mainly against more serious cases. As the case of Soddo Cotton District reveals this meant:

> Whenever the theft was discovered, it was reported to the *residenza* who ordered the arrest of the culprit; but the fact is it was not possible to arrest so many thousands of the natives and so the tendency was to eradicate the fraud by putting on trial those who committed serious theft and inflicting only corporal punishment on those who are guilty of petty thefts.[142]

Despite the severity of the measures, the political authorities were incapable of totally eradicating cotton theft. Nor did the package of coercive measures help the companies to overcome the chronic labour shortage or improve the level of production.[143]

With the setback of collective fields, advocacy for private fields began to gain momentum and for the supporters of this view the classic point of

[139] CD IAO AOI 1826, Pajella, Brevi notizie, p. 9. This same author suggests that ICAI intended to reduce the price further on the grounds that in the Belgian Congo the cotton was purchased for 60–70 cents per kg. But he does not support such a view; according to him, at the beginning cotton should be paid L.4–5 per kg and then be reduced very gradually to the current price once the cultivation becomes extensive and the peasant gets accustomed to it and discovers its economic importance.

[140] Ibid.

[141] Art. 2(d) of RD 7-1-1938, no. 443. This condition is incorporated in art. 4(c) or 5(c) of the contract of each cotton district.

[142] CD IAO AOI 1826, Pajella, Brevi notizie, p. 9.

[143] CD IAO AOI 1106, Rapporto schematico, p. 4.

reference was the Belgian Congo. The government was urged to 'foster cotoniculture by enacting, if necessary, measures similar to those in vigour in the Belgian Congo where in each cotton district every native family is compelled to cultivate cotton every year on a small patch of land'.[144] Basing their arguments on those of Edmond Leplae, the indefatigable propagandist for a similar system in the Congo, they justified the scheme for compulsory cotton cultivation on the alleged educational and welfare ground that it was the quickest way to improve the material well-being of the subject people who are otherwise lethargic to progress.[145] It was a sentiment equally shared by the Duke of Aosta. His early exposure to Belgian colonial experiment had left a strong impression on him.[146] The Duke had always maintained the view that the best policy for developing and expanding cotoniculture was to make it an autonomously functioning and, albeit under the control of a company, a family-based enterprise. In February 1941, when the failure of the collective fields appeared to prove his point, the Duke strongly reaffirmed this position claiming that 'he was more than ever convinced that only through the private fields can cotton production successfully expand'.[147] It is simply speculative to draw any conclusion about the success or failure if this scheme was put into practice but appeal to the Congo experiment, since itself was not conspicuous for its success, was hardly a reassuring example. What the debates among the officials make clear, however, is that the difficulties involved in the adoption of suitable systems were clearly far from simple.

Anyhow, Ethiopian resistance to cotton production was only one aspect of the problem. Even where there was substantial Ethiopian participation, as was the case in several areas of Qobbo Cotton District, there were well-known inherent weaknesses in the scheme.[148] Generally speaking, the Cotton Districts were, indeed, the most suitable cotton-growing areas but they had a significantly varying distribution of rainfall and temperature which involved several problems from the point of cotton production. An exception was Wuččale zone in Qobbo Cotton District, situated between 1,400 and 1,500 altitude which for its location,

[144] CD IAO AOI 1826, Pajella Brevi notizie, 11.

[145] According to Leplae 'To oblige the native to know and to develop one or another crop for export is the only way to improve the material situation of the entire population in a relatively short period . . . But the method of compulsion is . . . followed for educational purposes. It does not aim to keep the population under a regime of constraint indefinitely.' (E. Leplae, 'Cultures obligatories: Leurs résultats au Congo belge et dans d'autres pays tropicaux', *Revue des Questions Scientifiques* (Mar. 1934), 256–7; see also id., 'Transformation de l'agricolture indigène du Congo belge par les coltures obligatories', in *Technique agricole internationale*, ii (Rm, 1936).)

[146] Villa Santa *et al.*, *Amedeo*, 207.

[147] Ibid, 249.

[148] CD IAO AOI 1830, Pesarini to Maugini, pp. 1–3.

fertility of the soil, relative absence of malaria, and ample rainfall, gave repeatedly excellent results and, therefore, was considered as one of the most promising for early development of cotton. By-contrast, Alamaťa, the headquarters of the district, had rich soil, excellent temperature, and a vast area of level lands. But because of inadequate rainfall the outcome of the consecutive four years' experiments by both Cotetio and ECAI was an abject failure. The nearby zones of Zébul, Čärčär, and Qorbäta had similar situations and the technical officers had to eliminate these zones from their 1940–1 campaign. In districts with satisfactory rainfall, like Wäldia in Qobbo or Adama in Awash, the temperatures especially at night during the cotton-growing season were too low for cotton, often falling to near freezing-point.[149]

Location and insect pests also would have continued to handicap seriously the development of commercial production of cotton. Several districts were thinly inhabited, and quite inaccessible. Although absence of intensive Ethiopian cultivation put at the companies' disposal abundant land for extensive cultivation, it did also mean labour shortage. Attempts with forced resettlement schemes in such areas had to be stopped abruptly when the discontent of those resettled intensified. Most of them previously inhabited much more fertile, and healthy areas of the Ethiopian highlands and were reluctant to move out.[150] Despite the excellent possibilites for irrigation by pumping or gravity and great potential for cotton-growing, bad location, labour scarcity, and lack of good roads, made these areas quite inaccessible for commercial development of cotton, at least in the short run. With capital-intensive technologies these problems could have been mastered relatively easily. But this would require a huge investment in machinery, such as tractors, and road construction. The companies, however, operated within a stringent budget. But even if they had such resources, the cost of operation and maintenance of the machines would have probably made the scheme impractical.

Insects were the other most important agencies limiting cotton production on a large scale. With the introduction of U4 strains progress was made in overcoming serious Jassid attacks. In addition to Jassid, other reported pests included cotton-stainers, the capsid Helopeltis. Attempts were made to control insect pests by introducing strict quarantine regulations in the transport of cotton seeds, both into the country and within the country from one region to another. But these measures were not adequate to prevent the introduction of the pests nor was the government in a position to enforce them effectively. Therefore, even if the Italian

[149] Ibid.; Nicholson, 'General Survey', 10.

[150] Villa Santa *et al.*, *Amedeo*, 249.

government's scheme had not been disrupted by events following Ethiopian independence or even if the co-operation of the peasantry for the scheme was successfully enlisted, the magnitude of efforts required to grow cotton on a large scale cannot be underestimated. Indeed, it is difficult to draw any conclusion about the long-term success or failure of the scheme.

What was obvious was the fact that the 'insistent and paternal advice' with which the authorities intervened had made cotton production fairly unpopular. Yet unpopular or not the Cotton Districts, like the agricultural offices, laid down much in the way of infrastructures. With them began the transformation of rural life, for the cotton companies and agricultural offices' campaigns brought into rural areas for the first time agricultural experts and administrators. Numerous studies of the land and soils were undertaken and in some areas technological changes were promoted. For example, the use of new varieties of cotton and cereal crops, much improved agricultural tools, such as ploughs and other implements, were introduced. Even though they remained at the initial stage, a co-ordinated attempt was being made to introduce the Ethiopian peasantry to growing cash crops. The very fact that the cotton companies and agricultural offices tried to make use of traditional social and economic units of production was better than a plantation agriculture where the labourer worked far from home.

At the end of the occupation, only Täsänäy survived,[151] but cotton remained one of the most important crops grown in a number of other districts.[152] The most conspicuous legacy, however, was the spinning and weaving mill of Deré Dawa set up by Società Cotoniere Meridionali which still is the country's most important cotton-manufacturing plant. Finally, the work of the cotton companies and agricultural offices attempted to contribute to what was a very marked feature of European colonialism, i.e. the preservation of the pre-colonial social and even economic structures while at the same time trying to promote the production of large surpluses for the world market.

The expressed view of the cotton companies was that of raising the welfare level of the peasantry without involving any serious disruption of the existing agricultural structure. But in practice it became increasingly clear that the primary purpose of the scheme was to forcibly milk the rural agricultural and labour surplus in order to meet pressing metropolitan

[151] Nicholson, 'General Survey', 14. According to this report, in 1955 Täsänäy Cotton District comprised 40,000 acres. Of these, 10,000 acres were cultivated by 2,000 share-croppers consisting of about 6,000 individuals.

[152] Ibid. 10–14; Awash Valley became one of the key cotton plantation centres after 1960 (cf. Maknun Gamaledin Ashami, 'The Political Economy', CD IAO AOI 1106, Rapporto schematico, p. 6).

domestic needs. The scheme was largely formulated in the interest of the metropolitan textile industry at the expense of the indigenous population. Thus government's commitment to uplift the standard of life for the indigenous population had much rhetoric but little substance.

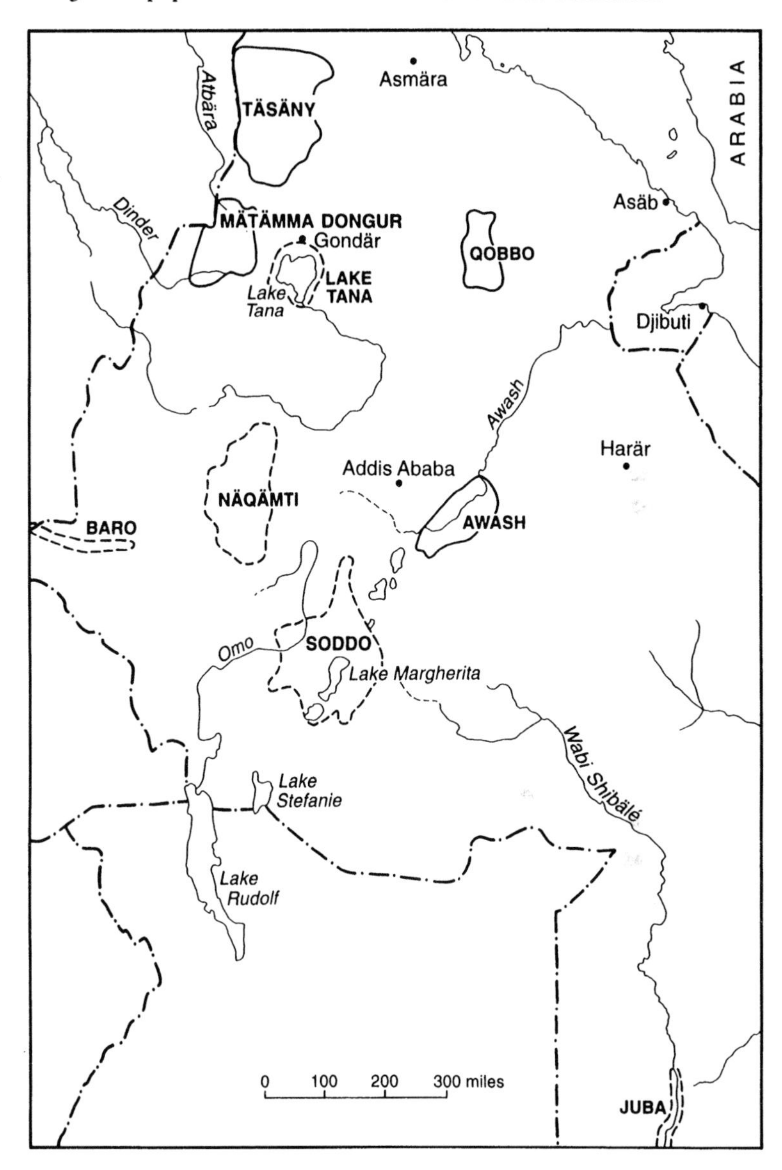


FIG. 5. Cotton-growing districts

# *Conclusion*

We have examined in some detail the nature, scope, and dynamics of Italian land-use policy in Ethiopia during the brief period of occupation. The policies, though devised as responses to pressures from the metropolitan's combined demands for a social outlet for its surplus population, reliable source of agricultural raw materials, and the need to consolidate the Empire as economically self-supporting, were far from being carefully integrated programmes demanded by later planning techniques. Yet they did maintain some continuity with Italian colonial tradition as manifested in its other colonies, both in terms of problems they addressed and the methods by which they attempted to tackle them. Emerging from this was the development of similar institutional frameworks which, in turn, were shaped by the presence of the same personalities.

As in Libya, the key criterion of Italian agricultural policy remained the demographic myth, with its stress on settlement. This was the cornerstone of Italian imperial policy, and constituted its distinctive feature *vis-à-vis* that of other colonial powers. In contrast to Libya, Ethiopia contained much more fertile land so it seemed that the conquest would make this dream real. Alongside these schemes, a bewildering variety of interests had to be accommodated, giving rise to a number of initiatives, differing in their orientation and methodology. These were individually justified by the diversity of climate and geo-ethnographic conditions, as well as economic factors in the various territories in which they operated.

Mussolini's vision of conquering Ethiopia to settle millions of Italians and solve Italy's economic problems was an attractive idea. It contributed considerably to the popularity of the Fascist regime. But once the country had been conquered the regime neither had clear ideas on how to implement this vision nor the necessary means and proper personnel. Despite his ministers' propaganda informing the Italians that the Duce personally guided everyday the course of imperial policy, in fact once the heroic deeds of conquest were over, he showed little interest in the development of Ethiopia. The popular exuberance of the Italians themselves also subsided once the excitement of easy victory was over.

The settlement of large numbers of Italian peasant families—the main economic argument to justify the conquest—proved unworkable. On 27 April 1940, less than a fortnight before Mussolini's entry into the European war, the then Minister of Italian Africa, Terruzzi, claimed that 3,550 families were settled on 113,760 ha of land. But it appears from other sources these were much inflated figures. The peasants actually transferred numbered no more than 400 and of these about 150 were joined by their families—a very far cry from the often proclaimed millions. Despite the tremendous publicity, by the end of 1939 the recruitment of peasant families willing to move to Ethiopia had become an arduous task.

As an outlet for emigration, Ethiopia was a total failure. In 1939 the total Italian population of IEA numbered about 130,000, of which about 60,000 were settled in Ethiopia. Yet the Italian population of New York was still ten times more than that of the entire Italian Empire. But the trend was that Italians 'were returning to Italy rather than coming out to the Empire'.[1]

The dream of the Empire as a land of opportunity also evaporated as the prospects of employment became poor, and many experienced extreme poverty, housing shortage, and spiralling living costs. At the end of 1936 there had been 146,000 Italian workers. By June 1939 this number dropped to 23,000. During this same year the number of Italian unemployed in Addis Ababa alone was estimated to be between 6,000 and 7,000. Many lived in great squalor in a camp on the outskirts of the town, often begging Ethiopians for food—thus making the so-called racial prestige policy a mockery.[2]

Despite the claims made for it, colonization did little or nothing to help solve Italy's pressing social problems. At best, perhaps the foundation of the Empire served as a comforting way to distract the Italian people from the reality of economic depression and dictatorship that surrounded them. For the handful of settlers the *coup de grâce* came with the collapse of Italian rule in 1941 with which they lost their *raison d'être*. Following the Anglo-Italian agreement effected in the midst of World War II, they were repatriated to Italy to start a new life.[3] With their return to a war-stricken

[1] FO371/23380/J1776/296/1, Bird to Halifax, AA 1 Apr. 1939.

[2] ACS SPDR 44/242/R/39, Farinacci to Presidente, Cremona 24 Apr. 1938; FO371/24635/J467/18/1, ACG Gibbs to PSSFA, AA 28 Dec. 1939.

[3] ECRE survived until 1946 in Eritrea where some of its members operated four farms at the outskirts of Asmära, 'biding their time in view of making the life of the occupying enemy difficult'. But soon it was exposed as 'the nest of Fascism', and, contrary to the plan, its members 'lived miserably, despised and hated by the anti-fascists and members of Free

TABLE 29. Italian population in Ethiopia, 1940

| Place | Male | Female | Total |
|---|---|---|---|
| Addis Ababa | — | — | 26,952 |
| Adwa | 200 | 30 | 230 |
| Ambo | — | — | 4,500 |
| Deré Dawa | 3,370 | 174 | 3,542 |
| Gondär | — | — | 2,000 |
| Härär | 3,018 | 436 | 3,453 |
| Jimma | 12,544 | 626 | 13,170 |
| Mäqälé | 92 | 43 | 105 |
| Koräm | 97 | 12 | 109 |
| Soddo | 170 | 3 | 173 |
| Yavéllo | 57 | 5 | 62 |
| TOTAL | 19,548 | 1,329 | 54,296 |

*Source: Popolo Fascista*, 25 Feb. 1939; *GI*, 25 Feb. 1939; 'Quadro demografico di Gimma', *CI*, 2 Mar. 1940. For a different but, somewhat inflated figure, see R. Ciferri and E. Bartolozzi, 'La produzionc cerealicola dell'AOI nel 1938,' *AC* 34/11 (1940), 441.

homeland the dream of Italian demographic settlement definitely faded, a myth that had besieged the minds of a section of its most influential intellectual and political élite since the early days of independence.

Another premiss on which the conquest of Ethiopia was justified was the commercial advantages and the profitable source of raw materials it was thought the country would provide. Italy's imports from her colonies in 1938 were little more than 2 per cent of her total imports and she had spent ten times as much on administering the Empire as the total volume of trade with it. Quite apart from the continuing military expense, which was kept secret, the cost of organizing and exploiting the Empire was over a thousand million lire in hard currency each year. This was more than one-tenth of the total foreign reserves available in 1938 and 1939.[4] For

Italy [Libera Italia] closely watched and harried by the British Intelligence and the Eritrean Police Force and in constant fear of being interned at any moment the enemy wishes or believes it appropriate.' (Cf. CD IAO 1921, Timò Mansueto, Attività dell'Ente Romagna, Rm 9 Feb. 1943.)

[4] F. Guarnieri, *Battaglie economiche tra le due grandi guerre*, ii (Milan, 1953), 404.

some Fascists this was an investment which would prove rewarding in the end. But for the Minister of Foreign Trade, Felice Guarnieri, such a rate of expenditure for more than a few years could not be afforded by the Italian State.[5]

Far from being self-sufficient as Mussolini had ordered, Ethiopia had to import most of the food for her Italian expatriates and industrial equipment. Italy exported to her East African Empire in 1938 goods to the value of over two thousand million lire, nearly twenty times as much as she received. Thus far from solving Italy's economic problems—apart perhaps from providing a secure market for some Italian products, such as textiles—it only aggravated them. Rather than strengthening national self-sufficiency and assisting Italy's balance of payments, the Empire, according to Guarnieri's own statement, was swallowing Italy itself.[6] The diversion of goods to Ethiopia that might have been sold elsewhere for foreign exchange and the use of ships more profitably employed on transatlantic routes made the deficit even worse. So exasperating was the whole economic condition that, whereas Italians had counted on abundant supplies of coffee from the Empire, 1938 saw the beginning of a serious coffee shortage in Italy.

Underlying such poor achievement was bad planning which can be understood by the speedy conquest that astonished not only the Italians but the entire world. The MC in Rome was unprepared for the many pressures coming from different quarters—largely unleashed by the multi-media Fascist propaganda machine: the unemployed of the immediate post-Depression era searching for work and land to cultivate; the social adventurers with their strong desire to see with their own eyes the mythical lands; demobilized masses of military recruits urgently needing settlement; a strong business lobby wanting to expand its enterprise; strategic and economic considerations calling for the Empire's self-sufficiency. The temptation was also great to show international opinion, hostile to the conquest, Italy's fitness to be a great power.

These multitudinous pressures were so overwhelming that any suggestions to delay action until the country was effectively pacified and the land tenure system was carefully studied, was felt to be an unaffordable luxury. The inevitable outcome of such pressures was the emergence of an uncoordinated and chaotic agricultural policy. Overnight the MC in Rome had to improvise a bold programme of development, which in practice was little more than a collection of projects lacking coherence

[5] Ibid. 198. [6] Mack Smith, *Mussolini* 121.

and realism. Its inherent flaw was the tremendous lack of knowledge about the physical, social, and economic conditions that prevailed in Ethiopia. There was a lack of hard facts for realistic prognoses and plans. The men who devised the programmes, did so on knowledge acquired either during a brief flying visit or by reading often unsubstantiated accounts. The programmes seemed deceptively easy to implement, yet in practice proved far from workable.

The first to realize their unworkability were the authorities in Ethiopia who had the task of translating the programmes into action. With their badly paid, numerically insufficient, and qualitatively poor personnel, they were burdened with the uneasy task of accommodating a wide range of conflicting metropolitan interests following the conquest: they had to provide the Empire with a solid organizational and administrative structure in a land that was not totally pacified; they had to provide assistance, hospitality, military escort, and means of transport to the representatives of agencies or groups of farmers who came to choose land for their prospective farms. The cost of the settlements' protection, the construction of roads linking each centre, and the erection of schools and hospitals and other social amenities, was an intolerable burden on the authorities' shoestring budget.

This situation was further complicated by the multiple, and largely hostile, traditional local interests which the administration had to keep reasonably satisfied. Overburdened by the enormity of such responsibilities within the country, the administration in Ethiopia repeatedly called on Rome to slow down, and if possible to halt, the agricultural programmes until such time that a comprehensive and realistic policy was worked out. But Rome was not in a position to fully grasp the situation in the colony. The colonial administration in Ethiopia thought Rome's plans had been conceived in the peaceful confines of the ivory tower of the MAI by armchair intellectuals, and offered only lukewarm support. The immediate bearing this had on colonization was to slow down its progress.

The centre of contention was understandably land. The first objective of the Italian administration was to constitute a group of domanial lands, under the jurisdiction of the state. The bulk of these lands was owned by rebels or active political exiles in addition to lands belonging to the Ethiopian state and the royal family. Lands seized in this way had various destinations: they were allocated to the settlers or kept for future needs of colonization or redistributed to Ethiopian allies or given in exchange to those Ethiopians displaced to make room for Italian settlers.

To accommodate the new settlers, the colonial administration in Ethiopia had to pursue policies that were largely drafted in Rome and which were also in clear contradiction to Rome's own avowedly stated policy of respecting the legitimate land interests of the Ethiopians. These were often moved off their lands to make room for new Italian farms. Even though the Ethiopians were given other lands in exchange, the land was normally relinquished reluctantly and under severe pressure. Nor was the process fast enough to satisfy the land rush that immediately followed occupation. Initially, the Italian regime was led to believe that it had physical power to repress any Ethiopian resistance to land alienation but as the economic costs of doing so became increasingly unacceptable, it proceeded fairly cautiously. The slow pace angered the land-hunters and led to violent disputes with the colonial administration.

The immediate impact of such a scarcity of land was the emergence of settlement enclaves, scattered and remote from each other. Dispersion in time and space made the sharing of experiences difficult and brought greater expense for the settler and intolerable cost for the government at a time when its overstretched human and material resources were desperately needed elsewhere. Attempts to create unified zones of colonization that were contiguous to each other, by means of a gradual elimination of interpolating Ethiopian properties through purchase, exchange, and other pressures resulted in further alienation of the Ethiopians, who charged the government with robbing them of their lands to benefit the White settlers.

Scarcity of land and hasty selection had led some demographic colonization agencies to settle in areas that were unhealthy, of high political risk, or with no possibility of expansion. Both the ECRE and ONC had to divert most of their operation to other areas on political grounds and for lack of free land to expand. In both cases, the cost to the agencies as well as to the central administration was enormous.

Demographic colonization was also undermined by its own internal weakness. The agencies were directed by remote bodies. In most cases, the leadership proved poor and incompetent. The farms were run autocratically and the peasants had little say in the expenditure of funds derived from their labour. The overwhelming majority of the colonists failed to possess those precious qualities and sentiments of racial prestige, social and economic responsibility, physical and technical fitness that constituted the ideal for their selection. Their behaviour stultified the workings of the racial prestige policy, forcing the government to have recourse to legal measures. And the economic resources were inadequate, so the demographic colonization was forced to function on a shoestring budget. The production, based on cereal crops, was threatened by competition from the Ethiopians, who, notwithstanding the statutory restrictions, nevertheless produced at much lower cost.

The trends within commercial farming were no more encouraging. With the exception of a few large companies, the commercial farms were reluctant to move their capital to Ethiopia. This can be attributed to the natural mistrust that these business interests had towards a new environment where the technical and economic potential was still unexplored, the climatic conditions unknown, the security of capital uncertain and profits only in the very long term. Notwithstanding this general reluctance, a number of companies or private entrepreneurs ventured, lured by attractive appearances: an agreeable climate, vegetation growing effortlessly with no or minimal irrigation, food crop prices that permitted fantastic profits even with the smallest yields, and above all, very cheap manpower. However, the enthusiasm even of those 'outstanding and honest farmers' was soon dampened once the stark reality of the enterprise revealed itself in all its severity and the enormous difficulties that they had to contend with became apparent: a government that lacked coherent policies; an intractable and amorphous bureaucracy which allocated land at an exasperatingly slow pace; an unsafe political environment; scarce and costly manpower. Most of their efforts were also frustrated by bad weather, drought, and continuous battle with marauding insects, against which their technical skill offered little remedy. Despite media reports to the contrary, archival materials clearly reveal the low level of experience and technical expertise of those men who found themselves located in a region totally new to them. The cumulative effect was dismal results and high operational costs that at the beginning of 1940 left most of the farms in a lamentable state. For their survival they depended upon a generous subsidy from the government on whose anxiety to bolster production they capitalized.

One of the most notable legacies of Italian agricultural policy was the introduction of racial competition and conflict over land and production. In the colonization zones, Ethiopians were under severe pressures to move off their lands to make room for Italian farmers. A host of mechanisms that gained their justification in often mistaken interpretation of local agrarian customs and traditions, were employed to perpetuate this harmful policy. Only complicated political and economic considerations prevented wholesale deportations from the colonization zones. Despite its obvious limitations, Ethiopian production helped relieve the tax burden on the Italians. Ethiopians were needed in the countryside as, without their labour, the colonists' farms were doomed to bankruptcy.

As labour became an intractable problem, the key issue remained not the terms on which the Ethiopians would be driven off their lands but how they would be permitted to settle. The government's policy quite deliberately and consciously aimed to force them into a system of integration with the settlers' farms through a variety of schemes such as sharecropping contracts, whereby Ethiopians were offered land in return for labour. And yet the Italians feared the challenges posed by the competition of Ethiopian farms whose success they felt would undermine the development of settler farming. Ethiopian agriculture was allowed to develop only as long as it was able to provide support for, or at least not compete with, settler farming. Despite attempts to control Ethiopian agriculture and direct it towards securing the economic domination of the Italian settlers, by 1940 it had become clear that the two economies could hardly coexist side by side. The suggestion was made to aggressively pursue the racial-segregation policy in rural areas, whereby Ethiopians would be allowed to farm only within their own reserved zone—land marginal to settler agriculture—and engage only in subsistence farming. This policy went against claims that the Italians attempted to upgrade the lifestyle of the Ethiopian population.

Certainly the Italians introduced modern agricultural techniques and offered technical advice to stimulate indigenous agricultural production. And yet these were only timid gestures meant largely to guarantee their own position by distributing a few crumbs. As its racial policy as well as a draft document on zone of colonization show, the Italian regime, like the rest of the European colonial powers, had little intention of granting genuine economic advancement to the Ethiopians. Of course, as it was dependent on their labour, the regime planned to integrate them into its own scheme of development. Otherwise, Ethiopians benefited from the Italian colonization only after Independence. Any economic benefit that may have accrued to the Ethiopians during the occupation resulted from accident and not design.

However, the real achievements during the period under consideration have to be seen in a perspective that incorporates some notion of the underdevelopment of the country before the occupation. When considered against this background and with full account taken of the shortage of trained agriculturalists and the short timespan and intractable political, social, and economic climate in which the regime operated, the transformation of the agricultural economy, particularly in areas solidly

TABLE 30. Major Italian companies in post-1941 Ethiopia

| Company name | Area | Size (ha) | Owner | Capital (Eth. $) | Crop |
|---|---|---|---|---|---|
| Arba Gou Gogu Plantation Co.[a] | Bakaka, Minneh, Gololch | 6,500 | Di Vassallo Giraudi and Porta | 1,000,000 | Coffee: 500,000 plants and annual produce 15,000 ql |
| *Avvocato* Ghila Farm | Mizan Täfäri | 5,005 | Ghila | * | * |
| Awasa Farm | Shashämäné | 800 | V. Pettinelli[b] | * | Coffee: 100,000 ql orchards |
| Čaffa | Däsé | 4,000 | Comm. Luigi Ertola[c] | * | Cereal, vegetable, and oilseeds |
| De Boi Industries | * | * | Ing. Casati | * | Forestry |
| Galanti Giuseppe Farm | * | 400 | Galanti Giuseppe | * | * |

| | | | | | |
|---|---|---|---|---|---|
| Maramalca Farm | * | * | Varafa[d] | * | * |
| SNIA Viscous Agricultural Farm | Lake Tana | unknown but very vast | * | * * | Eucalyptus and paper-mill |
| Tana Plantation Ltd. | Mizan Täfäri | 4,000 | Ing. Fuzzi[e] | 300,000 | Coffee: 300,000 plants produce 500 ql |
| Varafa Fruit and Vegetable Farm | Bishoftu | * | Varafa[f] | * | Fruit and vegetable |

*Notes:* * Indicates that no information is available

[a] Originally owned by the Belgian plantation company and then by SIA.

[b] Known formerly under the name of Mauro Rapetti, during the occupation he owned a farm in Wändo Källa [Sidamo] (cf. CD IAO GS 1822, p. 8).

[c] During occupation he owned farms in Eritrea.

[d] Former name Saboure Montanati.

[e] He was the ex-president of ECRE.

[f] Former name was Lebaró.

*Source:* CD IAO 3168, Fiumana to [Maugini], AA 1 Oct. 1958; CD IAO 4299, [Maugini], Appunti sull'agricoltura in Etiopia, FC 22 Jan. 1966.

controlled by the regime, was not insignificant. Whatever their motives, and notwithstanding the brutalities they committed, the Italians had laid down developmental infrastructures and initiated agricultural policies that had the effect of vigorously forcing Ethiopia into greater participation in the world capitalist system. They had done much to dismantle Ethiopian traditional structures and replace them with what they considered to be more modern institutions conducive to exploiting the country's rich agricultural potential. In doing so the Italians had somehow accelerated the process that was already in operation at the time of their occupation.

Perhaps one of the most important of such changes was the curbing of the power of the traditional élites who vetoed most of Haile Sellassie's earlier attempts at land reform: once powerful regional armies and their commanders were either eliminated or made effectively powerless because of their collaboration; the Church had been weakened. Most significantly, more than 4,000 miles of all-weather roads were built which linked the main regions of the Empire as never before. These roads can be used for the agricultural exploitation of the peripheral areas and marketing of the produce. The monetization of the economy was stimulated by the encouragement of the production of cash crops to satisfy the needs of soldiers and settler families. A limited consumer society emerged, particularly among those who collaborated with the Italians in commerce and government. These accomplishments greatly strengthened the restored regime which reinstated most of the agricultural changes that the Fascist regime introduced and made reforms that further eroded the power of the regional lords.

Considerable pioneering work was carried out in the field of agriculture. The Italians were the first to put forward an outline development plan which made provision for surveys, research as well as actual development work. Perhaps even today it would be difficult in Ethiopia to mention any major project, whether still under discussion or already implemented, which had not been investigated, sometimes in a fair amount of detail, during the period of occupation.

We know little concerning the extent to which the post-Independence planning in agriculture was inspired by the Italian experience. And yet although the Italian initiatives were halted in mid-stream, their ideas and the debates they generated had some influence in the long term. Of the many other Italian colonial developmental initiatives which have a place in the history of Ethiopian agriculture, commercial farming is worth mentioning. Retrospectively, the picture that emerges is that the policy and practice of achieving self-sufficiency through commercial farms within a short period did not materialize. To what extent, then, does the commercial farming and production for export of the independent Ethiopia owe to Italian economic policy? Italians laid down important developmental infrastructures but the direct bearing in the independent Ethiopian post-war cash-crop economy is much more limited specially when compared with the conspicuous role played by the British farmers in Kenya whose economic foundations served as the basis for the country's post-independence economy. Most of the Ethiopian exports derive from either peasant cultivations or from tenuously Italian-linked post-war developments. Even though negligible as a direct factor, the overall influence of Italian policy in post-liberation economy is considerable. Indeed, it had been either instrumental in facilitating post-1941 economic policies and efforts or many of the most successful industries of this period had Italian antecedents. Obvious examples are cotton farming, sugar[7] and other oleaginous plantations, forestry exploitation and their related industries—such as Deré Dawa textile mills, sugar-mills, oil-mills, flourmills, and sawmills.[8] With the sole exception in Eritrea where the Italian farming community remained largely unscathed by the post-occupation developments, almost none of the Italian farms in Ethiopia survived beyond the occupation. Yet many Italians played an important role in post-1941 cash-crop farming and ran concessions largely on thirty years' lease while a substantial number of others were employed as technicians by the Ethiopian entrepreneurs (see Table 30). In policy terms, Italian influences are visible particularly in the para-statal organizations, such as Ethiopian National Trading Company and Awash Valley Authority, which certainly took their cues from Fascist Italy's State monopolies that dominated the country's economy. However, it would be of great interest to explore further Italian economic impact on the economic policies pursued by Imperial Ethiopia in the 1960s and 1970s, as those very Italian ideas and notions of development based on commercialization of agriculture and encouragement of agrarian capitalism (import substitution policies) seem to have served as a driving engine of the country's development plans.

[7] Wänji sugar, set up on the same site as one of the CFA members' sugar-cane plantation, is a case in point.

[8] In a wider economic context, one has to add cases such as Deré Dawa cement factory and the country's largest hydro-electric power plant Qoqa, which began with the Italians and later was completed with the Italian money paid for war reparation. Of course, many of these were planned before the occupation and, in case of agricultural farms, most of them predated Italian presence.

# *Appendix 1*

## LABOUR CONTRACT BETWEEN ONC AND GUNTUTA INHABITANTS (TRANSLATION)

1. [ONC's] Holäta Farm grants to cultivate an irrigated [*size*] ha of land for the purpose of kitchen garden.
2. The Ethiopian abides to cultivate land granted to him according to the instruction to be given from time to time by the ONC and according to farming requirements.
3. Produce shall be divided equally in half save the right of ONC to buy at current market price native's share. In the same way, if the native agrees to sell to the ONC, the agency will buy at the day's market price.
4. Upon signing the contract, the agency will pay L.50 to the native in indemnity for expenses born out by demolition, transportation of materials, and rebuilding of his *tukul*. The agency will provide seeds necessary charging him with half of the expenses that will be recovered from the sale of produce at one or more time.
5. No irrigation claims shall be made against the agency.
6. The native shall be responsible for small implements necessary for cultivation.
7. The contract shall be in force for two years starting from the day in which the native takes over the new farm. Should there be need for the land, the agency agrees to resettle the native or give him and his family new work.
8. The agency has the right to terminate the contract at any time if the native refuses to do the required works or commits a breach of grave disciplinary rules. The termination will be communicated through the *Residente* who has the final judgement on the dispute.
9. The native may terminate the contract giving at least two months advance notice.
10. The native [*name*] declares that the subject [*name*] acts as his guarantor to ensure that the present contract is observed in its entirety.

# *Appendix 2*

## SALARY CONTRACTS BETWEEN ONC'S COLONISTS AT HOLÄTA AND ETHIOPIANS (TRANSLATION)[1]

1. Native [*name*] shall agree to work throughout the duration of the contract, except holidays, for the settler [*name*] according to the request and instructions of the colonist. The work shall be exclusively of agricultural nature and be carried out within the farm.
2. The settler [*name*] shall pay the native [*name*] L.150 per month, of which L.100 in foodstuff, computed at the day's Holäta market price, and L.50 in Italian lire. Equal treatment shall be provided to the members of the family, provided they are over 16 years of age; for male and female members of the native's families if less than 16 years of age, the salary will be assessed at one and a half, on the understanding that the colonist likes to engage them.
3. Within a month of the contract the native will be given a plot of 5000 sq m for building his house and cultivating what his family needs.
4. The native will not keep if large stock more than three, and if sheep more than eight. All will be registered and kept under strict veterinary control and the native will inform the farm of any change in size by the way of birth, purchase, death or sale. The colonist indicates from time to time which grazing land to use.
9. The contract is valid for a year from 1 March 1939 and, unless the native is dismissed by two months' advance notice, i.e. 1 January of each year, is tacitly renewable.
10. The colonist reserves the right to revoke the whole contract if the native refuses to do the work required from him or for breaches of disciplinary matters. Sudden dismissal has to be communicated through the *residenza* of Holäta who will settle the dispute without further appeal.
11. The native presents as his guarantor to the execution of the present contract [*name*] who is considered fit by the *residenza* for this purpose.

[1] ACS ONC 7, Nasi to ONC, Contratto di salariato fra colonizzatori ed indigeni, AA 20 May 1939.

# *Appendix 3*

## SHARECROPPING CONTRACTS BETWEEN THE ETHIOPIANS AND ITALIAN FARMERS

GG OF IEA

DSAE

No. 154722 AA 19 April 1938

A. SHARECROPPING CONTRACTS WITH THE NATIVES

..... Sharecropping contracts with the natives may take the following forms:

*Case No. 1.* The Italian farmer shall farm unbroken land with a machine and, for each half *gasha* tilled (ca. 20 ha), he shall give (the native) 2 harrows and 6 sickles. The native, for his part, shall supply the seed and carry out sowing, harrowing, weeding, harvesting and stacking works, and transport the sheaves to the threshing centre. The produce shall be apportioned between the Italian farmer and the native cultivator in the ratio of 44 and 56 per cent respectively. Both sharecroppers shall pay that same proportion of tithe to the Government at the threshing floor.

*Case No. 2.* The input of the Italian farmer shall involve tilling an already cultivated land, supplying 2 harrows and 6 sickles per each 0.5 *gasha*-land farmed, and mechanical threshing. And the native's input shall be as in no. 1. In this case the share of the produce will be 33 per cent to the Italian farmer and 67 per cent to the native cultivator. The tithe to the government will be paid at the threshing floor by the two parties in the same proportion.

*Case No. 3.* The Italian farmer's contribution shall consist of supply of 2 harrows and 6 sickles for each 0.5 *gasha*-land tilled and mechanical threshing. The native then shall provide the seed, and do sowing, weeding, harvesting, piling the sheaves and transporting them to the threshing centre. The produce will be partitioned between them in the ratio of 17 per cent by Italian farmer and 83 per cent by the Ethiopian cultivator. The payment of tithe will be as indicated above.

*Case No. 4.* The input of the Italian farmer shall be supplying 2 harrows and 6 sickles per 0.5 *gasha*-land farmed. The input by the native cultivator shall involve supply of seeds, tilling, sowing, harrowing, weeding, harvesting, piling the sheaves and transporting them to the threshing floor, and threshing. The produce shall be shared in the ratio of 5 per cent by the Italian and 95 per cent by the native cultivator. Tithe payment will be as above.

*Case No. 5.* Where the Italian farmer provides the seeds, these shall be repaid with 22 per cent interest at harvest.

All sharecropping contracts must take place with the full agreement of the native contractor and be signed in the presence of the chief, then ratified by the local political authority. . . .

B. A SAMPLE OF SHARECROPPING FORM[1]

No. ______________

*Name of the Concessionaire*: CONCESSIONAIRES BOIDI BROTHERS

*The Native* [*name*] ______________
*resident at* [*address*] ______________
*residency of* [*name*] ______________
agrees to the norms that are set out by the GG of IEA in the DGG No. 154722 of 19 April 1938.

______________

*Case No. (of the contract)*: ______________ *Household No.* ______________
Total Plot Size ______________
Farmed Area ______________
*Date* ______________
*Signature: Native* ______________ *Chief* ______________
*Resident* ______________

[1] CD IAO AOI 1936, Cerulli to MAI, Contratti di compartecipazione con gli indigeni, AA 22 Apr. 1938.

# *Appendix 4*

## COTTON-GROWING AGREEMENTS BETWEEN COTTON COMPANIES AND THE ETHIOPIANS

COTETIO

ADDIS ABABA

DATE______

*GENERAL HEADQUARTERS FOR*

SEED DISTRIBUTION AND COLLECTION

CENTRE______

COTTON-GROWING AGREEMENTS WITH THE INDIGINOUS FARMERS

COTETIO has handed over______kg of cotton seeds to (Mr)______. (Mr)______ declares to have received it for distribution to native farmers of his district according to the following regulations.
The purpose of handover is as follows:

1. All the seeds received shall be distributed without any interference to each person and sown on dates and according to the instructions that will be emanated by the officials of the Company.
2. All the works related to cotton farming—such as thinning out, hoeing, pruning—shall be carried out exactly according to the rules that will be communicated in due course by the officers of the Company.
3. Also the picking shall take place in the day and according to the order that will be issued by the officers of the Company.
4. All the cotton crop without any interference shall be taken and deposited in the storage centre that will be established in due course by the Company.
5. Upon collecting the crop, the Company shall pay the farmers or their chief the price that the Government will issue according to justice.
6. Each farmer may request and receive from the Company clean cotton necessary for his family needs within the conditions that will be issued by the Government.
7. Should the farmers fail to follow the terms and conditions issued by the Company's officers, the Company shall carry out the necessary works directly and deduct the relative expenses from the price of the crop.

Mr_____in his capacity as_____accepts willingly *per se* and on behalf of his dependents all the terms and conditions as described above. The sowing will begin accordingly and may God bless the crop.

Mr_____
(Signature)_____

Having seen the contract, we will abide to carry it out accordingly.

(Signature)_____ (Signature)_____

(Signature)_____ Agent for COTETIO

R. Residency of_____

(Signature) [*The Resident*]_____

# *Appendix 5*

## COTTON CAMPAIGN (TRANSLATION)

ETHIOPIAN COTTON COMPANY (COTETIO)
ADDIS ABABA

*Basic Cotton-Farming Rules*

To sow cotton prepare the soil over which no cotton plants have been before.

Sow after the first rains so that the land be ready to receive the seed and this can germinate quickly.

Make with plough as many furrows as possible each distant from the other two steps. Put the seed in small holes at the bottom of each furrow within a small distance from each other. Put only four grains in each hole and immediately cover them with soil.

So while you are passing with plough, your son can follow you depositing it inside the ground; four grains in each separated by a small step and cover them with soil using his feet.

The seed is expensive and difficult to get and yields a very good cotton. Thus take care not to misuse it.

Do not sell it and [if you do], you will be punished. If anyone asks you to have some, do not give it. Your friend can take it from us.

The seed yields ten times more than the cotton you already have in your plot.

Remember that for a land with 100 steps length and 100 steps width you need six *qunna* seeds.

Destroy all old plants of cotton of previous years. It is bad and may harbour diseases that can kill even the good plants that you obtain from our seed stock.

When the plants are as high as two hands leave only two in each hole and destroy the rest.

Use hoe to work the land and prevent grass from growing. Grass-infested cotton-field gives bad and small yield.

Do not be hasty to harvest the cotton, nor should you wait until it falls down on the ground: both early and fallen cotton will fetch you low price.

As it can be stained by smoke, do not store the cotton that you picked up in a *tukul* where there is too much smoke; nor store it on the floor as it can get dirty.

As dirty cotton that is adulterated with leaves, straw, pebbles, and soil will fetch you a low price, try to bring it to us immediately after you picked it.

Do not mix the cotton produced using our seeds with the old cotton you have in your plot. We will notice it easily and tender reduced price accordingly.

Be aware that the cotton you produced can be sold to nobody but us. Nor can you keep it at home for yourself.

We will purchase your cotton for a fair price as fixed by the government.

We will also give you clean cotton sufficient enough for the needs of your family.

Once the cotton is picked, pull out all cotton plants and burn them. For the following year you will be provided with far improved seeds.

You will sow this in a different land in the same manner that you have been instructed before.

COTETIO

# Bibliography

## UNPUBLISHED SOURCES

### 1. Archival

[ACS] GRA: 30/29/36, 32/47/1, 36/31/11, 36/31:12–13, 36/32/1, 38/33/3, 40/33:1–15, 43/34/22, 43/34/26, 44/36:1–2, 45/41/2, 45/41:4–5, 46/41/9, 46/41:14–16, 46/41/19.

[ACS] MCP B3/27, B5/52, B7/74, B18/261, B18 bis/267, B236, B240, B241.

[ACS] ONC AOI 1, 1/1, 1/2, 1/2 Bis, 1/3, 1/6, 2, 3, 3/A:1–2, 3/B, 3/B:1–3, 3/D, 4, 4/2, 4/4, 4/9, 4/16, 4/18, 4/20:1–2, 4/20/4, 4/20/20, 4/25, 4/33, 4/39, 4/A, 4/E, 5, 5:1–3, 5/A, 5/A/1, 6, 6/9, 6/4/D, 7, 7/7, 8/A, 9, 15, 15/15, 17, 20/4, 21, 21/1, 22, 22:A-D, 22/G, 26, 29, 33, 38, Misc.

[ACS] PNF: 1/J, 3/H, B11, B26.

[ACS] PS: 38/C/J4/F/1939, 426/A5G/194.

[ACS] SPDR: 44/242/R/39, 87/W/R/1, 87/W/R/1/LA.

AMAR: f4.

ASMAE PAPERS: 2/2, 2/2/1, 2/4/1, 3/1/1, 3/6/5/1, 6/1/1, 7/5/1, 7/7/1, 8/1/1, 8/2/1, 8/4/1, 12/8/84, 14/4/1, 14/9/1, 16/6/1, 18/13/84.

ASMAI PAPERS: 3/13, 4/14, 4/88, 11/78/2, 13/145, 19/162, 51/1:1–14, 51/22, 54/7/20, 54/8:21–22, 54/21/172, 256/52, PB:3–6, B/232, B/245, B/247, B/287, III/P4

ATdR: 24:79–80, 89–100, 101, 103–4, 108–9, 116, 119–20, 124, 124 bis, 148, 148 bis.

CD IAO AOI: 345, 606, 788–91, 793–4, 1038, 1097–8, 1106, 1111, 1120–1, 1232, 1247, 1323, 1325, 1328–9, 1335–8, 1636, 1774–6, 1778, 1792–3, 1799–1803, 1806, 1808, 1813, 1817–22, 1830, 1832, 1837–40, 1845–8, 1850, 1852–4, 1916, 1918, 1921–4, 1926, 1929–30, 1935–7, 1965, 1982, 1984–6, 1990, 1992, 2004, 2064, 2214, 2919, 3017, 3025, 3034, 3168, 4299, 4414.

PUBLIC RECORD OFFICE: FO371/20167/J5632/45/1. FO371/20209/J8291/4321/1. FO371/22020: J40/40/1, J297/40/1, J368–71/40/1, J383/40/1, J395/40/1, J622/40/1, J641/40/1, J657/40/1, J755/40/1. FO371/22021: J1214/40/1, J1216/40/1, J1224/40/1, J1321/40/1, J1501/40/1, J1747/40/1, J1804/40/1, J2363/40/1, J2376/40/1, J2447/40/1, J2512/40/1, J2677/40/1, J2728/40/1, J2923/40/1, J2926/40/1, J3016/40/1, J3287/40/1, J3439/40/1. FO371/23376/J574/41/1, FO371/23377/J1835/41/1. FO371/23380: J296/296/1, J575/296/1, J1324/296/1, J1776/296/1, J1981/296/1, J1992/296/1, J2053/296/1, J2209/296/1, J2292/296/1, J4188/296/1, J4499/296/1, 4710/296/1, J4814/296/1. FO371/24635: J5/5/1, J376/18/1, J412/18/1, J466–7/18/1, J887/18/1. FO371: 24643/J247/18/1; 35641/J280/2801.

## 2. *Theses and Other Papers*

ABDEL RAHIM, A. W., 'An Economic History of the Gezira Scheme 1900–1956', Ph.D. thesis (Manchester, 1968).

MAKNUN G. ASHAMI, 'The Political Economy of the Afar Region of Ethiopia: A Dynamic Periphery', Ph.D. thesis (Cambridge, 1985).

BAUER, F., 'Land, Leadership and Legitimacy among the Inderta Tigray of Ethiopia', Ph.D. thesis (University of Rochester, 1972).

CASTAGNO, A. A., 'The Development of Expansionist Concept in Italy, 1861–1896', Ph.D. thesis (Columbia University, 1957).

'Fetha Nägäst' [Law of Kings], trans. Paulos Tzadua (AA, HSIU mimeograph).

LOMBARDI, P., 'La colonizzazione agraria in Libia durante il periodo fascista', paper presented at Seminario Sulla Libia: Storia e Rivoluzione, Rm, 27–9 Jan. 1981.

MCCLELLAN, CHARLES, W., 'Reaction to Ethiopian Expansionism: The Case of Darassa, 1895–1935', Ph.D. thesis (Michigan State University, 1978).

MESFIN WOLDE MARIAM, 'Some Aspects of Land Ownership in Ethiopia', paper presented at the Seminar of Ethiopian Studies, HSIU, 1965.

SBACCHI, A., 'Italian Colonialism in Ethiopia', Ph.D. thesis (University of Illinois at Chicago-Circle, 1975).

STELLA, GIAN CARLO, 'Qualche nota su organizzazioni agricole fasciste', Ravenna 3 Aug. 1985 [typed]. Material provided during his correspondence with the author.

### PUBLISHED SOURCES

## 1. *Official Publications*

CDDD, XV, Libro Verde: Ethiopia, Rm: Tip. CD, 1890.

Circolare 25-11-1937, no. 163647, Riscossione della decima, in *GU* 2/23, 1 Dec. 1937.

DGG 4-7-1936, no. 82, Confisca dei beni di Ras Nasibù e di Blattenghietà Uolde Mariam, in *GU* 2/1, 1 Jan. 1937.

DGG 17-7-1936, no. 84, Restituzione beni a Ras Hailù, in *GU* 2/8 (suppl.), 26 Apr. 1938.

DGG 30-7-1936, no. 135, Confisca beni del suddito coloniale Blattengheità Herui, in *GU* 2/8, (suppl.), 26 Apr. 1938.

DGG 4-10-1936, no. 95, Norme che evitano, nell 'attuale periodo di sviluppo industriale e commerciale dei territori dell'AOI, ingiustifcati accaparramenti di terreni, in *GU* 2/20, 16 Oct. 1937.

DGG 31-8-1937, no. 656, Concessione taglio macchia di sottobosco per far carbone, in *GU* 2/22, 16 Nov. 1937.

DGG 27-9-1937, no. 748, Ripristino in tutti i nuovi territori dell'AOI dei tributi erariali tradizionali, in *GU* 2/21, 1 Nov. 1937.

DGG 29-9-1937, no. 693, Disciplina del commercio e consumo dei cereali, in *GU* 2/21, 1 Nov. 1937.

DGG 15-10-1937, no. 738, Confisca beni di sudditi indigeni, in *GU* 2/22, 16 Nov. 1937.

DGG 25-10-1937, no. 751, Confisca beni di sudditi indigeni, in *GU* 2/23, 1 Dec. 1937.

DGG 25-10-1937, no. 752, Confisca beni di sudditi indigeni, in *GU* 2/23, 1 Dec. 1937.

DGG 13-11-1937, no. 805, Abrogazione dell'art. 1 del DGG 4-10-1936, no. 95, riflettente le norme intese ad evitare ingiustificati accapparramenti di terreno, in *GU* 2/23, 1 Dec. 1937.

DGG 11-1-1938, no. 15, Confisca dei beni dei sudditi Degiac Auraris Dammenà e Ascalè Mariam, in *BUGA* 3/2–3, 2 Mar. 1937.

DGG 10-2-1938, no. 94, Esenzione dai diritti di confine del petrolio, della benzina e dei residui della distillazione degli olii minerali destinati esclusivamente alle attività agricole dell'AOI, in *GU* 3/10, 16 May 1938.

DGG 3-3-1938 no. 155, Abrogazione dei decreti vicereali no. 95 del ottobre 1936, no. 805 del 13-11-1937, ed emanazione di nuove norme disciplinanti la materia delle alienazioni del terreni, sia a scopo agricolo che a scopo edilizio, in *GU* 3/10, 16 May 1938; *BUGA* 3/2–3, Feb.–Mar. 1938.

DGG 5-4-1938, no. 858, Istituzione a Guder di una scuola agraria per indigeni, in *GU* 3/17, 1 Sept. 1938.

DGG 25-5-1938, no. 651, Concorsi a premio tra I coltivatori indigeni residenti nei territori del GAA, nei Commandi Settori di Ambò, Debra Berhan e del Commissariato di G. di Adama, in *GU* 3/15, 1 Aug. 1938.

DGG 23-7-1938, no. 828, Prezzi dei carburanti agricoli, in *GU* 3/15, 1 Aug. 1938.

DGG 7-11-1938, no. 1300, Permesso concesso alla ditta Eugenio Bertolani per l'utilizzazione della massa legnosa ritirabile dalle formazioni boschive demaniali site nella zona Metchà Coritchà, in *GU* 4/5, 1 Mar. 1939

DGG 24-1-1939, no. 56, Permesso concesso alla signora Germana Dentyu ved[ova]. Prasso, per l'utilizzazione della massa legnosa ritirabile dalle formazioni boschive demaniali site sul versante nord dei monti Roggè, Arfingiò Dagà e, Deballé in *GU* 4/6, 16 Mar. 1939.

DGG 1-2-1939, no. 86, Norme per la disciplina del commercio e del consumo dei cereali, in *GU* 4/6 (suppl.), 22 Mar. 1939.

DGG 15-3-1939, no. 230, Divieto d'importazione di cotone in bioccoli e norme per l'importazione di seme di cotone nei territori dell'AOI, in *GU* 4/9, 1 May 1939.

DGG 22-4-1939, no. 373, Concessione agricolo-pastorale nella zona di Mencherrè (Lago Ascianghi) accordata a S. E. Rodolfo Graziani Marchese di Neghelli, in *GUGS* 4/16, 9 Aug. 1939.

DGG 10-5-1939, no. 417, Autorizzazione concessa alla ditta Cocciarfico Quirino per l'utilizzazione e la vendita della massa legnosa ritirabile dalla zona boschiva demaniale di Arera, in *GUGS* 4/16, 9 Aug. 1939.

DGG 30-6-1939, no. 573, Revoca confisca beni del Deggiac Auraris Demmenà e della Uoizerò Ascalè Mariam, in *BUGA* 4/15, 15 Aug. 1939.

DGG 18-7-1939, no. 665, Decreto col quale il sig. Petrulio Antonio è autorizzato all'utilizzazione e alla vendita della massa legnosa ritirabile dalle zone boscate demaniali di Colbà, Malca Dagà, Ciollè, Monte Harro e Monte Colbà in territorio di Guder, in *GUGS* 4/33, 6 Dec. 1939.

DGG 4-9-1939, no. 803, Norme per la salvaguardia e la disciplina delle piantaggioni di caffè nel territorio del G Hr, in *GUGS* 4/23, 27 Sept. 1939.

DGG 16-10-1939, no. 987, Istituzione presso il GGS della gestione di EDR per la valorizzazione agraria di una superficie di circa 2,000 ettari compresa tra I torrenti Bore ed Affolè, in *GUGS* 4/26 (suppl.), 19 Oct. 1939.

DGG 13-11-1939, no. 1103, Concessione di avvaloramento forestale in località Acachi al sig. Bernardo Giannotti ed approvazione del relativo disciplinare, in *GUGS* 5/17, 14 Feb. 1940.

DGG 23-11-1939, no. 1174, Autorizzazione alla ditta Loretelli Rodolfo per l'utilizzazione e vendita della massa legnosa ritirabile dalle zone boscate demaniali Giallò-Gabù-Matused approvazione del relativo disciplinare, in *GUGS* 5/7, 14 Feb. 1940.

DGG 5-12-1939, no. 1225, Istituzione di un ente assistenziale per gli indigeni nel GSc, in *GUGS* 5/7, 14 Feb. 1940.

DGG 22-1-1940, no. 89, Istituzione presso ogni GAOI di una commissione allo scopo di determinare il demanio disponibile per la colonizzazione, in *GUGS* 5/ 9, 28 Feb. 1940.

DGG 23-1-1940, no. 98, Provvedimenti per l'inceremento dell'autotrazione a gassogeno in AOI, in *GUGS* 5/15, 10 Apr. 1940.

DGG 13-2-1940, no. 193, Estensione delle attività dell'ente assistenziale UMA in AOI, in *GUGS* 5/13, 27 May 1940.

DGG AOI 21-3-1940, no. 387, Modifiche apportate alle norme del DGG 15-3-1939, no. 230, che regola e disciplina la coltivazione del cotone nell'AOI, in *GUGS* 5/27, 3 July 1940.

DGov Am 25-10-1937, Determinazione dei salari massimi per opera e lavoratori indigeni, in *BUGA* 2/11–12, Nov.–Dec. 1937.

DGov Am 5-6-1937, no. 46144, Indemaniamento del territorio dell'Acefer (circoscrizione dei commissariati di GGn e del Gojjam), in *BUGA* 3/4, 16 Apr. 1938.

DGov Am 8-6-1937, no. 46204, Indemaniamento dei territori del Ginfranchrerà, Uorchemder, Gianorà, Tseghede, Gusquam (circoscrizione dei commissariati di GGn e del Semien), in *BUGA* 3/4, 16 Apr. 1938.

DGov Am 9-2-1938, no. 73412, AF, Indemaniamento del territorio dell'Uogherà (zona di Dabat), in *BUGA* 2/2–3, Feb.–Mar. 1938.

DGov Am 11-3-1938, no. 134918 AP, Assegnazione ai due vescovi copti del territorio dell'Amara di L.6000 mensili, in *BUGA* 3/2–3, 2 Mar. 1938.

DGov Am 17-3-1938, no. 53303 AF, Indemaniamento dell'Uogherà (zona Amba Ghiorghis), in *BUGA* 3/2–3, Feb.–Mar. 1938.

DGov Am 18-7-1938, no. 46, Concorsi a premio da assegnarsi agli agricoltori indigeni che avranno ottenuto nel prossimo raccolto una maggiore produzione unitaria di grano ed altri cereali, in *BUGA* 3/15, 1 Aug. 1938.

DGov Am 31-12-1938, no. 206152, Accordata l'esenzione dal pagamento della decima agli imprenditori di Dessiè e all'ECRE per l'annata agraria 1938–9, in *BUGA* 4/3, 14 Feb. 1939.

DGov Am 21-1-1939, no. 226420, Istituzione di una commissione consultiva per esprimere il parere sull'accoglimento delle domande per il trasferimento delle famiglie di lavoratori nel territorio dell'Am, in *BUGA* 4/3–14, Feb.-Mar. 1939.

DGov Am 7-7-1939, no. 385, Nomina del 1°. Segretario di G dr. Dodaro Ugo a Reg. la sezione della colonizzazione presso la direzione della CL, in *BUGA* 4/15, 15 Aug. 1939.

DGov Hr 13-7-1938, no. 439, Noleggio di macchine ed attrezzi agricoli dell'amministrazione a favore degli agricoltori, in *GU* 3/17, 1 Sept. 1938.

DGov Sc 9-3-1937, no. 76, Fissazione della retribuzione da corresponders ai manovali indigeni, in *GU* 2/3, 1 Apr. 1937.

DGov Sc 9-3-1939, no. 223, Concessione accordata al sig. Marini Antonio di un lotto di terreno sito in località Uriel per sfruttamento agricolo, in *GUGS* 4/15, 2 Aug. 1939.

DGov Sc 10-3-1939, no. 224, Concessione accordata al sig. Liri Clemente di un lotto di terreno sito sulla destra della strada AA-Moggio, per sfruttamento agricolo, in *GUGS* 4/15, 2 Aug. 1939.

DGov Sc 10-3-1939, no. 225, Concessione, a scopo agricolo, di un lotto di terreno sito nei pressi della strada AA-Gm al sig. Carlo Cataldo, in *GUGS* 4/24, 4 Oct. 1939.

DGov Sc 12-6-1939, no. 513, Permesso accordato al sign. Riccini Alviero per lo sfruttamento della massa legnosa ritirabile dalla zona boscata demaniale di Guder, in *GUGS* 4/26, 18 Oct. 1939.

DGov Sc 17-7-1939, no. 145, Concessione agricola accordata al sig. Tedeschi dott. Aldo, in *GUGS* 4/29, 8 Nov. 1939.

DGov Sc 19-9-1939, no. 264, Concessione agricola ai sigg. Barbieri Orlando e Peta Francesco, in *GUGS* 5/11 (suppl.), 9 Mar. 1940.

DGov Sc 1-6-1940, no. 88, Concessione al sig. Pettini Cosimo per la durata di anni 9 di un lotto di terreno, in *GUGS* 5/25, 19 June 1940.

DGov Sc 13-2-1940, no. 201, Concessione di anticipazioni di frumento da semina e di carburanti agricoli ad agricoltori nazionali che svolgono la loro attività nel territorio del G Sc in *GUGS* 5/25, 19 June 1940.

DIM 18-5-1938, Distretti cotonieri: agevolazioni in materia fiscale e doganale, in *BUGA* 3/16, 16 Feb. 1938.

DM 25-3-1937 Estensione di alcuni articoli di regolamento fondiario per la colonia Er ai territori dei Governi della Sm, dei GS e Hr, in *BUGA* 2/3–4, Mar.–Apr. 1937.

DM 16-5-1938, IDC di: Cobbò; Auasc; Metemma–Dongur; in *GU* 3/17, 1 Sept. 1938.

DM 1-12-1938, IDC del Basso Giuba, in *GU* 4/4, 16 Feb. 1939.

DM 1-12-1938, IDC di Lechenti la cui organizzazione e la gestione vengono affidate alla ICAI, in *GU* 4/12, 16 June 1939.

DM 1-12-1938, IDC di Soddu la cui organizzazione e la gestione vengono affidate alla ICAI, in *GU* 4/12, 16 June 1939.

DM 31-1-1940, IDC dell'Uollega, in *GUGS* 5/29, 17 July 1940.

DM 31-1-1940, Concessione alla Cotetio di alcune esenzioni e riduzioni fiscali e doganali sui tributi coloniali, in *GUGS* 5/35 (suppl.), 17 July 1940.

DM 31-5-1940, Revoca della concessione del distretto cotoniero di Lekemti [*sic*] alla ICAI e passaggio dello stesso alla Cotetio in *GUGS* 5/35 (suppl.), 31 Aug. 1940.

DReg Gov Am 21-6-1937, Concorsi a premio per la coltivazione del grano e delle patate agli agricoltori indigeni, in *BUGA* 2/3–4, May–June 1937.

Istituzione e statuto della sezione agraria di istituto tecnico superiore specializzato nell'agricoltura coloniale presso il R. Istituto Agronomico per l'AI', *AC* 34/2 (1940), 45–8.

Legge 10-6-1937, no. 1029, Conversione in legge del RDL 19-12-1936, no. 2467, che conferisce un diritto di preferenza nella concessione delle terre dell'AOI a coloro che hanno ivi partecipato alle operazioni militari in qualità di combattenti, in *GU* 2/15, 1 Aug. 1937.

Legge 23-12-1937, no. 2644, Odinamento del servizio catastale e tecnico erariale nell'AOI, in *REAI* 26/4, (1938).

Legge 3-6-1938, no. 965, Istituzione di una commissione per studi fondiari nell'AOI, in *REAI* 26/8, (1938).

Legge 25-8-1940, no. 1415, Costituzione dell'ente di colonizzazione per gli italiani all'estero, in *GUGS* 5/48, 27 Nov. 1940.

MAE, *L'Avvaloramento e la Colonizzazione*, 1/2 of *L'Italia in Africa* (Rm: Soc. Abete, 1970).

MAI, 'La Valorizzazione e Colonizzazione Agraria', *AAI* 2/3 (1939), 179–316.

MAI *La Costruzione dell'Impero*, 3/1 of *AAI* (Milan: Mondadori, 1940).

MAI *Progetti di Ordinamento fondiario per l'AOI* (Rm: Ufficio Studi MAI, 1942).

'Ordinamento dei servizi dell'Agricoltura nell'AI', *AC* 33/4 (Apr. 1939), 209–21.

RD 3-6-1938, no. 965, Istituzione di una commissione per studi fondiari nell'AOI, in *GU* 3/16, 16 Aug. 1938.

RD 29-7-1938, no. 2221, Ordinamento dei servizi dell'agricoltura nell'AI, in *BUGA* 4/7–1, 15 Apr. 1939.

RDL 7-10-1937, no. 2513, Costituzione dell'ECAI con sede in Rm, in *GU* 3/5, 1 Mar. 1938; *GURI* 41, 19 Feb. 1938.

RDL 6-12-1937, no. 2325, Costituzione dell'ECPE, in *REAI* 26/2 (1938).

RDL 6-12-1937, no. 2300, Costituzione dell'ECRE, in *REAI* 26/2 (1938).

RDL 6-12-1937, no. 2314, Costituzione dell'ECVE, in *REAI* 26/2 (1938).

RDL 7-1-1938, no. 443, Istituzione di distretti cotonieri nell'AOI, in *GU* 3/13, 1 July 1938; *GURI*, 10 May 1938.

RDL 13-5-1940, no. 823 Modificazione dell'art. 7 del RDL dicembre 1937, no. 2314, Costitutivo dell'ECVE, in *GUGS* 5/34, 21 Aug. 1940.

Relazione della Commissione di Competenza per l'Agricoltura, 'L'avvaloramento agrario dell'Impero', *REAI* 25/10 (1937), 1561–76.

Relazione Generale della Regia Commissione d'Inchiesta sulla Colonia Eritrea, Rm: Tip. delle Mantellate, 1891.
UA Am 'Aspetti generali e zootecnici del Lago Tana', *AC* 32/6 (1938), 252–72.
UA Gn, 'L'Acefer: Notizie di indole generale', *AC* 32/2 (1938), 55–9.

## 2. *Articles*

AMBROSINI, GASPARE, 'Le caratteristiche della colonizzazione italiana in Africa', *Studi di Storia e Diritto*, 1 (1944), 317–35.

'ANDAMENTO [L'] e lo sviluppo della cotonicoltura dell'AOI: Una detagliata relazione al Duce del presidente della Cotetio', *Rassegna d'Oltremare* (Sept. 1938).

ANDERSON, DAVID, and DAVID, THROUP, 'Africans and Agricultural Production in Colonial Kenya: The Myth of the War as a Watershed', *JAH* 26/4 (1985), 327–45.

ANJORIN, A. O., 'European Attempts to Develop Cotton Cultivation in West Africa, 1850–1910', *ODU* 3/1 (1960), 3–15.

'AOI nel quadro dell'autarchia nazionale', *II* (Apr. 1940).

'Attività colonizzatrice a Puglie d'Etiopia', *EI* (May 1940).

BALDRATI, I., 'Lo sviluppo dell'agricoltura in Eritrea nei cinquanta anni di occupazione italiana', *REC* 7/1 (1933), 43–53.

BARTOLOZZI, ENRICO, 'Il commercio del caffè nell'AOI', *AC* 32/7 (1938), 316–19.

—— 'L'aratro abissiono', *AC* 32/11 (1938), 538–45.

—— 'Missione di studio in AOI', *AC* 33/3 (1939), 126–31.

BASSETT, THOMAS J., 'The Development of Cotton in Northern Ivory Coast, 1910–1965', *JAH* 29/2 (1988), 267–84.

BONINSEGNI, SERGIO, 'Esperimenti di coltivazione di frumenti italiani Ad Ugri nel 1937', *AC* 32/5, (1938), 205–10.

CAPANNA, ALBERTO, 'Prospettive autarchiche dell'AI: I cereali minori', *EI* (Mar. 1940).

—— 'Economic Problems and Reconstruction in Italy', *International Labour Review*, 62 (June 1965), 607–52.

CARACCIOLO, MATTIA M., 'Prospettive ed orientamenti di politica fondiaria nell'Impero', *REAI* 26/10 (1938), 1571–5.

CAROSELLI, FRANCESCO S., 'Aspetti economici dell'avvaloramento agrario nell'Impero', *AC* 35/2 (1941), 47–54.

CASTELLANI, ETTORE, 'Prima ricognizione fitopatologica in AOI', *AC* 33/3 (1939), 143–8.

—— 'Problemi fitopatologici dell'Impero: Osservazioni ed orientamenti', *Georgofili*, 6/5 (1939), 545–9; *AC* 34/1 (1940): 5–15.

—— 'Ruggine e granicoltura nell'Africa tropicale montana', *Georgofili*, 7/8 (1942), 74–8.

CERULLI, ENRICO, 'La colonizzazione del Harar', *AAI* 6/1 (1943), 63–79.

CESARINI, PAOLO, 'Vita nell'Africa nostra: Pattuglie del Grano nel Gm', *GP*, 9 May 1940.

CHIAROMONTE, A., 'L'VIII Congresso Internazionale di Agricoltura Tropicale e Subtropicale di Tripoli', *AC* 33/4 (1939), 221–4.

CHIUDERI, ARRIGO, 'Alcune considerazioni su una campagna granaria nel GS', *Georgofili*, 7/7 (1940), 116–20.

—— 'L'agricoltura indigena nel GS ed i mezzi per farla progredire', *AC* 36/10 (1942), 269–74.

CICCARONE, ANTONIO, 'Malattie delle piante segnalate nel 1939 nell'AOI', *AC* 34/9 (1940), 388–90.

—— 'Note sulla biologia della "nebbia del frumento" (*Erysiphe Graminis* D.C.) nello Sc', *AC* 35/6 (1941), 232–8.

CIFERRI, RAFFAELE, 'Frumenti e granicoltura indigena in Etiopia': *Georgofili*, 6/5 (1939), 233–42; *AC* 33/6 (1939), 337–49.

—— 'Problemi del caffè nell'AOI': *Georgofili*, 6/6 (1940), 11–20; *AC* 34/4 (1940), 135–44.

—— and BARTOLOZZI, ENRICO, 'La produzione cerealicola dell'AOI nel 1938', *AC* 34/11 (1940), 441–50; 12 (1940), 502–15.

—— and GIGLIOLI, G. R., 'Considerazioni pratiche sul problema della produzione frumentaria nell'AOI', *AC* 33/12 (1939), 660–5.

COHEN, JOHN, 'Peasant and Feudalism in Africa: The Case of Ethiopia', *JMAS* 8/1 (1974): 155–7.

'[La] Colonizzazione in AOI e in libia', *stampa*, 13 June 1939.

COLUCCI, MASSIMO, 'Premesse per la colonizzazione dell'Impero', *RDA* 18 (1939), 150–63.

'Concessioni terriere nel Harar assegnate al 31 gennaio 1940', *Consulente coloniale*, 10 June 1940.

CONFORTI, E., 'Cenni sulla regione dei Guraghé', *AC* 33/7 (1939), 415–26.

—— 'La regione dei Guraghé', *AC* 35/6 (1941), 239–50.

'Consistenza numerica degli agricoltori nell'Impero', *EI* (May 1940).

COOKEY, S. J. S., 'The Concession Policy in the French Congo and the British Reaction, 1898–1906', *JAH* 7/2 (1966), 263–78.

CORDELL, DENNIS D., and GREGORY, JOEL W., 'Labour Reservoirs and Population: French Colonial Strategies in Koudougou, Upper Volta, 1914 to 1939', *JAH* 23/1 (1982), 205–24.

DAGNE HAILE GABRIEIL, 'The Gebezenna Charter, 1894', *JES* 9/2 (1972), 67–80.

DAVID, MASSIMO, 'L'Impero al terzo anno: La prima Casa Pugliese', *GP*, 14 Jan. 1939.

DE BENEDICTIS, ANTONIO, 'L'autarchia alimentare dell'Impero', *Georgofili*, 6/3 (1937), 468–93.

—— 'L'autarchia alimentare dell'Impero', *AC* 32/1 (1938), 1–12.

—— 'Il problema del miglioramento del bestiame bovino indigeno nell'AOI', *AC* 32/4 (1938), 148–57.

DE BIASE, L. 'Le regioni del Mens e del Marabetie', *AAI* 5/4 (1942), 931–47.

Dei Gaslini, M., 'Lettera dal GS, il caffè: Produzione, consorzio, mercato, tutelà', *EI*, Apr. 1940.

De Michelis, Giuseppe, 'La Valorizzazione agricola dell'Impero', *REAI* 26/1 (1938), 5–11.

Di Crollalanza, Arnaldo, 'La valorizzazione agricola dell'Impero', *REAI* 25/4 (1937), 489–95.

—— 'L'avvaloramento agricolo dell'Impero', *REAI* 26/5 (1938), 712–17.

—— 'Relazione sui programmi di colonizzazione demografica nell'Impero da parte dell'ONC', *REAI* 26/5 (1938), 739–63.

—— 'L'avvaloramento agricolo dell'Impero', *REAI* 27/11 (1939), 1195–1207.

Di Lauro, Raffaele, 'Panorama politico-economico dei GS', *REAI* 26/7 (1937): 1082–7; 26/8 (1938): 1272–8; 26/12 (1938): 1886–94.

—— 'L'Impero può bastare a se stesso?', *L'Autarchia Alimentare*, 2/2 (1939), 8–10.

—— 'Il bracciantato indigeno e la colonizzazione dell'Impero', *EI* (Mar. 1939), 105–8.

'Disposizioni transitorie per il conseguimento del titolo di perito agrario coloniale da parte di licenziati dei già corso medio-superiori di agricoltura coloniale dell'Istituto Agricolo Coloniale di Firenze', *AC* 34/4 (1940): 133–5.

'Distribuzione di aratri ai nativi dell'Eritrea', *EI* (Sept. 1938).

Duly, L. C., 'The Failure of the British Land Policy at the Cape, 1812–1828', *JAH* 6/3 (1965): 351–71.

Falorsi, Giorgio, 'L'Istituto Agricolo Coloniale', *AC* 33/3 (1939), 166–93.

Ferrari, Angelo, 'Gli strumenti dell'autarchia alimentare nell'Impero', *Autarchia Alimentare* (Sept. 1938), 13–20.

Fossa, Davide, 'L'intervento del partito nel governo dell'Impero', *REAI* 27/3 (1939), 255–62.

Furfaro, Domenico, and Bianco, Gianna, 'L'ideologia dell'imperialismo fascista nella "Rivista delle Colonie Italiane" ', in F. Bozzi (ed.), *Miscellanea di Storia delle Esplorazioni*, iv. 223–54 (Geneva, 1979).

Gaetani, Livio, 'Politica agraria dell'Impero', *AC* 32/3 (1938), 97–104.

Gebre-Weld Ingida Werq, 'Ethiopia's Traditional System of Land Tenure and Taxation', *EO* 5/4 (1962), 302–39.

Gennari, Giulio, 'La colonizzazione agraria di popolamento nell'economia corporativa dell' Impero: Osservazioni e proposte di un legionaro', *Georgofili*, 6/2 (1936), 502–20.

Gennari, Giulio, 'L'agricoltura nell'AOI', *REAI* 25/12 (1937), 1889–1925.

Giaccardi, Alberto, 'La colonizzazione dell'Impero', *REAI* 27/1 (1939), 9–18.

Giannoccaro, Giambattista, 'Prime tappe dell'Ente "Puglia d'Etiopia' in AOI', *AI* 1/1 (Nov. 1938): 25–8.

—— 'L'ECPE', *AI* 7 (May 1939): 19.

[Giglio, C.], 'Necessità di colonizzazione demografica nell'Impero', *REAI* 26/9 (1938), 1410–17.

[GIGLIO, C.], 'Rapporti della colonizzazione demografica con la colonizzazione capitalistica e l'agricoltura indigena', *REAI* 26/10, (1938), 1561–5.

——'Importanza dell'ambiente fisico-agrologico-economico nella colonizzazione demografica', *REAI* 27/1 (1939), 28–38.

——'Il finanziamento degli enti di colonizzazione demografica in AOI', *REAI* 27/2 (1939), 154–61.

——'Le prime realizzazioni della colonizzazione demografica in AOI', *REAI* 27/3 (1939), 259–62.

——'Da colono al proprietario in AOI con la colonizzazione demografica', *REAI* 27/3 (1939), 401–5.

——'Di alcuni problemi della colonizzazione demografica', *REAI* 27/12 (1939), 1872–8.

GIGLIOLI, GUIDO R., 'Impressioni sull'economia agraria del territorio Borana', *AC* 30/9 (1936), 331–9;

——'L'opera dell'istituto per la sperimentazione agraria nell'AI', 33/3 (1939), 163–6.

GOGLIA, LUIGI, 'Note sul razzismo coloniale fascista', *Storia Contemporanea*, 6 (1988), 1223–66.

GOSIOTTO, ADOLFO, 'La proprierà terriera ecclesiastica nel Tigrai', *RDA* 18 (1939), 417–21.

GUIDI, GUIDI, 'Una visita all'Ente Romagna d'Etiopia', *Impero Illustrato* (14–20 May 1940).

GUIDOTTI, ROLANDO, and DALLARI, GIACCHINO, 'Bassopiano occidentale oltre il Setit e territorio del Tana: Studio agrologico e rilevamento economico agrario di larga massima', *AC* 31/6 (1937), 209–19.

HAILE M. LAREBO, 'The EOC and Politics in the Twentieth Century', *Northeast African Studies*, 9/3 (1987): 1–15; 10/1, (1988): 1–22.

Iniziativa agricola nel territorio dei Borana e Hararino, *EI* (July 1938).

JEWSIEWICKI, PAR B., 'Le Colonat agricole Européen au Congo-Belge, 1910–1960: Questions politiques et économiques', *JAH* 20/4 (1979) 559–70.

LAMA, ERNESTO, 'Premesse ed aspetti di politica economica coloniale', *REAI* 25/15 (1937), 1939–57.

LEPLAE, EDMONDO, 'La colonizzazione agricola italiana dell'Etiopia', *AC* 30/8 (1936), 283–6.

'Lettera dall'Impero: Conoscere l'Impero', *Critica Fascista* (1 Sept. 1939).

LONSDALE, JOHN, and BERMAN, BRUCE, 'Coping with Contradictions: The Development of Colonial State in Kenya', *JAH* 20/3 (1980), 487–505.

LUSIGNANI, G. B., 'Iniziative cerealicole dell'UA Am', *Autarchia Alimentare* (June 1938): 37.

——'La visita di Attilio Terruzzi a Gn', *Autarchia alimentare*, 1/3 (15 Aug. 1938), 32–6.

——'L'ECRE', *Autarchia alimentare*, 1/7 (15 Dec. 1938), 35–8.

MACDONALD, J. S., 'Italy's Rural Social Structure and Emigration', *Occidente*, 22 (Sept.-Oct. 1956), 437–56.

MAHTEME SELLASSIE WOLDE MASKAL, 'The Land System of Ethiopia', *EO* 1/9 (Oct. 1957), 283–301.

MALAPARTE, CURZIO, 'Visita ai Pionieri della Romagna d'Etiopia', *CS*, 1 Aug. 1939.

MANETTI, CARLO, 'Scuole indigene professionali a tipo agrario in AOI', *REAI* 27/4 (1939), 431–6.

MANGANO, GUIDO, 'La colonizzazione agraria dell'AOI', *AC* 31/5 (1937), 161–70; *AC* 32/6 (1938): 230–9.

—— 'Programma di massima di attività dell'ECAI', *AC* 32/9 (1938), 414–18.

—— 'La politica del cotone', *Georgofili*, 6/5 (1939), 266–78.

MARKAKIS, JOHN, Review of *Land Tenure among the Amhara of Ethiopia: The Dynamics of Cognatic Descent*, by Allan Hoben, *JMAH* 12/2 (1974), 341–2.

—— 'Italian Conquest and Colonization', *JAH* 28/1 (1987), 168–9.

MASSA, LUIGI, 'La dura in Eritrea e la sua coltivazione a Tessenei' *AC* 30/6 (1936), 205–13.

MASSI, ERNESTO, 'Economia dell'Impero', *RISS* 48/11 (1940), 424–54.

MAUGINI, ARMANDO, 'La valorizzazione agricola della colonia Eritrea', *REC* 19/3–4 (Mar.–Apr. 1931), 365–79.

—— 'L'agricoltura nelle colonie: Esperienze e nuovi doveri', *AC* 30/11 (1936), 401–17.

—— 'Guida-questionario per lo studio generale della produzione terriera nei territori dell'AOI', *AC* 31/6 (1937), 286–94.

—— 'Impressioni sull'agricoltura dell'Impero', *AC* 33/3 (1939), 114–19.

—— 'Programmi autarchici nell'AI', *AC* 33/12 (1939), 653–5.

—— 'L'opera dell'Italia a favore dell'economia indigena nelle colonie dell'AO', *Georgofili*, 6/1 (1935), 272–92.

—— 'Primi orizzonti della valorizzazione agricola dell'Impero', *Autarchia Alimentare* (June 1938), 17–20.

—— 'Appunti sulle praspettive agricole dell'Impero', *AC* 35/11 (1941), 401–14.

MAZZEI, JACOPO, 'Le conseguenze economiche del possesso coloniale per la madrepatria', *REAI* 25/8 (1937), 1084–98.

MAZZOCCHI ALEMANNI, N. 'Orientamenti nella colonizzazione demografica dell'Impero: Prime realizzazioni dell'ONC', *Georgofili*, 6/4, (1939), 93–115; see also *AC* 32/4 (Apr. 1938), 158–69; *AC* 32/5 (May 1938), 199–204.

MAZZONI, GUGLIELMO, 'Le colture stagionali della tradizione rurale scioana', *AC* 37/9–10 (1943), 225–31.

—— 'Prima impostazione del problema dei frumenti nelle terre alte dell'AOI', *Georgofili*, 7/7 (1941), 99–105; see also *AC* 35/4 (1941), 136–43.

MININNI, MATTIA, 'Prospettive e orientamenti di politica fondiaria nell'Impero', *REAI* 27/3 (1939), 267–73.

MISSAGLIA, ALDO, 'La colonizzazione demografica dell'Impero: Storia di Bari d'Etiopia', *PI*, 3 Apr. 1940.

MONDAINI, GENNARO, 'I problemi del lavoro nell'Impero', *REAI* 25/5 (1937), 747–52; *REAI* 25/7 (1937), 942–7.

MONDAINI, GENNARO, 'La partecipazione degli indigeni all 'onere tributario dell'AOI', *REAI* 27/11 (1939), 1213–22.

—— 'Tradizione e innovazione nell'incipiente ordinamento fondiario dell'AOI', *Georgofili*, 7/7 (1941), 355–86.

MONTEFOSCHI, MAURIZIO, 'I centri agricoli di Oletta' e Biscioftù: Successo di un esperimento', *IC*, May 1939.

MORENO, MARTINO M., 'Politica di razza e politica coloniale Italiana', *AAI* 2/2 (1937), 450–67.

—— 'La politica indigena Italiana in AOI', *AAI* 5/1 (1942), 64–77.

—— 'Politica religiosa e paganesimo in Etiopia', *AAI* 5/3: 641–55.

—— 'Il regime terriero abissino nel GS', *REAI* 25/10 (Oct. 1937), 1496–1508.

MORESCHINI, TULLIO, 'Il problema della disponibilità delle terre per la colonizzazione nello Sc: Prime esperienze', *Georgofili*, 7/7 (1941), 105–10; *AC* 35/5 (1941), 194–8.

NADEL, S. F., 'Land Tenure on the Eritrean Plateau', *Africa*, 16/1 (1946): 1–23; 16/2, 99–109.

NANNINI, SERGIO, 'Il commissariato per le migrazioni e la colonizzazione in AOI', *REAI* 25/4 (1937), 507–8.

NASI, G., 'L'opera dell'Italia in Etiopia', *Italia e Africa* (Rm: n.d.) 11–21.

NASTRUCCI, MARIO, 'Lavorazioni del terreno nell'Impero', *AC* 32/5 (1938), 225–8.

NEWMAN, POLSON, 'Abyssinia under Italian Rule', *Daily Telegraph* (13 Aug. 1937).

NICHOLAS, GILDAS, 'Peasant Rebellions in the Socio-Political Context of Today's Ethiopia', *Pan African Journal*, 7/3 (Fall 1974), 236–62.

NICHOLSON, EDWARD G., 'A General Survey of Cotton Production in Ethiopia and Eritrea', *Empire Cotton Review*, 32/1 (1955), 1–17.

'Notiziario agricolo commerciale: Impero Etiopico', *AC* 30/5 (1936), 195–6.

'Notiziario agricolo commerciale: AOI', *AC* 30/9 (1936), 354–5; 31/4 (1937), 158–9; 32/1, (1938), 44; 32/2: 90; 32/3: 135–40; 32/5: 235; 32/9: 477; *AC* 33/11 (1939), 648; 34/1 (1940), 38–40; 34/6: 217; 34/11: 482; 35/1 (1941), 42.

'Opera di colonizzazione nell'Am', *II* (Dec. 1938).

ORIO, RICCARDO, 'Colonizzazione cerealicola nell'AOI vista da un Argentino', *AC* 31/1 (1937), 1–14.

PACHAI, BRIDGLAL, 'Land Policies in Malawi: An Examination of the Colonial Legacy', *JAH* 14/4 (1973): 681–98.

PAJELLA, ACHILLE, 'Contributo allo studio dell'economia agraria nel territorio dei GS', *AC* 32/4 (1938), 170–84.

PALMER, ROBIN, 'Working Conditions and Worker Responses on Nysaland tea estates, 1930–1953', *JAH* 27 (1986), 105–26.

PANKHURST, RICHARD, 'Italian Settlement Policy in Eritrea and its Repercussions, 1889–1896', in Jeffrey Butler (ed.), *Boston University Papers on African History*, (Boston, Mass.: 1964), i. 121–56.

—— 'Fascist Racial Policies in Ethiopia 1922–1940; *EO* 12 (1969), 270–86.

—— 'The Ethiopian Patriots: The Lone Struggle, 1936–1940', *EO* 13/1 (1970): 40–55.

—— 'The Ethiopian Patriots and the Collapse of Italian Rule in East Africa, 1940–41', *EO* 13/2 (1970), 92–127.

—— 'A Page of Ethiopian History: Italian Settlement Plans during the Fascist Occupation of 1936–1941', *EO* 13/2 (1970), 145–56.

—— 'A Chapter in Ethiopia's Commercial History: Developments during the Fascist Occupation of Ethiopia, 1936–1941', *EO* 14/1 (1971), 47–67.

—— 'Economic Verdict on the Italian Occupation of Ethiopia', *EO* 14/1 (1971), 68–82.

—— 'Italian and "Native" Labour during Italian Fascist Occupation of Ethiopia, 1935–41', *Ghana Social Science Journal,* 2/2 (Nov. 1972), 42–74.

'Perchè il Pesce di Biscioftù Costa Sedici Lire?', *CI*, 14 July 1938.

PERINETTI, ANTONIO, 'Risultati delle esperienze eseguite sui grani dell'AI', *AC* 37/8 (1943), 209–13.

PESCE, GIOVANNI; 'Gli obbligi e le facoltà del concessionario', *Azione coloniale*, 6 July 1939.

PIANI, G., 'Attività autarchica nell'Hararino', *Autarchia Alimentare* (June 1938), 38.

—— 'La visita di Attilio Terruzzi nell'Hararino' *Autarchia Alimentare* (15 July 1939), 32–40.

—— 'L'agricoltura indigena nel G Hr e i mezzi per farla progredire', *AC* 33/7 (1939), 401–8.

PICCIALUTI, TITO, 'Progetto per la formazione di una legione di lavoratori agricoli per l'AOI', *REAI* 26/2 (1938), 192–203.

PICCIOLI, ANGELO, 'La coltura industriale del caffè nella colonia Eritrea', *Autarcha Alimentare* (Aug. 1938), 21–5.

PICCOLI, GUALFARDO, 'Le iniziative dei servizi agrari dell'Am nel campo zootecnico', *Georgofili*, 7/7 (1941), 111–16.

PIERRUCCI, VINCENZO, 'Impressioni agrarie sull'Aussa', *AC* 34/4 (1940), 158–66.

PINI, GIUSEPPE, 'La terra Ricchezza d'Impero: Romagna d'Etiopia', *Azione coloniale*, 22 June 1939.

PIRRÒ, C., 'La politica autarchica e le sue ripercussioni sui traffici commerciali', *EI* (July 1938).

PISTOLESE, GENNARO E., 'Problemi dell'Impero', *EI* (July 1938).

—— 'Le attività agricole dell'AOI', *EI* (Feb. 1939).

—— 'L'Impero nell'economia di guerra' *EI* (May 1940).

POMILIO, MARCO, 'I problemi attuali dell'Impero nel pensiero di Attilio Terruzzi', *Autarchia Alimentare* (Aug. 1938): 7–8; see also *Alimentazione Italiana*, 13–31 Aug. 1938.

PRINZI, DANIELE 'Regime fondiario e colonizzazione in AOI', *REAI* 25/10 (1937), 1513–28.

—— 'La manodopera indigena nella colonizzazione nazionale', *REAI* 25/11 (1937), 1713–19.

QUARANTA, FERDINANDO, 'Note ed Appunti', *Autarchia Alimentare* (July 1938), 56–7.

'Rassegna agraria commerciale', *AC* 31/6 (1937), 297–8; 33/12 (1939), 687; 35/5 (1940), 211–14; 36/3 (1942), 113.

ROSSI-DORIA, M., 'Land Tenure System and Class in Southern Italy', *American Historical Review*, 64 (Oct. 1958), 46–53.

SAITTA, ACHILLE, 'Il sistema tributario del vecchio Impero Abissino', *REAI* 27/3 (1939), 282–9.

SALOME GABRE EGZIABHER, 'The Ethiopian Patriots 1936–1941', *EO* 12 (1969), 63–91.

SANTAGATA, FERNANDO, 'Il sistema economico degli Arussi', *REAI* 27/9 (1939), 1069–81.

'Semine nel Gm', *II* (Dec. 1939).

SLATER, HENRY, 'Land, Labour and Capital in Natal: The Natal Land and Colonisation Company, 1860–1948', *JAH* 16/2 (1975), 257–83.

'Sviluppo della sericoltura nell'Impero', *EI* (Sept. 1939).

'Sviluppo dell'agricoltura indigena in Eritrea', *EI* (May 1940).

SUTTON, INEZ, 'Labour in Commercial Agriculture in Ghana in the Late Nineteenth and Early Twentieth Centuries', *JAH* 24/3 (1983), 461–83.

'Territori dell'Impero e la loro valorizzaione agraria', *Georgofili*, 6/3 (1937), 340–61.

TOSH, JOHN, 'Lango Agriculture during the Early Colonial Period: Land and Labour in a Cash-Crop Economy', *JAH* 19/3 (1978), 415–39.

TRIULZI, ALESSANDRO, Review of books on *Italian Colonialism And Ethiopia*, *JAH* 23/2 (1982), 237–43.

VAUGHAN, MEGAN, 'Food Production and Family Labour in Southern Malawi: The Shire Highlands and Upper Shire Valley in the Early Colonial Period', *JAH* 23/3 (1982), 351–64.

VICINELLI, PAOLO, 'L'Istituto del "Distretto Cotoniero" nell'AOI', *AC* 33/1 (1939): 31–43.

VILLARI, GERARDO, 'I "gultì" della regione di Axum', *REAI* 26/9 (1938): 1430–44.

### 3. *Periodicals*

The following periodicals were consulted:

*Agenzia Le Colonie*, *Azione Coloniale*, *Colonie*, *Consulente Coloniale* (*CC*), *Corriere Eritreo* (*CE*), *Corriere dell'Impero* (*CI*), *Corriero Mercantile* (*CM*), *Corriere Padano*, *Corriere della Sera* (*CS*), *Daily Telegraph*, *Gazzetta del Mezzogiorno* (*GM*), *Gazzetta del Popolo* (*GP*), *Gazzetta di Venezia* (*GV*), *Giornale di Agricoltura della Domenica*, *Giornale di Genova*, *Giornale d'Italia* (*GI*), *Impero e Autarchia*, *Impero del Lavoro* (*IL*), *Italia Coloniale* (*IC*), *Italia d'Oltremare* (*IOM*), *Lavoro Agricolo Fascista*, *Lavoro Fascista*, *Lunedì dell'Impero* (*LI*), *Mattino*, *Messaggero*, *Nazione*, *Pattuglia*, *Piccolo*, *Popolo Fascista*, *Popolo d'Italia* (*PI*), *Popolo di Roma* (*PR*), *Popolo di Sicilia*, *Regime Fascista*, *Resto del Carlino*, *Sera*, *Sole milano*, *Somalia Fascista*, *La Stampa*, *The Sunday Times*, *The Times*, *Tribuna*.

*4. Books*

ADDIS HIWET, *Ethiopia from autocracy to Revolution* (London: Review of African Political Economy, 1975).

ALEMANNI, MAZZOCCHI, *Colonizzazione demografica* (Rm: Colombo, 1938).

AMBAYE ZEKARIAS, *Land Tenure in Eritrea (Ethiopia)* (AA, Addis Printing Press, 1966).

AMBROSIO, VINCENZO, *Tre anni fra I Galla e I Sidama 1937–1940: Lettere di un funzionario coloniale e testimonianze della sua morte sul campo* (Rm: MAI, 1942).

ARDEMANNI, ERNESTO, *Tre pagine gloriose nella storia militare-civile-religiosa della colonia Eritrea* (Rm.: Tip. R. Lastrucci, 1901).

ANNARATONE, CARLO, *In Abissinia* (Rm: Enrico Uoghera, 1914).

ANTINORI, O., *Viaggio Fra i Bogos* (Rm: Presso. La Societa Geografica Italiana, 1887).

BAHRU ZEWDE, *A History of Modern Ethiopia, 1855–1974* (London: James Curry, 1991).

BALDRATI, I., *Mostra delle attività economiche della colonia Eritrea* (As: Coloniale Fioretti, 1932).

Banco di Roma di AA, *Configurazione economica dell'Impero Etiopico: Notizie raccolte dal Banco di Roma di AA* (AA: Banco di Roma, 1936).

BARAVELLI, G. C., [Missiroli, Mario], *The Last Stronghold of Slavery: What Abyssinia is* (Rm: Soc. Ed. di Novissima, 1935).

BARDI, ADELMO, *Dall'Etiopia selvaggia all'Impero d'Italia* (San Remo: G. Gandolfi, 1936).

BARKER, A. J., *The Civilising Mission: The Italo-Ethiopian War* (London: Cassell, 1968).

BARTOLOZZI, ENRICO, *Case rurali nell'AOI* (Fl: Istituto Agronomo, 1940).

BASSI, CARLO, *Come finanziare il potenziamento dell'Impero Italiano d'Etiopia* (Perugia: Economica, 1936).

BATTISTELLA, GIACOMO, *Il credito agrario e fondiario in Africa* (Rm: Soc. An. Arte della Stampa, 1941).

BENT, T., *The Sacred City of the Ethiopians* (London: Longmans, Green & Co., 1896).

BERHANOU ABBEBE, *Évolution de la propriété foncière au Choa* (Paris: Imp Nationale, 1971).

BERKLEY, G. F. H., *The Campaign of Adlowa and the Rise of Menelik* (London: Archibald Constable & Co. Ltd., 1935).

BERRETA, A., *Con Amedeo d'Aosta in AOI in pace e guerra* (Milan; Casa ed. Ceschina, 1952).

BERTOLA, ARNALDO, *Il regime dei culti nell'AI* (Bologna: Cappelli, 1939).

BIANCHI, G., *Alla terra dei Galla* (Milan: Treves, 1884).

BIRKBY, CAREL, *It's a Long Way to Addis* (London: Frederick Muller Ltd., 1942).

BOTTEGO, VITTORIO, *Il Giuba esplorato* (Rm: E. Loescher & Co., 1895).

BRIANI, V., *L'emigrazione italiana ieri e oggi* (Rm: La Navicella, 1957).

BROTTO, ENRICO, *Il regime delle terre nel GHr. (Studio del Consigliere del Governo)* (AA: Servizio Tipografico GG AOI, 1939).

BURNS, EMILE, *Abyssinia and Italy* (London: Victor Gollancz Ltd., 1935).

[CAGNASSI, I.] *I nostri errori: Tredici anni in Eritrea* (Turin: Casanova, 1898).

CARNEVARI, E., *La guerra Italiana: Retroscena della disfatta* (Rm: Tosi, 1948).

CERULLI, ENRICO, *Etiopia schiavista* (Rm: Nuova Antologia, 1935).

CFA, Relazione di una missione di agricoltori in AOI (Fl: IACI, 1937).

—— *Gli agricoltori per la valorizzazione dell'Impero* (Rm: Ramo degli Agricoltori, 1938).

CIANO, G., *Ciano's Diary, 1937–1938* (London: Methuen, 1952); *Diari 1937–1938* (Bologna: Cappelli, 1948).

CIASCA, R., *Storia coloniale dell'Italia contemporanea* (Milan: Hoepli, 1940).

CICCARONE, A., *Il problema delle ruggini nei grani d'Etiopia* (Fl: IAC, 1947).

CIFERRI, R., *Frumenti e granicoltura indigena in Etiopia* (Fl: IAC, 1939).

CIPRIANI, L., *Un assurdo etnico: L'Impero etiopico* (Fl: R. Bemporadi & F., 1935).

—— *Abitazioni indigene dell'AOI* (Naples: della Mostra d'Oltre Mare, 1940).

COHEN, JOHN M., and WEINTRAUB, DOV., *Land and Peasants in Imperial Ethiopia: The Social Background to a Revolution* (Assen: Van Gorcum & Comp. BV, 1975).

COLETTI, F., *Dell'immigrazione italiana,* ii (Rm: H. Hoepli, 1911).

COLUCCI, M., *Il regime della proprietà fondiaria nell'AI* (Rocca San Casciano: Cappelli, 1942).

—— *Proprietà individuale e proprietà collettiva delle colonie* (Rm: Universitaria, 1939).

CONFORTI, EMILIO, *Impressioni agrarie su alcuni itinerari dell'altopiano Etiopico* (Relazione e Monografie Agrario-Coloniali, 65; Fl: RIA AI, 1941).

CONTI ROSSINI, CARLO, *Principi di diritto consuetudinario dell'Eritrea* (Rm: Tip dell'Unione Ed., 1916).

—— *Il regime fondiario indigeno in Etiopia e i mezzi di accertamento della proprietà in Sindacato Tecnici Agricoltori* (*Agricoltura ed Impero*, 2; Rm: Stabilmento Tipografico, 1937).

CORTESE, GUIDO, *Problemi dell'Impero* (Rm: Princiana, 1938).

CRISPI, F., *La prima guerra d'Africa* (Milan: Treves, 1914).

—— *Scritti e discorsi politici, 1848–1890* (Turin: Casa ed. Nazionale, n.d.).

CRUMMEY, DONALD, *Priests and Politicians: Protestant and Catholic Missions in Orthodox Ethiopia, 1830–1868* (Oxford: Clarendon Press, 1972).

DÄMSÉ WÄLDÄ AMANUÉL, *Bä-Iteyop̌eya agärachen lay ammest amät läwärrärän assäqaqi yähäzän engurguro* (AA: Täsfa Maryam Printing Press, 1951 EC).

DANIEL TEFERRA, *Social History and Theoretical Analyses of the Economy of Ethiopia* (African Studies, 4; Lewiston, NY.: Edwin Mellen Press, 1990).

DE FELICE, RENZO, *Mussolini il Duce: Gli anni del consenso, 1926–1936* (Turin: Einaudi, 1974).

DEL BOCA, ANGELO, *Gli Italiani in Africa Orientale*, i. *Dall'unità alla marcia di Roma*; ii. *La conquista dell'Impero*; iii. *La caduta dell'Impero*; iv. *Nostalgia delle colonie*; 4 vols. (Bari: Laterza, 1976–84).

DEL THEI, PRESENTI F., *Clima, acqua, terreno, dove e cosa si produce e si alleva in AOI* (Venice La Borsa del Libro, 1938).
DE MARCO, ROLAND, *The Italianization of African Natives: Government Native Education in the Italian Colonies, 1890–1937* (Colombia University, 1942).
DESSALEGN RAHMATO, *Agrarian reform in Ethiopia* (Trenton, NJ: Red Sea Press, 1985).
DIEL, L., *'Behold our new Empire'—Mussolini* (London: Hurst & Blackett Ltd. 1939).
DI LAURO, RAFFAELE, *Tre anni a Gondar* (Milan: Mondadori, 1936).
—— *Le terre del Lago Tana: possibilità economiche attuali del nord ovest etiopico* (Rm: Soc. Ital. Arti Grafiche, 1936).
DONHAM, DONALD, and JAMES, WENDY (eds.), *The Southern Marches of Imperial Ethiopia* (Cambridge: Cambridge University Press, 1986).
DORE, G., *La democrazia italiana e L'immigrazione in America* (Brescia: Morcelliana, 1964).
DUCHESNE-FOURNET, J., *Mission en Ethiopie 1901–1903* (Paris: J. Masson et Cie Editeure, 1908–9).
FANTOLI, AMILCARE, *Elementi preliminari del clima dell'Etiopia* (Fl: Sansoni, 1940).
FARAGO, LADISLAO, *Abyssinia on the Eve* (London: Wyman & Sons Ltd., 1935).
FARINA, G. G., *Follie delle follie* (Rm: Staderini, 1945).
FEDERZONI, LUIGI, *Venti anni di azione coloniale* (Milan: Mondadori, 1926).
—— *AO: Il post al sole* (Bologna: Zanichelli, 1936).
FOSSA, DAVIDE, *Lavoro Italiano nell'Impero* (Milan: Mondadori, 1938).
FRANCHETTI, LEOPALDO, *Sulla colonizzazione agricola dell'altopiano Etiopico: Memoria dell'On. Franchetti Deputato al Parlamento Italiano* (Rm: MAE, 1890).
—— *L'Italia e la sua colonia africana* (Città di Costello: Lapi, 1891). *Mezzogiorno e Colonie* (Fl: La Nuova Italia, 1950).
GAITSKELL, A., *Gezira: A Story of Development in Sudan* (London: Faber and Faber, 1959).
GÄRIMA TÄFÄRRA, *Gondäré Bägashaw* (AA: Tasfa Gabre Sellassie Printing Press, 1949 EC).
GEBRE HIWET BAYKEDAGN, *Berhan yehun: Aše Menilikenna Iteyop̌eya* (As: 1912).
GENNARI, GIUGLIO, *L'agricoltura nell'AOI* (Rm: REAI, 1938).
GIGLIO, CARLO, *La colonizzazione demografica dell'Impero* (Rm: *REAI*, 1939).
GLASS, D. V., and EVERSLEY, D. E. C. (eds), *Population in History* (Chicago: E. Arnold, 1965).
GRAZIANI, RODOLFO, *Fronte sud* (Milan: Mondadori: 1936).
GUARIGLIA, RAFFAELE, *Ricordi, 1922–1946* (Naples: Scientifiche Italiane: 1949).
GUARNIERI, F., *Battaglie economiche tra le due grandi guerre*, ii (Milan: Garzanti, 1953).
GUEBRE SELLASSIE, *Chronique du regne de Menelik II, roi des rois d'Éthiopie* (Paris: Maisonneuve Frères, 1930).
HAILE SELLASSIE (Emperor), *The Autobiography of Emperor Haile Sellassie I: My*

*Life and Ethiopia's Progress, 1892–1937*, ed. and trans. E. Ullendorff (Oxford: Oxford University Press, 1976).

HARLOW, VINCENT, CHILVER, E. M., and SMITH, ALISON, *History of East Africa*, ii (Oxford: Clarendon Press, 1965).

HESS, ROBERT L., *Italian Colonialism in Somalia* (Chicago: University of Chicago Press, 1966).

HIWET HIDARU, *Yachi Qän tärässach* (AA: Artistic Printing Press, 1975).

HOBEN, ALLAN, *Land Tenure among the Amhara of Ethiopia: The Dynamics of Cognatic Descent* (Chicago: Chicago University Press, 1973).

HUNTINGFORD, G. W. B., *The Land Charters of Northern Ethiopia* (AA: HSIU & Oxford University Press, 1965).

Impresa Cotoniera AI, *Primo esperimento di cotonicultura, maggio XVI–marzo XVIII. Soddu (GS)* (Milan: Pozzi, 1939).

IAC, *Main Features of Italy's Actions in Ethiopia, 1936–1941* (Fl: L'Arte della Stampa, 1946).

KÄBBÄDÄ TÄSÄMMA, *Yätarik Mastawäsha* (AA: Artistic Printing Press, 1970).

LAFFREDO, SILVA, *Ricerche chimiche sui sederimenti del Lago Tana, in Missione di Studio al Lago Tara*, iii/3 (Rm: Reale Accademia d'Italia, 1940).

LAGUZZI, OLINTO, Pioneri dell'Impero fascista: 'Raimondo franchetti "Il Lawrence italiaro"!' (Lavagna: Tip. Artigianelli, 1932).

LEJEAN, G., *Voyage aux deux Nils, 1860–1864* (Paris: Hachette, 1865).

LESSONA, ALESSANDRO, Scritti e discorsi coloniali (Milan: Arle e Storia, 1935).

—— *AI nel primo anno dell'Impero* (Rm: Edizioni della 'REAI', 1937).

—— *Verso l'Impero: Memorie per la storia politica del conflitto Italo-Etiopico* (Fl: Sansoni, 1939).

—— *Memorie* (Fl: Sansoni, 1958).

LEVI, CARLO, *Cristo si è fermato ad Eboli* (Milan: Mondadori, 1970).

LEWIS, HERBERT S., *A Galla Monarchy: Jimma Abba Jiffar: Ethiopia 1830–1932* (Madison, University of Wisconsin Press, 1965).

LONFERINI, BRUNO, *I Sidamo: Un antico popolo Cuscita* (Bologna: Nigrizia, 1971).

LOVATO, ARMANDO, *Per un'associazione delle famiglie numerose* (Bologna: Officine Grafiche Combattenti, 1936).

MCCANN, JAMES C., *From Poverty to Famine in Northeast Ethiopia: A Rural History, 1900–1935* (Philadelphia: University of Pennsylvania Press, 1987).

—— *A Great Agrarian Cycle? A History of Agricultural Productivity and Demographic Change in Highland Ethiopia, 1900–1981* (Working Papers in African Studies, 131, Boston, Mass: Boston University, 1988).

MCCLELLAN, CHARLES W., *State Transformation and National Integration: Gedeo and the Ethiopian Empire, 1895–1935* (East Lansing, Mich.: Michigan State University, African Studies Center, 1988).

MACK SMITH, DENIS, *Mussolini's Roman Empire* (Harmondsworth: Penguin Books, 1977).

MAHTEME SELLASSIE WOLDE MASKAL, Zekrä Nägär (AA: Näšannät Printing Press, 1942, EC).

MALLASALEM ANNELY, *Balläfut ammest yämäkära amätat Fashestoch bä-Iteyopeya* (AA, 1947 EC).

MANETTI, CARLO, *Panorama economico agrario dell'AOI e dell'Abisinnia* (Fl: Arte della Stampa, 1936).

MANTEL-NIECKO, JOANNA, *The Role of Land Tenure in the System of Ethiopian Imperial Goverment in Modern Times* (Warsaw: Wydanictwa Universytetu Warszawskiego, 1980).

MANZOTTI, F., *La polemica sull'emigrazione nell'Italia unita fino alla prima guerra mondiale* (Milan: Soc. Ed. Dante Alighieri, 1962).

MARCUS, HAROLD, *Haile Sellassie I: The Formative Years, 1892–1936* (Berkeley, Calif.: University of California Press, 1987).

—— *The Life and Times of Menelik II: Ethiopia 1844–1913* (Oxford: Clarendon Press, 1974).

MARESCALCHI, GIANNINO, *Eritrea* (Milan: Bietti, 1935).

MARKAKIS, JOHN, *Ethiopia: Anatomy of a Political Polity* (Oxford: Clarendon Press: 1974).

MAROI, FULVIO, *Aspetti di diritto agrario nelle terre dell'AOI* (Città di Castello: Unione Artigrafiche, 1941).

MARTINELLI, R., *Sud. Rapporto di un viaggio in Eritrea e in Etiopia* (Fl: Vallecchi, 1930).

MARTINI, F., MARINELLI, O., *et al.*, *L'Eritrea economica* (Novara: Istituto Geagrafico De Agostini, 1913).

MASOTTI, PIER MARCELLO, *Ricordi d'Etiopia di un funzionario coloniale* (Milan: Pan Milano, 1981).

MASSA, L., and SACCARDO, D., *Attraverso il territorio dei Galla e Sidama* (*Note agrarie*) (Relazioni e monografie agrio-coloniali, 55, Fl: RIA AI, 1939).

MAUGINI, ARMANDO, *Flora ed economia agraria degli indigeni* (Rm: MC, 1931).

——*Appunti sulle prospettive agricole* (Fl: IAC, 1941).

MELLI, B. *La Colonia Eritrea* (Parma: Leo Battei, 1899).

MESFIN WOLDE MARIAM, *Rural Vulnerability to Famine in Ethiopia 1958–1977* (London: Intermediate Technology Publications, 1986).

MIEGE, J. L., *L'imperialismo coloniale italiano dal 1870 ai giorni nostri* (Milan: Société d'edition d'enseignement Supérieure, 1976).

MOCKLER, ANTONY, *Haile Selassie's War* (Oxford: Oxford University Press, 1984).

MONDAINI, G., *Legislazione coloniale Italiana*, i–ii (Milan: Istituto Di Politica Internazionale 1941).

MONELLI, P., *Mussolini: An Intimate Life* (London: Thames and Hudson, 1953).

MONTEGAZZI, V., *La guerra in Africa* (Fl: Successori Le Monnier, 1896).

NEWMAN, E. L. POLSON, *The New Abyssinia* (London: Rich and Cowan Ltd., 1938).

NUNNO, SETTIMO, *Nella terra del Negus* (Milan: Pontificio Istituto Missioni Estere, 1967).

OMODEO, A., PEGLION, V. and VALENTI, G., *La colonia Eritre: Condizioni e problemi* (Rm: Tip. Nazionale del G. Bertero, 1913).

*Ottavo Congresso di Agricoltura Tropicale e Subtropicale, Tripoli*, Carocci Buzzi (ed.) (Rm: Federazione Internazionale Tecnici Agricoli, 1941).

OXAAL, IVAR, BARNET, TONY, and BOOTH, DAVID (eds.), *Beyond the Sociology of Development* (London: Routledge and Kegan Paul, 1975).

PACE, B., *L'Impero e la collaborazione internazionale in Africa* (Rm: Istituto Nazionale della Cultura Fascista, 1938).

PANETTA, RINALDO, *Culqualber fine dell'Impero* (Rm: Volpe, 1965).

PANKHURST, RICHARD, *State and Land in Ethiopian History* (AA: HSIUP, 1966).

—— *Economic History of Ethiopia, 1800–1935* (AA: HSIUP, 1968).

PERHAM, MARGERY, *The Government of Ethiopia* (London: Faber and Faber Ltd, 1969).

PIEROTTI, FRANCESCO, *Vita in Etiopia, 1940–1941* (San Casciano: F. Cappelli, 1959).

PIRONTI DI CAMPAGNA, GIUSEPPE, *La difesa dell'Impero* (Rm: La Libreria Cattolica Internazionale, 1937).

POGGIALI, CIRO, *Albori dell'Impero: L'Etiopia come è e come sarà* (Milan: Treves, 1938).

—— *Diario AOI. 15 giugno 1936–4 ottobre 1937* (*Gli appunti segreti dell'inviato del 'Corriere della Sera'*) (Milan: Longanesi & Co., 1971).

POLLERA, ALBERTO, *Il regime della proprietà terriera in Etiopia e nella colonia Eritrea* (MC Monografie e Rapporti Coloniali, Rm: G. Bertero e C., 1913).

—— *Lo stato Etiopico e la sua Chiesa* (Rm: Soc. D'Arte Illustrata, 1926).

QUARANTA, FERNANDO, *Ethiopia: An Empire in the Making* (London: P. S. King & Son, Ltd., 1939).

RAINERO, ROMAN, *I primi tentativi di colonizzazione agricola e di popolamento dell'Eritrea* (*1890–1895*) (Milan: Marzorati, 1960).

*Relazione di una missione di agricoltori in AOI* (Relazioni e Monografie Agrario Coloniali; Fl: IAC, 1937).

RIVERA, VINCENZO, *Prospettive di colonizzazione dell'AOI* (Rm: Libreria di Scienze e Lettere, 1939).

RIZZUTI, ANTONIO, *Le concessioni agricole nelle colonie Italiane* (Rm: Foro Italiano, 1939).

ROBERTSON, ESMONDE M., *Mussolini as Empire Builder: Europe and Africa, 1932–1936* (London: Macmillan, 1977).

ROBERTSON, A. F., *People and the State: An Anthropology of Planned Development* (Cambridge: CUP, 1984).

RUBENSON, SVEN, *The Survival of Ethiopian Independence* (Goran Rystad and Sven Tägil Series in Lund Studies in International History, 7; London: Heinemann, 1976).

S.A.Co. Ital. Importatori Caffè, *Relazione sui lavori svolti della missione inviata in AOI per lo studio dei problemi inerenti al caffè dell'Impero* (Turin: Fedetto, n.d.).

SALIS SEROTOLI, RENZO, *L'ordinamento fondiario Eritreo* (Padua: CEDAM, 1932).

SANTAGATA, FERNANDO, *L'Harar: Territorio di pace e di civiltà* (Milan: Garzanti, 1940).

SANTI, NOVA, *Elementi di dommatica della colonizzazione* (Fl: Poligrafica Universitaria, 1937).

SAVONA, V., and STRANIERO, M., *Canti dell'Italia Fascista* (Milan: Garzanti, 1979).

SBACCHI, ALBERTO, *Ethiopia under Mussolini: Fascism and the Colonial Experience* (London: Zed Books Ltd., 1985).

SCASSELLATI-SFORZOLINI, *La Società Agricola Italo-Somala* (Fl: IACI, 1926).

SEGRÈ, CLAUDIO G., *Fourth Shore: The Italian Colonization of Libya*, (Series in Studies in Imperialism, ed, Robin L. Winks; Chicago: The University of Chicago Press, 1974).

SPERONI, GIGI, *Amedeo duca d'Aosta: La resa dell'Amba Alagi e la morte in prigionia nei documenti segreti inglesi* (Milan: Rusconi Libri S.P.A., 1984).

STARACE, ACHILLE, *La marcia su Gondar: Della colonna Celere AO e le sue successive operazioni nella Etiopia occidentale* (Milan: Mondadori, 1936).

STEER, GEORGE L., *Sealed and Delivered* (London: Hodder & Stoughton Ltd., 1942).

—— *Caesar in Abyssinia* (London: Hodder & Staughton Ltd., 1936).

TADDÄSSÄ MÄČA, *Ťequr Änbäsa bä-meerab Ityop̌eya* (AA: A. Corriere Eritreo officina Grafica, 1943 EC).

TADDÄSSÄ TAMRAT, *Church and State in Ethiopia, 1270–1527* (Oxford: Clarendon Press, 1972).

TADDÄSSÄ ZÄWÄLDÄ, *Qärin gärrämaw. Yä-arbäña tarik* (AA: Berhanenna Sälam Haile Sellassie Press, 1960 EC).

TADDIA, IRMA, *L'Eritrea—Colonia, 1890–1952* (Milan: Franco Angeli, 1986).

TALBOT, I. D., *Agricultural Innovation in Colonial Africa: Kenya and the Great Depression* (African Studies, 18; Lewiston, Me.: Edwin Mellen Press, 1990).

TEKESTE NEGASH, *Italian Colonialism in Eritrea, 1882–1941: Policies, Praxis and Impact* (Acta Universitatis Upsaliensis Studia Historica Upsaliensia, 148; Uppsala, 1987).

—— *No Medicine for the Bite of a White Snake: Notes on Nationalism and Resistance in Eritrea, 1890–1940* (Uppsala: University of Uppsala, 1986).

TODARO, U., *La conquista della terra (1930–1943)* (Rassegna dell'ONC: L'opera, 1940).

TRAVERSI, L., *Let Marefià* (Milan: Alpes, 1931).

VEROI, G. P., *I servizi bancari nell'Impero e l'organizzazione del Banco di Roma* (Rm: Istituto Poligrafico dello Stato, 1938).

VILLA SANTA, N., SCAGLIONE, ATTIGLIO, *et al.*, *Amedeo Duca D'Aosta* (Rm: Istituto del Nastro Azzurro fra Combattenti Decorati al Valone Militare, 1954).

WALKER, C. W., *The Abyssinian at Home* (London: Sheldon Press, 1933).

WAUGH, E., *Waugh in Abyssinia* (London: Longmans, Green & Co., 1936).

YOLANDE, MARA, *The Church of Ethiopia: The National Church in the Making* (As: Il Poligrafico, 1972).

YUDELMAN, MONTAGUE, 'Imperialism and the Transfer of Agricultural Techniques', in Peter Duignan and L. H. Gann (eds.), *The Economics of Colonialism* (Colonialism In Africa 1870–1960, 4; Cambridge: CUP, 1975), 329–57.

ZERVOS, ADRIEN, *L'Empire d'Ethiopie* (Alexandria: Impr. del'École Professionelle des Frères, 1936).

ZEWDE GABRE-SELLASSIE, *Yohannes IV of Ethiopia: A Political Biography* (Oxford: Clarendon Press, 1975).

# Glossary

*Abba* — (Eccles.) father; Reverend.

*Abbat* — Father; ancestor.

*Abun/ä* — (Eccles.) title of a Bishop; the Head of the EOC.

*Addi* — Country, village [*Tegreña*].

*Aläqa* — Leader; (Eccles.) head of a *Däbr*.

*Anťafi* — (Eccles.) upholsterer.

*Aqabé Säat* — (Lit. Custodian of hour); (eccles.) title of the Abbot of Däbrä Häyeq.

*Aqabit* — (Eccles.) church-keeper, custodian.

*Asrat* — (Lit. tithe); part of a tax on produce.

*Aťabi* — (Eccles.) washer; cleaner of church utensils.

*Balabbat* — (Lit. Indigenous or one with the father); a local low-level official who mediated between the people and the government.

*Bétä-Kehnät* — (Lit. the house of clergy) Eccles. order.

*Bétä-Mängest* — (Lit. the house of government); royal household; public domain.

Capo — (Lit. head, chief); title associated with Mussolini; foreman.

*Čat* — A narcotic widely grown in the south-eastern regions.

*Čeguraf-gottet* — See *shäna*.

*Čeqa* — Mud.

*Čeqa-Shum* — (Lit. Chief mud); a minor official appointed for a fixed period from among the elders of *rest*-owning peasantry to act as intermediary between the government and the peasants of the area.

*Čisäña* — See *ťisäña*.

*Däbr/-ä* — (Eccles.) abbey or an endowed Church.

*Däbtära* — (Eccles.) learned but unordained cleric who acts as chorister, poet, astrologer.

*Däga* — Highland [2,400–4,600 m].

*Dähena Märét* — Semi-fertile land.

*Däjach* — See *däjazmach*.

*Däjazmach* — (Lit. Commander of the Gate); a nobleman's title equivalent to count.

*Dässa* — See *Shäna*.

*Dawulla* — Weight measurement unit consisting of 20 *qunna* or about 100 kg.

| | |
|---|---|
| Duce | (Lit. Leader); title associated with Mussolini. |
| *Ekkul Arash* | (Lit. tiller of half); sharecropping system whereby the produce is divided in half between the parties. |
| *Ente* | (Ital.) agency, corporation. |
| *Erbo Arash* | (Lit. tiller of a quarter); sharecropping system where the produce is divided on a rate of one quarter. |
| *Fitawurari* | (Mil. lit. leader of the vanguard); title equivalent to viscount. |
| *Gäbäz* | (Eccles.) administrator of a church. |
| *Gäbbar* | Tax-payer; peasant, farmer. |
| *Gädam* | (Eccles.) monastery. |
| *Ganä-gäb* | Lit. pot- or royal-bound; land produce directly administered by the central government. |
| *Gasha* | A land unit of between 30–40 ha or one *qällad*. |
| *Gäťar church* | Lowest rank of Parish church. The lowest unit of church organization. Rural Church. |
| *Geber* | Tax; part of land tax traditionally paid in kind for the *gult-gäz* [*gult*-holder]. |
| *Gebbi* | Royal court, palace. |
| *Gult* | Land granted for a particular purpose or service; fief; benefice; administrative unit. |
| *Gult-gäž* | (Lit. *gult*-ruler); official holding *gult* rights over an estate; administrator. |
| *Gundo* | Capacity measurement used mainly for honey [*Gundo Mar*] consisting of 60 *berelé* or [500 × 60 gm]. |
| *Hudad* | Public domain; land temporarily set aside and worked by statute labour for the benefit of a governor or the state. |
| *Kahenat* | Clergy. |
| *Liqä Kahnat* | High Priest/Arch-Priest. Bishop Representative at local church. |
| *Madärya* | Land allocated to State officials in lieu of salary. |
| *Mad-bét* | (Lit. dining-room/hall) royal kitchen. |
| *Mäggabi* | (Lit. one who administers; feeds) quartermaster; (eccles.) administrator of a *gädam*. |
| *Maläfeya märét.* | Fertile, beautiful land. |
| *Mämher* | Teacher; (eccles.) head of a monastery. |
| *Mäqomiya* | a piece of grazing land given to a tenant farmer for his cattle. |
| *Mär* | See *gundo*. |
| *Märét* | Land; soil. |
| *Mäťefo märét* | Unproductive land. |
| *Mofär zämach* | non-resident or mobile ploughman/farmer. |
| *Näfťäña* | (Lit. Gun-carrier); northern soldier-settler in southern Ethiopia. |

*Negus* King.

*Qäl[l]ad* Rope used in land survey. ca. 150 cubit; land measured by qälad.

*Qés Gäbäz* (Eccles.) a priest *Gäbäz*.

*Qolla* Lowland [below 1,400 m].

*Qunna* Grain measurement unit consisting of about 5 litres.

*Qurť geber* Fixed tribute/tax.

*Ras* (Mil. lit. Head) a nobleman's title equivalent to duke.

*Rest* Inherited land; use of such rights.

*Restä-gult* A hereditary *gult*.

*Sämon märét* (Eccles. lit. Weekly land); land burdened with the obligation to provide weekly mass.

*[Yä] Säqäla märèt* Government land; non-hereditary land given to a state official.

*Shäna* Village or communal land tenure system often involving periodical distribution.

*Sisso Arash* (Lit. Tiller of one-third); sharecropping system where the produce is divided between the parties on a rate of one third.

*Ťämaj* A yoker of oxen; plough man.

*Ťéf* An indigenous cereal crop widely grown in highland areas.

*Tekläña* (Lit. the one who is planted); settler.

*ťisäña* Tenant farmer.

*Wärägänu* Royal herdsmen.

*Wäynä-däga* Midlands [1,400–2,400 m].

*Yä-* Preposition meaning 'Of; belonging to'; e.g. *yä-sämon märét* = land of *sämon*; *yä-bétä mängest* = belonging to royal household, state.

# Index